A Generalized Framework of Linear Multivariable Control

A Generalized Framework of Linear Multivariable Control

Liansheng Tan

Butterworth-Heinemann
An imprint of Elsevier
elsevier.com

Butterworth-Heinemann is an imprint of Elsevier
The Boulevard, Langford Lane, Kidlington, Oxford OX5 1GB, United Kingdom
50 Hampshire Street, 5th Floor, Cambridge, MA 02139, United States

Notices
Knowledge and best practice in this field are constantly changing. As new research and experience broaden our understanding, changes in research methods, professional practices, or medical treatment may become necessary.

Practitioners and researchers must always rely on their own experience and knowledge in evaluating and using any information, methods, compounds, or experiments described herein. In using such information or methods they should be mindful of their own safety and the safety of others, including parties for whom they have a professional responsibility.

To the fullest extent of the law, neither the Publisher nor the authors, contributors, or editors, assume any liability for any injury and/or damage to persons or property as a matter of products liability, negligence or otherwise, or from any use or operation of any methods, products, instructions, or ideas contained in the material herein.

Library of Congress Cataloging-in-Publication Data
A catalog record for this book is available from the Library of Congress

British Library Cataloguing-in-Publication Data
A catalogue record for this book is available from the British Library

ISBN: 978-0-08-101946-7

For information on all Butterworth-Heinemann publications
visit our website at https://www.store.elsevier.com/

Working together
to grow libraries in
developing countries

www.elsevier.com • www.bookaid.org

Publisher: Joe Hayton
Acquisition Editor: Sonnini R. Yura
Editorial Project Manager: Ana Claudia Abad Garcia
Production Project Manager: Kiruthika Govindaraju
Cover Designer: Greg Harris

Typeset by SPi Global, India

Contents

Introduction

1

In this book several contributions to the theory of generalized linear multivariable control are made that provide extensions to some results in *polynomial matrix description* (PMD) theory, behavioral theory, and $\mathcal{H}_\infty$ control theory. We highlight the application of linear multivariable control in congestion control of the Internet as well. The remaining parts of this book are organized as follows.

In Chapter 2, we outline the basic components of vector algebra, matrix algebra, linear differential equation, matrix differential equation, and Laplace transform.

Chapter 3 presents a brief introduction of the generalized inverse of matrix, which is needed in the following expositions. This introduction includes the left inverse and right inverse, the Moore-Penrose inverse, the minimization approach to solve an algebraic matrix equation, the full rank decomposition theorem, the least square solution to an algebraic matrix equation, and the singular value decomposition.

The polynomial matrix fraction description (PMFD) is a mathematically efficient and useful representation for a matrix of rational functions. With application to the transfer function of a multiinput, multioutput linear state equation, the PFD can inherit the structural features that, for example, permit natural generalization of minimal realization having established for single-input, single-output state equations. The main aim of Chapter 4 is to lay out a basic introduction to the polynomial fraction description, which is a key mathematical component in the current exposition of the generalized linear multivariable control theory. The main components covered by this chapter are right (left) polynomial fractions, column and row degrees, minimal realization, poles and zeros, and state feedback.

Chapter 5 gives an introduction of the mostly important and basic elements in the stability theory. We surveyed the following theoretical concepts, terms, and statements: internal stability (IS), Lyapunov stability, and input-output stability.

Chapter 6 surveys the fundamental approaches to control system analysis, including the PMD theory of linear multivariable control systems, behavioral approach in systems theory, and the chain-scattering representations (CSR).

The finite and infinite frequency structures of a rational matrix are fundamental to system analysis and design. The classical methods of determining them are not stable in numerical computations, because the methods are based on unimodular matrix transformations that result in an extraordinarily large number of polynomial manipulations. Van Dooren et al. [1] presented a method for determining the Smith-McMillan form of a rational matrix from its Laurent expansion about a particular finite point $s_0 \in C$. Subsequently, Verghese and Kailath [2] and Pugh et al. [3] extended

A Generalized Framework of Linear Multivariable Control. http://dx.doi.org/10.1016/B978-0-08-101946-7.00001-9

this theory to produce a method of determining the infinite pole and zero structures of a rational matrix from its Laurent expansion about the point at infinity. The difficulty inherent in these approaches is the computation of the Laurent expansion of the original rational matrix, one is forced to calculate all the coefficient matrices in the Laurent expansion; furthermore there is no explicit stop criterion for recursion available in these approaches.

In Chapter 7 a novel method is developed that determines the finite and infinite frequency structure of any rational matrix. For a polynomial matrix, a natural relationship between the rank information of the Toeplitz matrices and the number of the corresponding *irreducible elementary divisors* (IREDs) in its Smith form is established. This relationship proves to be fundamental and efficient to the study of the finite frequency structure of a polynomial matrix. For a rational matrix, this technique can be employed to find its infinite frequency structure by examining the finite frequency structure of the *dual* of its *companion polynomial matrix*. It can also be extended to find the finite frequency structure of a rational matrix via a PMFD. It is neat and numerically stable when compared with the cumbersome and unstable classical procedures based on elementary transformations with unimodular matrices. Compared to the methods of Van Dooren et al. [1], Verghese and Kailath [2], and Pugh et al. [3], our approach, which is based on analyzing the *nullity* of the Toeplitz matrices of the derivatives of the polynomial matrices rather than the Laurent expansion of the original rational matrices, will be more straightforward and much simpler, for it is easier and more direct to obtain the derivatives of a polynomial matrix than to obtain its Laurent expansion. The special Toeplitz matrices, which are based on the information of the *Smith zeros* [4, 5] of the system, are relatively easy to compute due to the fact that several numerical algorithms [6] have been proposed for finding the locations of the Smith zeros. Moreover, the procedure will terminate after a minimal number of steps, which thus represents another numerical refinement.

In Chapters 8–10, several contributions are presented concerning the solution of regular PMD.

Chapter 8 considers regular PMDs or *linear nonhomogeneous matrix differential equations* (LNHMDEs), which are described by

$$A(\rho)\beta(t) = B(\rho)u(t), \quad t \geq 0, \tag{1.1}$$

where $\rho := d/dt$ is the differential operator, $A(\rho) = A_q\rho^q + A_{q-1}\rho^{q-1} + \cdots + A_1\rho + A_0 \in R[\rho]^{r \times r}$, $\text{rank}_{R[\rho]}A(\rho) = r$, $A_i \in R^{r \times r}$, $i = 0, 1, 2, \ldots, q$, $q \geq 1$, $B(\rho) = B_l\rho^l + B_{l-1}\rho^{l-1} + \cdots + B_l\rho + B_0 \in R[\rho]^{r \times m}$, $B_j \in R^{r \times m}$, $j = 0, 1, 2, \ldots, l$, $l \geq 0$, $\beta(t) : [0, +\infty) \longrightarrow R^r$ is the *pseudo-state* of the PMDs, $u(t) : [0, +\infty) \longrightarrow R^m$ is a p times piecewise continuously differentiable function called the input of the PMD. Its homogeneous case

$$A(\rho)\beta(t) = 0, \quad t \geq 0 \tag{1.2}$$

is called the homogeneous PMD or the *linear homogeneous matrix differential equations* (LHMDEs). Both the regular *generalized state space systems* (GSSSs), which are described by

$$E\dot{x} = Ax(t) + Bu(t),$$

where E is a singular matrix, rank $(\rho E - A) = r$, and the (regular) state space systems, which are described by

$$\dot{x} = Ax(t) + Bu(t)$$

are special cases of the regular PMDs (Eq. 1.1).

Regarding the solutions of the GSSSs, there have been many discussions [7–11]. In [12], the impulsive solution to the LHMDEs was presented in a closed form. For both the regular PMDs either in homogeneous cases or in nonhomogeneous cases, Vardulakis [13] developed their solutions under the assumption that both the initial conditions of the state and the input are zero. However, as we will see later, in some cases, the initial conditions of the state might result from a random disturbance entering the system, and a feedback controller is called for. Since the precise value of the initial conditions of the state is unpredictable and the control is likely to depend on those initial conditions of the state, so the assumption of zero initial conditions is somewhat stronger than necessary.

In this book, based on any resolvent decomposition of $A(s)$ we will present a complete solution to Eq. (1.1) that displays the impulse response and the slow response created not only by the initial conditions of $\beta(t)$, but also by the initial condition of $u(t)$. Also a reformulation to the solution of the regular PMDs in terms of the *regular derivatives* of $u(t)$ is given. By defining the *slow state* (*smooth state*) and the *fast state* (*impulsive state*) of the solution components, it is shown that the system behaviors of the regular PMDs can be decomposed into the *slow response* (*smooth response*) and the *fast response* (*impulsive response*) completely. This approach is conveniently applied to discuss the impulse free initial conditions of Eq. (1.1).

So far there have been many discussions about the impulse free initial condition either to the generalized state space systems [14, 15] or to the regular PMDs [13, 16]. However, the common concern in the known results is the impulse created by the appropriate initial conditions of the state alone. To the regular PMDs, a different analysis concerning this issue will be carried out in this book that considers both the impulse created by the initial conditions of $u(t)$ and that which is created by the initial conditions of $\beta(t)$.

The solution of the above PMDs is generally proposed on the basis of any resolvent decomposition of $A(s)$ given by

$$A^{-1}(s) = C(sI_n - J)^{-1}Z + C_\infty(sJ_\infty - I_\varepsilon)^{-1}Z_\infty, \tag{1.3}$$

where (C, J) is the finite Jordan pair of $A(s)$, and (C_∞, J_∞) is the infinite Jordan pair [13, 17] of $A(s)$. Such resolvent decompositions are not unique, but play an important role in formulating the solution of the regular PMDs. The difficulties in the problem of obtaining the solution of the PMD are specific to the particular resolvent decomposition used. From a computation point of view, when the matrices in Eq. (1.3) are of minimal dimensions, it is obviously easiest to obtain the inverse matrix of $A(s)$.

For Eq. (1.1) Gohberg et al. [17] proposed a particular resolvent decomposition, and the solution of it was formulated according to this decomposition. However, the impulsive property of the system at $t = 0$ was not fully considered in this work. One advantage of this resolvent decomposition however is that it is constructive, due to the fact that the matrices Z, Z_∞ in Eq. (1.3) are formulated in terms of the finite and the infinite Jordan pairs. On the other hand in this resolvent decomposition, there appears certain redundant information, in that the dimensions of the infinite Jordan pair are much larger than is actually necessary, which in turn brings some inconvenience in computing the inverse matrix of $A(s)$. In fact some of the *infinite elementary divisors* [18] that correspond to the infinite poles of $A(s)$ and actually contribute nothing to the solution. Thus the resolvent decomposition can be proposed in a simpler form that has the advantage of giving a more precise insight into the system structure and bringing some convenience in actual computation. Vardulakis [13] obtained a general solution of Eq. (1.1) under the assumption that the initial conditions of $\xi(t)$ are zero. The main idea of this approach is to find the minimal realizations of the strictly proper part and the polynomial part of $A^{-1}(s)$ and then to obtain the required resolvent decomposition. Although this procedure does give good insight into the system structure, the realization approach is not so straightforward by itself and it is consequently more difficult to be applied in actual computation. Furthermore, in this procedure, no explicit formula for Z and Z_∞ is available. Apparently, differences arise between the solution of Gohberg et al. [17] and that of Vardulakis [13] due to the fact that these two solutions are expressed through two different resolvent decompositions. Although it is found that the redundant information contained in the solution of Gohberg et al. [17] can be decoupled, an overly large resolvent decomposition definitely brings some inconvenience to actual computation.

One of the main purposes of Chapter 9 is to present a resolvent decomposition, which is a refinement of both results obtained by Gohberg et al. [17] and Vardulakis [13]. It is formulated in terms of the notions of the finite Jordan pairs, infinite Jordan pairs and the generalized infinite Jordan pairs that were defined by Gohberg et al. [17] and Vardulakis [13]. We make clear the issue of *infinite Jordan pair* noted by Gohberg et al. [17] and Vardulakis [13]. This refined resolvent decomposition captures the essential feature of the system structure. Further, the redundant information that is included in the resolvent decomposition of Gohberg et al. [17] is deleted through a certain transformation, thus the resulting resolvent decomposition inherits the advantages of both the results of Gohberg et al. [17] and of Vardulakis [13]. This refined resolvent decomposition facilitates computation of the inverse matrix of $A(s)$ due to that the dimensions of the matrices used are of minimal.

Based on this proposed resolvent decomposition, a complete solution of a PMD follows that reflects the detailed structure of the *zero state response* and the *zero input response* of the system. The complete impulsive properties of the system are also displayed in our solution. Such impulsive properties are not completely displayed in the solution of Gohberg et al. [17]. Although for the homogeneous case this problem is considered in Vardulakis [13], such complete impulsive properties of the system for the general nonhomogeneous regular PMD is not available from Vardulakis [13].

The known solutions are all based on the *resolvent decomposition* [13, 17] of the regular polynomial matrix $A(s)$, which is formulated in terms of the finite Jordan pairs and the infinite Jordan pairs of $A(s)$. On the one hand, such treatments have the immediate advantage that they separate the system behavior into the slow (smooth) response and the fast (impulsive) response, which may provide a deep insight into the system structure. On the other hand, such treatments bring some inconvenience for the actual computation since the classical methods of determining the finite Jordan pairs or the infinite Jordan pairs have to transform the polynomial matrix $A(s)$ into its Smith-McMillan form or its Smith-McMillan form at infinity. Such transformations are well known not to be stable in numerical computation terms, for they result in an extraordinarily large number of polynomial manipulations.

In Chapter 10 a novel approach via linearization is also presented to determine the complete solution of regular PMDs that takes into account not only the initial conditions on $\beta(t)$, but also the initial conditions on $u(t)$. One kind of linearization [17] of the regular polynomial matrix $A(s)$ is the so-called *generalized companion matrix*, which is in fact a regular matrix pencil. The Weierstrass canonical form of this matrix pencil can easily be obtained by certain constant matrix transformations [5]. In this book certain additional properties of this companion form are established, and a special *resolvent decomposition* of $A(s)$ is proposed that is based on the Weierstrass canonical form of this generalized companion matrix. The solution of the regular PMD is then formulated from this resolvent decomposition. An obvious advantage of the approach adopted here is that it immediately avoids the polynomial matrix transformations necessary to obtain the finite and infinite Jordan pairs of $A(s)$, and only requires the constant matrix transformation to obtain the Weierstrass canonical form of the generalized companion form, which is less sensitive than the former in computational terms. Since numerically efficient algorithms to generate the canonical form of a matrix pencil are well developed [19, 20], the formula proposed here is more attractive in computational terms than the previously known results.

One of the fundamental problems treated in this book originates from classical network theory, where a circuit representation called the *chain matrix* [21] has been widely used to deal with the cascade connection of circuits arising in analysis and synthesis problems. Recently, Kimura [22] has developed the CSR, which was subsequently used to provide a unified framework for H^∞ control theory. The CSR is in fact an alternative way of representing a plant. Compared to the usual transfer function formulation, it has some remarkable properties. One is its *cascade structure*, which enables *feedback* to be represented as a matrix multiplication. The other is the *symmetry (duality)* between the CSR and its inverse called the dual *chain-scattering representation* (DCSR). Due to these characteristic features, it has successfully been used in several areas of control system design [22–24].

Consider a plant P with two kinds of inputs (w, u) and two kinds of outputs (z, y) represented by

$$\begin{bmatrix} z \\ y \end{bmatrix} = P \begin{bmatrix} w \\ u \end{bmatrix} = \begin{bmatrix} P_{11} & P_{12} \\ P_{21} & P_{22} \end{bmatrix} \begin{bmatrix} w \\ u \end{bmatrix}, \tag{1.4}$$

where P_{ij} ($i = 1, 2$; $j = 1, 2$) are all rational matrices with dimensions $m_i \times k_j$ ($i = 1, 2$; $j = 1, 2$). In order to compute the CSR and DCSR of Eq. (1.4), P_{21} and P_{12} are generally assumed to be invertible [22]. However, for general plants, if neither of P_{12} and P_{21} is invertible, one cannot use CSR and DCSR directly. Although the approach was extended by Kimura [22] to the case in which P_{12} is full column rank and P_{21} is full row rank by augmenting the plant, a systematic treatment is still needed in the general setting when neither of these conditions is satisfied.

In Chapter 11, we consider general plants, and therefore make no assumption about the rank of P_{12} and P_{21}. From an input-output consistency point of view, the conditions under which the CSR exist are developed. These conditions essentially relax those assumptions that were generally put on the rank of P_{21} and P_{12} and make it possible to extend the applications of the CSR approach from the regular cases of the 1-block case, the 2-block case and the 4-block case [25] to the general case. Based on this, the chain-scattering matrices are formulated into general parameterized forms by using the matrix generalized inverse. Subsequently the cascade structure property and the symmetry are investigated.

Recently, behavioral theory [26, 27] has received broad acceptance as an approach for modeling dynamical systems. One of the main features of the behavioral approach is that it does not use the conventional input-output structure in describing systems. Instead, a mathematical model is used to represent the systems in which the collection of time trajectories of the relevant variables are viewed as the *behavior* of the dynamical systems. This approach has been shown [26, 28] to be powerful in system modeling and analysis. In contrast to this, the classical theories such as Kalman's state-space description and Rosenbrock's PMD take the input-output representation as their starting point. In many control contexts it has proven to be very convenient to adopt the classical input-state-output framework. It is often found that the system models, in many situations, can easily be formulated into the input-state-output models such as the state space descriptions and the PMDs. Based on such input-state-output representations, the action of the controller can usually be explained in a very natural manner and the control aims can usually be attained very effectively.

As far as the issue of system modeling is concerned, the computer-aided procedure, i.e., the automated modeling technology, has been well developed as a practical approach over recent years. If the *physical description* of a system is known, the automated modeling approach can be applied to find a set of equations to describe the *dynamical behavior* of the given system. It is seen that, in many cases, such as in an electrical circuit or, more generally, in an interconnection of blocks, such a physical description is more conveniently specified though the *frequency domain behavior of* the system. Consequently in these cases, a more general theoretic problem arises, i.e., if the *frequency domain behavior description* of a system is known, what is the corresponding dynamical behavior? In other words, what is the input-output or input-state-output structure in the time domain that generates the given frequency domain description. It turns out that this question can be interpreted through the notion of *realization of behavior* which we shall introduce in Chapter 12. In fact, as we shall see later, realization of behavior, in many cases, amounts to introduction of latent variables in the time domain. From this point of view, realization of behavior can be understood as a converse procedure to the latent variable elimination theorem [27]. It should be emphasized that realization of behavior also generalizes the notion of transfer function matrix realization in the classical control theory framework, as the behavior equation is a more general description than the transfer function matrix. As a special case, the behavior equations determine a transfer matrix of the system if they represent an input-output system [27, 29], i.e., when the matrices describing the system satisfy a full rank condition.

More recently in [30], a realization approach was suggested that reduces high-order linear differential equations to the first-order system representations by using the method of "linearization." From the point of view of *realization* in a physical sense one is, however, forced to start from the system frequency behavior description rather than from the high-order linear differential equations in time domain.

The main aim of Chapter 12 is to present one new notion of *realization of behavior*. Further to the results of [31]. The input-output structure of the GCSRs and the DGCSRs are thus clarified by using this approach. Subsequently the corresponding autoregressive-moving-average (ARMA) representations are proposed and are proved to be realizations of behavior for any GCSR and for any DGCSR. Once these ARMA representations are proposed, one can further find the corresponding first-order system representations by using the method of Rosenthal and Schumacher [30] or other well-developed realization approaches such as Kailath [32]. These results are interesting in that they provide a good insight into the natural relationship between the (frequency) behavior of any GCSR and any DGCSR and the (dynamical) behavior of the corresponding ARMA representations. Since no numerical computation is involved in this approach, realization of behavior is particularly accessible for the situations in which the coefficients are symbolic rather than numerical.

The process of designing a control system generally involves many steps. There are many issues, most importantly modeling, which need to be considered before a controller is designed. The task in control system design is, therefore, not merely to design control systems for the known plants. It also involves investigating practical models. It is thus important to realize at the outset that some practical problems [33, 34] arising in engineering practice often provide rich information of the system *input signals*.

Modern $\mathcal{H}_\infty$ and $\mathcal{H}_2$ control theory are powerful tools for control system design. Zames [35] originally formulated the $\mathcal{H}_\infty$ optimal control problem in an input-output setting. The more recent $\mathcal{H}_\infty$ control theories ([36, 37], only document two instances) took the standard control feedback configuration (*model*), on which some standing assumptions are made, as their starting point. Such approaches, capture all the essential features of the general problem and the proposed theories are thus accessible to engineering practice on one hand, but involve some sacrifice of generality on the other hand. Such sacrifice of generality essentially result from the standing assumptions that are made on the system (*plant*) to achieve system *well-posedness* (WP) and IS. In some circumstances, these assumptions are not necessarily satisfied by the actual plants [38]. Such sacrifice of generality may also result from the impact of the input signal on the system control design not being included in the consideration.

In this case, it is interesting to consider the relaxations of the general requirements of WP and IS from the point of view of the impact of the input signal information on the system control design, to treat this issue explicitly and to give quantitative and qualitative results about it. This subject is addressed in this book. The input signal information is considered in analyzing the issues that are related to system WP and IS.

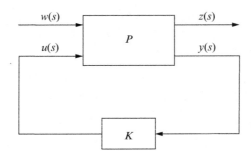

Fig. 1.1 Standard control system.

Consider the standard feedback configuration given in Fig. 1.1, where P is the plant and K is the *controller*. The plant accounts for all system components except for the controller. The signal w contains all external inputs, including disturbances, sensor noise, and commands. In some circumstances, the input signal w belongs to a set that can be characterized to some degree [39]. The output z is an error signal, y is the vector of measured variables, and u is the control input. All these signals $z, y, w,$ and u are *real rational* vectors in the complex variable $s = \lambda + j\omega$. The equations that describe the feedback configuration of Fig. 1.1 are

$$\begin{bmatrix} z \\ y \end{bmatrix} = P \begin{bmatrix} w \\ u \end{bmatrix},$$

$$u = Ky, \tag{1.5}$$

where P and K are *real rational* transfer matrices in the complex variable $s = \lambda + j\omega$. This means that only finite-dimensional linear time-invariant systems are considered. We write

$$P = \begin{bmatrix} P_{11} & P_{12} \\ P_{21} & P_{22} \end{bmatrix}, \tag{1.6}$$

where the partitioning is consistent with the dimensions of z, y and w, u. These will be denoted respectively by l_1, l_2 and m_1, m_2.

Roughly speaking, $\mathcal{H}_\infty$ control problem consists of finding a controller that makes the closed-loop system well-posed, have IS and minimizes the $\mathcal{H}_\infty$ norm of the closed-loop transfer function. The basic requirements in this problem are WP and IS [36, 39, 40]. The so-called WP requirement is that the matrix $(I - P_{22}K)$ (or $(I - KP_{22})$) be invertible, while IS guarantees bounded signals z, u, y for all bounded input signals w.

There are various reasons to consider relaxation of these two requirements in some circumstances; in other words, it is necessary to generalize these two classical notions to include some nonregular cases:

- The known approaches in $\mathcal{H}_\infty$ control theory either in the state space framework [37] or in the frequency domain framework [36] make a number of technical assumptions as to the plant, including certain rank conditions on P_{21} and P_{12}. These conditions, simplify the problems and make the approaches be more accessible in engineering practice on the one hand and limit the applicability of the standard problem on the other hand, due to the fact that these conditions are indeed irrelevant for control system design [38]. It was seen [38] that, in some actual situations, such as in the minimum sensitivity problem [35], the usual rank assumptions on P_{21} and P_{12} are not satisfied.

- In some practical circumstances, the *input signal w* belongs to a set that can be characterized to some degree [39]. It is frequently known in advance or is measurable on-line. It will be shown later in this book that the known information of the input signal opens possibility of relaxing the standing requirements of WP and IS.

- The $\mathcal{H}_\infty$ or $\mathcal{H}_2$ control problem in fact amounts to the constrained minimization problems. The constraints come from a WP requirement and an IS requirement and the object we seek to minimize is the $\mathcal{H}_\infty$ norm or $\mathcal{H}_2$ norm of the closed-loop transfer function. If these two requirements are relaxed, it will definitely enable us to search the appropriate controllers in a wider area; the resulting controllers might thus be more flexible in accomplishing the control aims. This need is indeed seen from the following situations. The fact that P_{12} and P_{21} do not need to have full rank at ∞, as usually required, allows the study of the celebrated minimum-sensitivity problem [35]. Also by relaxing the usual requirement that the entire feedback system be stable, the relaxed requirement that K stabilize P_{22} allows the inclusion unstable weighting filters (including such that have poles on the imaginary axis) in the mixed sensitivity problem [41].

- For some plants, there is no controller that can make the closed-loop systems have IS. There exist, however, controllers to make the input-output transfer function $T_{z,w}$ (from w to z) stable. For such plants, one may still be interested in designing a controller to achieve this input-output stability and also to accomplish other control aims. In this situation, one thus has to relax the IS requirement.

The main aim of Chapter 13 is to present a certain generalization to the classical concepts of WP and IS. The input consistency and output uniqueness of the closed-loop system in the standard control feedback configurations are investigated. Based on this, a number of notions are introduced such as *fully internal well-posedness* (FIWP), *externally internal well-posedness* (EIWP), and *externally internal stability* (EIS), which characterize the rich input-output and stability features of the general control systems in a general setting. It is shown that, FIWP is equivalent to the classical assumption WP, which has been widely adopted in control system designs, while EIWP and EIS generalize the notions of WP and IS. Some conditions are established to verify EIWP by using the approach of the generalized matrix inverses. Natural links between EIWP and WP, EIS and IS are pointed out. This approach also leads to a generalization of the *linear fractional transformation* (LFT). The *generalized linear fractional transformations* (GLFTs) are formulated into some general parameterized forms by using the generalized inverse of matrices. Finally, on the basis of these notions of EIWP, EIS, and GLFT, the extended $\mathcal{H}_\infty$ control problem is defined in a general setting.

The nonstandard H_∞ control problem is the case where the direct feedthroughs from the input to the error and from the exogenous signal to the output are not necessarily of full rank. This problem is reformulated based on the generalized chain-scattering representation (GCSR) in Chapter 14. The GCSR approach leads naturally to a generalization of the homographic transformation. The state-space realization for this generalized homographic transformation and a number of fundamental cascade structures of the H_∞ control systems are further studied in a unified framework of GCSR. Certain sufficient conditions for the solvability of the nonstandard H_∞ control problem are therefore established via a (J, J')-lossless factorization of GCSR. These results present extensions to Kimura's results on the CSR approach to the H_∞ control in the standard case.

The working mechanism of today's Internet is a feedback system, with the responsibility of managing the allocation of bandwidth resources between competing traffic flows and control the congestion. Congestion control mechanisms in the Internet represent one of the largest deployed artificial feedback systems [42]. The main aim of Chapter 15 is to give an outline of the Internet congestion control from the perspective of linear multivariable control system. We bring this subject into a control domain by looking at the transferring relations between the dynamics of individual flow rates and the dynamics of aggregate flow rates, the transferring relations between link price and flow aggregate price from a multivariable control point of views including the Smith-McMillan forms. Finally we look at the issue of feedback control design to relate the TCP source flow rate with the aggregate flow price, which is in a proportional (P) controller structure. These analyses in the flow level then constitute a theoretical basis for TCP protocol design in the packet level.

We now summarize the key components of the current book: Chapters 2 through 5 covers the mathematical preliminaries and the basic background in linear control systems. Chapter 6 presents an introduction to such fundamental theories on control system structure and behavior as the PMD theory, the CSR approaches, the behavioral theory, and $\mathcal{H}_\infty$ control theory. Chapter 7 gives a novel method to determine the finite and infinite frequency structure of a rational matrix. Chapters 8–10 are devoted to the resolvent decomposition of a regular polynomial matrix and the solution of regular PMDs. A generalization to the CSR for general plants is presented in Chapter 11. A new notion realization of behavior and its applications are introduced in Chapter 12. Some related extensions to system well-posed-ness and internal stability in $\mathcal{H}_\infty$ control theory are represented in Chapter 13. Chapter 14 defines and discusses the nonstandard H_∞ control problem from a GCSR approach perspective and its solution. Chapter 14 presents applications of linear multivariable control in Internet congestion control. Finally Chapter 16 gives conclusions and some topics for further research.

Mathematical preliminaries

![2](chapter number 2)

In this chapter, we briefly discuss some basic topics in linear algebra that may be needed in the sequel. This chapter can be skimmed very quickly and used mainly as a quick reference. There has been a huge number of reference resources for matrix and linear algebra, to mention a few [43–50].

2.1 Vector algebra

If one measures a physical quantity, one will obtain a number or a sequence of numbers in a particular order. This is the so-called scalar, which is a single number quantity, and one can use it to gauge the magnitude or the value of the measurement. Scalars can be compared only if they have the same units. For example, we can compare a speed in km/h with another speed in km/h but we cannot compare a speed with a density because they are in different units.

An array is defined as a sequence of scalar numbers in a certain order. A rectangular table, which consists of an array of arrays, that are arranged in terms of a certain rule, is called a matrix. A matrix consists of m rows and n columns. For instance, the following is a 3×4 real matrix:

$$A = \begin{bmatrix} -1.5 & 0 & 13 & 7 \\ 6 & 13 & 7.8 & 0 \\ 1 & 4 & 6 & 76 \end{bmatrix}.$$

A special case of a matrix where it has only one row or one column is called a vector. Before we present the definition of a vector, let us take a look from how we distinct an array and a vector from computational point of view. A vector is usually used as a quantity that requires both magnitude and direction in space. Examples of vector can be displacement, velocity, acceleration, force, momentum, electric field, the traffic load in links of Internet, and so on. Vectors can be compared if they have the same physical units and geometrical dimensions. For instance, you can compare two-dimensional force with another two-dimensional force but you cannot compare two-dimensional force with three-dimensional force. Similarly, you cannot compare force with velocity because they have different physical units. You cannot compare the routing matrix with a traffic matrix in Internet because they are of different philosophy.

The total number of elements in a vector is called the dimension or the size of the vector. Since a vector can have a number of elements, we say that the space where the vector lies is a multidimensional space with the specific dimensions.

A Generalized Framework of Linear Multivariable Control. http://dx.doi.org/10.1016/B978-0-08-101946-7.00002-0

The magnitude of a vector is sometimes called the length of a vector, or norm of a vector. The direction of a vector in space is measured from another vector (i.e., standard basis vector), represented by the cosine angle between the two vectors.

Unlike ordinary arrays, vectors are special and can be viewed from both an algebraic and a geometric point of view. Algebraically, a vector is just an array of scalar elements. From a computational point of view, a vector is represented as a two-dimensional array while an ordinary array is represented as a one-dimensional array.

There are rich implications for vectors in geometry. A vector can also be represented as a point in space. When a vector is represented as a point in space, we can also view that vector as an arrow starting at the origin of a coordinate system pointing toward the destination point. Because the coordinate system does not change, we only need to draw the points without the arrows. In a geometrical diagram, a vector is usually plotted as an arrow in space, which can be translated along the line containing it. Subsequently, the vector can also be applied to any point in space if the magnitude and direction of both do not change.

2.2 Matrix algebra

2.2.1 Matrix properties

Determinant: Each $n \times n$ square matrix A has an associated scalar number, which is termed as the determinant of that matrix and is noted either by $\det(A)$ or by $|A|$. This value determines whether the matrix has an inverse or not.

When the matrix order is 1, That is, $A_{1 \times 1} = [a]$ then we calculate its determinant as $|A| = a$.

When the matrix order is 2, that is

$$A_{2 \times 2} = \begin{bmatrix} a_1 & b_1 \\ a_2 & b_2 \end{bmatrix},$$

then we have

$$|A| = \begin{vmatrix} a_1 & b_1 \\ a_2 & b_2 \end{vmatrix} = a_1 b_2 - a_2 b_1.$$

When the matrix order is 3,

$$A_{3 \times 3} = \begin{bmatrix} a_1 & b_1 & c_1 \\ a_2 & b_2 & c_2 \\ a_3 & b_3 & c_3 \end{bmatrix},$$

then we compute the determinant by the following formula:

$$|A| = \begin{vmatrix} a_1 & b_1 & c_1 \\ a_2 & b_2 & c_2 \\ a_3 & b_3 & c_3 \end{vmatrix} = a_1 \begin{vmatrix} b_2 & c_2 \\ b_3 & c_3 \end{vmatrix} - a_2 \begin{vmatrix} b_1 & c_1 \\ b_3 & c_3 \end{vmatrix} + a_3 \begin{vmatrix} b_1 & c_1 \\ b_2 & c_2 \end{vmatrix}.$$

In general, when we have the following general order of matrix $A_{n \times n} = [a_{ij}]$, then we define the following components:

- The minor of $[a_{ij}]$ is the determinant of an $(n-1) \times (n-1)$ submatrix denoted by M_{ij}. The submatrix M_{ij} is then obtained from the matrix $A_{n \times n}$ by deleting the row and column containing the element a_{ij}.
- The signed minor is called cofactor of a_{ij} denoted by A_{ij}. The cofactor is a scalar sign being determined by $A_{ij} = (-1)^{i+j} \det(M_{ij})$. This sign follows the following order:

$$\begin{bmatrix} + & - & + & - & + & \cdots \\ - & + & - & + & - & \cdots \\ + & - & + & - & + & \cdots \\ \vdots & \vdots & \vdots & \vdots & \vdots & \ddots \end{bmatrix};$$

- The determinant of matrix $A_{n \times n} = [a_{ij}]$ is the summed product of the first column entries with its cofactor. That is,

$$|A| = a_{11}A_{11} + a_{12}A_{12} + \cdots + a_{1n}A_{1n}.$$

Matrix trace: Trace of a square matrix is the summation of matrix diagonal entries $Tr(A) = \sum_{j=1}^{n} a_{jj}$. Trace of a matrix is also called spur of a square matrix. For square matrices A and B, it is true that

$$Tr(A) = Tr(A^T),$$
$$Tr(A + B) = Tr(A) + Tr(B),$$
$$Tr(\alpha \times A) = \alpha \times Tr(A),$$

where A^T denotes the transpose. The trace is also invariant under a similarity transformation, i.e., if

$$A_1 = BAB^{-1},$$

we have

$$Tr(A_1) = Tr(A).$$

The trace of a product of two square matrices is independent of the order of the multiplication. That is, we have

$$Tr(AB) = Tr(BA).$$

Therefore, the trace of the commutator of A and B is given by

$$Tr([A, B]) = Tr(AB) - Tr(BA) = 0.$$

The trace of a product of three or more square matrices, on the other hand, is invariant only under cyclic permutations of the order of multiplication of the matrices, by a similar argument. The product of a symmetric and an antisymmetric matrix has zero trace, i.e.,

$$Tr(A_S B_A) = 0.$$

Matrix rank: The dimension of row space of a matrix is equal to the dimension of the column space of that matrix. The common dimension of the row space and column space of matrix A is defined as the rank of matrix A, denoted by $\text{rank}(A) = R(A)$. It is seen that the rank of a matrix is the number of linearly independent vectors and equal to the dimension of space spanned by those vectors. To put it in another way, the rank of A is the number of basis vectors we can form from matrix A, because performing an elementary row operation does not change the row space. To compute the rank of a matrix, we usually perform manipulations on the matrix to bring it into the so-called reduced row echelon form (RREF, also called row canonical form) and subsequently the nonzero rows of the RREF matrix will form the basis for the row space.

A matrix is in RREF if it satisfies the following conditions:

- It is in row echelon form (REF).
- Every leading coefficient is 1 and is the only nonzero entry in its column.

The RREF of a matrix may be computed by Gauss-Jordan elimination. Unlike the REF, the RREF of a matrix is unique and does not depend on the algorithm used to compute it.

The following matrix is an example of one in RREF:

$$\begin{bmatrix} 1 & 0 & a_1 & 0 & b_1 \\ 0 & 1 & a_2 & 0 & b_2 \\ 0 & 0 & 0 & 1 & b_3 \end{bmatrix}.$$

Note that this does not always mean that the left of the matrix will be an identity matrix, as this example shows.

For matrices with integer coefficients, the Hermite normal form is an REF that may be calculated using Euclidean division and without introducing any rational number or denominator. On the other hand, the reduced echelon form of a matrix with integer coefficients generally contains noninteger entries.

We assume that A is an $m \times n$ matrix, and we define the linear map f by $f(x) = Ax$. The rank of an $m \times n$ matrix is a nonnegative integer and cannot be greater than either m or n, that is,

$$\text{rank}(A) \leq \min(m, n).$$

A matrix that has rank $\min(m, n)$ is said to have full rank; otherwise, the matrix is rank deficient. Note that only a zero matrix has rank zero. f is injective (or "one-to-one") if and only if A has rank n (in this case, we say that A has full column rank), while f is surjective (or "onto") if and only if A has rank m (in this case, we say that A has full row rank). If A is a square matrix (i.e., $m = n$), then A is invertible if and only if A has rank n (i.e., A has full rank). If B is any $n \times k$ matrix, then

$$\text{rank}(AB) \leq \min(\text{rank}(A), \text{rank}(B)).$$

If B is an $n \times k$ matrix of rank n, then

$$\text{rank}(AB) = \text{rank}(A).$$

If C is an $l \times m$ matrix of rank m, then

$$\text{rank}(CA) = \text{rank}(A).$$

The rank of A is equal to r if and only if there exists an invertible $m \times m$ matrix X and an invertible $n \times n$ matrix Y such that

$$XAY = \begin{bmatrix} I_r & 0 \\ 0 & 0 \end{bmatrix},$$

where I_r denotes the $r \times r$ identity matrix.

Sylvesters rank inequality: if A is an $m \times n$ matrix and B is $n \times k$, then

$$\text{rank}(A) + \text{rank}(B) - n \leq \text{rank}(AB).$$

This is a special case of the next inequality. The inequality due to Frobenius: if AB, ABC, and BC are defined, then

$$\text{rank}(AB) + \text{rank}(BC) \leq \text{rank}(B) + \text{rank}(ABC).$$

Subadditivity:

$$\text{rank}(A + B) \leq \text{rank}(A) + \text{rank}(B),$$

when A and B are of the same dimension. As a consequence, a rank-k matrix can be written as the sum of k rank-1 matrices, but not fewer. The rank of a matrix plus the nullity of the matrix equals to the number of columns of the matrix. (This is the well-known rank-nullity theorem.) If A is a matrix over the real numbers then the rank of A and the rank of its corresponding Gram matrix are equal. Thus, for real matrices

$$\text{rank}(A^T A) = \text{rank}(AA^T) = \text{rank}(A) = \text{rank}(A^T).$$

This can be shown by proving equality of their null spaces. The null space of the Gram matrix is given by vectors x for which $A^T A x = 0$. If this condition is fulfilled, we also have $0 = x^T A^T A x = |Ax|^2$. If A is a matrix over the complex numbers and A^* denotes the conjugate transpose of A, then one has

$$\text{rank}(A) = \text{rank}(\overline{A}) = \text{rank}(A^T) = \text{rank}(A^*) = \text{rank}(A^* A) = \text{rank}(AA^*).$$

2.2.2 Basic matrix operations

Matrix transpose: Transpose of a matrix is obtained by interchanging all rows with all columns of the matrix. If the matrix size is $m \times n$, the transpose matrix size is $n \times m$.

Matrix addition: If two matrices are of the same order, we then take the summation of any pair of elements, which are at the same position in a peer-to-peer manner in these two matrices to yield the summation. Namely, for two matrices A and B, then the sum of the two matrices $C = A + B$ is a matrix whose entry is equal to the sum of the corresponding entries $c_{ij} = a_{ij} + b_{ij}$.

Matrix subtraction: When we have two matrices, A and B then the matrix subtraction or matrix difference $C = A - B$ is a matrix whose entry is formed by subtracting the corresponding entry of B from each entry of A, that is $c_{ij} = a_{ij} - b_{ij}$.

Matrix subtraction works only when the two matrices have the same size and the result is also the same size as the original matrices.

Matrix multiplication: One of the most important matrix operations is matrix multiplication or matrix product. The multiplication of two matrices A and B is a matrix $C = A \times B$ whose element c_{ij} consists of the vector inner product of the ith row of matrix A and the jth column of matrix B, that is $c_{ij} = \sum_{k=1}^{m} a_{ik} b_{kj}$. Matrix multiplication can be done only when the number of column of A is equal to the number of rows of B. If the size of matrix A is $m \times n$ and the size of matrix B is $n \times p$, then the result of matrix multiplication is a matrix size m by p, or in short $A_{m \times n} B_{n \times p} = C_{m \times p}$.

Matrix scalar multiple: When we multiply a real number k with a matrix A, we have the scalar multiple of A. The scalar multiple matrix $B = kA$ has the same size of the matrix A and it is obtained by multiplying each element of A by k. That is $b_{ij} = ka_{ij}$.

Hadamard product: Matrix element-wise product is also called Hadamard product or direct product. It is a direct element by element multiplication. If matrix $C = AB$ is the direct product two matrices A and B, then the element of Hadamard product is simply $c_{ij} = a_{ij} \times b_{ij}$. The direct product has the same size as the original matrices.

Horizontal concatenation: It is often useful to think of a matrix composed of two or more submatrices or a block of matrices. A matrix can be partitioned into smaller matrices by setting a horizontal line or a vertical line. For example

$$A = \begin{pmatrix} 1 & 2 & 3 & 4 \\ 5 & 6 & 7 & 8 \\ 9 & 0 & 1 & 2 \end{pmatrix} = \begin{bmatrix} A_{11} & A_{12} \\ A_{21} & A_{22} \end{bmatrix}.$$

Matrix horizontal concatenation is an operation to join two sub matrices horizontally into one matrix. This is the reverse operation of partitioning.

Vertical concatenation: Matrix vertical concatenation is an operation to join two sub matrices vertically into one matrix.

Elementary row operation: In linear algebra, there are three elementary row operations. The same operations can also be used for column (simply by changing the word row into column). The elementary row operations can be applied to a rectangular matrix size $m \times n$:

1. interchanging two rows of the matrix;
2. multiplying a row of the matrix by a scalar;
3. add a scalar multiple of a row to another row.

Applying the above three elementary row operations to a matrix will produce a row equivalent matrix. When we view a matrix as the augmented matrix of a linear system, the three elementary row operations are equivalent to interchanging two equations, multiplying an equation by a nonzero constant and adding a scalar multiple of one equation to another equation. Two linear systems are equivalent if they produce the same set of solutions. Since a matrix can be seen as a linear system, applying the above three elementary row operations does not change the solutions of that matrix. The three elementary row operations can be put into three elementary matrices. Elementary matrix is a matrix formed by performing a single elementary row operation on an identity matrix. Multiplying the elementary matrix to a matrix will produce the row equivalent matrix based on the corresponding elementary row operation.

Elementary Matrix Type 1: Interchanging two rows of the matrix

$$r_{12} = E_1 = \begin{bmatrix} 0 & 1 & 0 \\ 1 & 0 & 0 \\ 0 & 0 & 1 \end{bmatrix}, \ r_{13} = E_1 = \begin{bmatrix} 0 & 0 & 1 \\ 0 & 1 & 0 \\ 1 & 0 & 0 \end{bmatrix}, \ r_{23} = E_1 = \begin{bmatrix} 1 & 0 & 0 \\ 0 & 0 & 1 \\ 0 & 1 & 0 \end{bmatrix}.$$

Elementary Matrix Type 2: Multiplying a row of the matrix by a scalar

$$r_1(k) = E_2 = \begin{bmatrix} k & 0 & 0 \\ 0 & 1 & 0 \\ 0 & 0 & 1 \end{bmatrix}, \ r_2(k) = E_2 = \begin{bmatrix} 1 & 0 & 0 \\ 0 & k & 0 \\ 0 & 0 & 1 \end{bmatrix}, \ r_3(k) = E_2 = \begin{bmatrix} 1 & 0 & 0 \\ 0 & 1 & 0 \\ 0 & 0 & k \end{bmatrix}.$$

Elementary Matrix Type 3: Add a scalar multiple of a row to another row

$$r_{12}(k) = E_3 = \begin{bmatrix} 1 & k & 0 \\ 0 & 1 & 0 \\ 0 & 0 & 1 \end{bmatrix}, \ r_{13}(k) = E_3 = \begin{bmatrix} 1 & 0 & k \\ 0 & 1 & 0 \\ 0 & 0 & 1 \end{bmatrix}, \ r_{23}(k) = E_3 = \begin{bmatrix} 1 & 0 & 0 \\ 0 & 1 & k \\ 0 & 0 & 1 \end{bmatrix},$$

$$r_{21}(k) = E_3 = \begin{bmatrix} 1 & 0 & 0 \\ k & 1 & 0 \\ 0 & 0 & 1 \end{bmatrix}, \ r_{31}(k) = E_3 = \begin{bmatrix} 1 & 0 & 0 \\ 0 & 1 & 0 \\ k & 0 & 1 \end{bmatrix}, \ r_{32}(k) = E_3 = \begin{bmatrix} 1 & 0 & 0 \\ 0 & 1 & 0 \\ 0 & k & 1 \end{bmatrix}.$$

Reduced row echelon form: There is a standard form of a row equivalent matrix and if we do a sequence of row elementary operations to reach this standard form we may gain the solution of the linear system. The standard form is called RREF of a matrix, or matrix RREF in short. An $m \times n$ matrix is called to be in an RREF when it satisfies the following conditions:

1. All zero rows, if any, are at the bottom of the matrix.
2. Reading from left to right, the first non zero entry in each row that does not consist entirely of zeros is a 1, called the leading entry of its row.
3. If two successive rows do not consist entirely of zeros, the second row starts with more zeros than the first (the leading entry of second row is to the right of the leading entry of first row).
4. All other elements of the column in which the leading entry 1 occurs are zeros.

When only the first three conditions are satisfied, the matrix is called in Row Echelon Form (REF). Using Reduced Row Echelon Form of a matrix we can calculate matrix inverse, rank of matrix, and solve simultaneous linear equations.

Finding inverse using RREF (Gauss-Jordan): We can use the three elementary row operations to find the row equivalent of RREF of a matrix. Using the Gauss-Jordan method, you can obtain matrix inverse. Gauss-Jordan method is based on the following theorem: A square matrix is invertible if and only if it is row equivalent to the identity matrix. The method to find inverse using the Gauss-Jordan method is as follows:

• we concatenate the original matrix with the identity matrix;
• perform the row elementary operations to reach RREF;

- if the left part of the matrix RREF is equal to an identity matrix, then the left part is the inverse matrix;
- if the left part of the matrix RREF is not equal to an identity matrix, then we conclude the original matrix is singular. It has no inverse.

Finding a matrix's rank using RREF: Another application of elementary row operations to find the row equivalent of RREF of the matrix is to find matrix rank. Similar to trace and determinant, the rank of a matrix is a scalar number showing the number of linearly independent vectors in a matrix, or the order of the largest square submatrix of the original matrix whose determinant is nonzero. To compute the rank of a matrix through elementary row operations, simply perform the elementary row operations until the matrix reach the RREF. Then the number of nonzero rows indicates the rank of the matrix.

Singular matrix: A square matrix is singular if the matrix has no inverse. To determine whether a matrix is singular or not, we simply compute the determinant of the matrix. If the determinant is zero, then the matrix is singular.

2.3   Matrix inverse

When we are dealing with ordinary number, when we say $ab = 1$ then we can obtain $b = 1/a$ as long as $b \neq 0$. We write it as $b = a^{-1}$ or $aa^{-1} = a^{-1}a = 1$. Matrix inverse is similar to division operation in ordinary numbers. Suppose we have matrix multiplication such that the result of matrix product is an identity matrix $AB = I$. If such matrix B exists, then that matrix is unique and we can write $B = A^{-1}$ or we can also write $AA^{-1} = I = A^{-1}A$.

Matrix inverse exists only for a square matrix (that is a matrix having the same number of rows and columns). Unfortunately, matrix inverse does not always exist. Thus, we term that a square matrix is singular if that matrix does not have an inverse, it is called nonregular matrix as well. Because matrix inverse is a very important operation, in linear algebra, there are many ways to compute matrix inverse:

1. The simplest way to find matrix inverse for a small matrix (order 2 or 3) is to use Cramers rule that employs the determinant of the original matrix. Recall that a square matrix is singular (i.e., no inverse) if and only if the determinant is zero. The matrix inverse can be computed by scaling the adjoint of the original matrix with the determinant. That is,

$$A^{-1} = \frac{adj(A)}{|A|}.$$

The adjoint of a matrix is the one whose element is the cofactor of the original matrix. For a 2 by 2 matrix

$$A = \begin{bmatrix} a & b \\ c & d \end{bmatrix},$$

we have

$$A^{-1} = \frac{1}{ad - bc} \begin{bmatrix} d & -b \\ -c & a \end{bmatrix}.$$

The determinant is set by multiplying the diagonal elements minus the product of the off-diagonal elements and the adjoint is set by reversing the diagonal elements and taking the minus sign of the off diagonal elements.

2. For a matrix of medium size (order 4–10), the usage of elementary row operations to produce RREF matrix using Gaussian Elimination or Gauss-Jordan is useful.

3. Other techniques to compute matrix inverse of medium to large size are to use numerical methods such as LU decomposition, singular value decomposition (SVD), or the Monte Carlo method.

4. For a large square matrix (order more than 100), numerical techniques such as the Gauss-Siedel, Jacobi method, Newton's method, the Cayley-Hamilton method, eigenvalue decomposition and Cholesky decomposition are used to calculate the matrix inverse accurately or approximately.

5. For a partitioned block matrix with 2×2 blocks, we have some specific method to compute its inverse, as outlined in the sequel.

We briefly introduce the following specific methods for computing the inverse of a matrix:

Gaussian elimination: Gauss-Jordan elimination is an algorithm that can be used to determine whether a given matrix is invertible and to find the inverse. An alternative is the *LU* decomposition, which generates upper and lower triangular matrices that are easier to invert.

Newton's method: Newton's method as used for a multiplicative inverse algorithm may be convenient, if it is possible to find a suitable starting seed such that

$$X_{k+1} = 2X_k - X_k A X_k.$$

Newton's method is particularly useful when dealing with families of related matrices that behave enough like the sequence manufactured for the homotopic above: sometimes a good starting point for refining an approximation for the new inverse can be the already obtained inverse of a previous matrix that nearly matches the current matrix, e.g., the pair of sequences of inverse matrices used in obtaining matrix square roots by Denman-Beavers iteration; this may need more than one pass of the iteration at each new matrix, if they are not close enough together for just one to be enough. Newton's method is also useful for making up corrections to an Gauss-Jordan algorithm that has been contaminated by small errors due to imperfect computer arithmetic.

Cayley-Hamilton method: The Cayley-Hamilton theorem allows us to represent the inverse of A in terms of $\det(A)$, traces and powers of A. Particularly, for a matrix A, we have its inverse formulated as follows:

$$\mathbf{A}^{-1} = \frac{1}{\det(\mathbf{A})} \sum_{s=0}^{n-1} \mathbf{A}^s \sum_{k_1,k_2,\ldots,k_{n-1}} \prod_{l=1}^{n-1} \frac{(-1)^{k_l+1}}{l^{k_l} k_l!} tr(\mathbf{A}^l)^{k_l},$$

where n is the dimension of A, and $tr(A)$ is the trace of matrix A given by the sum of the main diagonal. The sum is taken over s and the sets of all $k_l \geq 0$ satisfying the following linear Diophantine equation:

$$s + \sum_{l=1}^{n-1} l k_l = n - 1.$$

The formula can be rewritten in terms of the complete Bell polynomials of arguments $t_l = -(l-1)!\,\mathrm{tr}(A^l)$ as

$$\mathbf{A}^{-1} = \frac{1}{\det(\mathbf{A})} \sum_{s=0}^{n-1} \mathbf{A}^s \frac{(-1)^{n-1}}{(n-1-s)!} B_{n-1-s}(t_1, t_2, \ldots, t_{n-1-s}).$$

Eigenvalue decomposition: If matrix A can be eigenvalue decomposed and if none of its eigenvalues are zero, then A is invertible and its inverse is given by

$$\mathbf{A}^{-1} = \mathbf{Q}\mathbf{\Lambda}^{-1}\mathbf{Q}^{-1},$$

where Q is the square $(N \times N)$ matrix whose ith column is the eigenvector q_i of A and $\Lambda = [\Lambda_{ii}]$ is the diagonal matrix whose diagonal elements are the corresponding eigenvalues, i.e., $\Lambda_{ii} = \lambda_i$. Furthermore, because Λ is a diagonal matrix, its inverse is easy to calculate in the following manner:

$$\Lambda^{-1} = \frac{1}{\lambda_i}.$$

Cholesky decomposition: If matrix A is positive definite, then its inverse can be obtained as

$$\mathbf{A}^{-1} = (\mathbf{L}^*)^{-1}\mathbf{L}^{-1},$$

where L is the lower triangular Cholesky decomposition of A, and L^* denotes the conjugate transpose of L.

Block-wise inversion: Matrices can also be inverted blockwise by using the following analytic inversion formula:

$$\begin{bmatrix} \mathbf{A} & \mathbf{B} \\ \mathbf{C} & \mathbf{D} \end{bmatrix}^{-1} = \begin{bmatrix} \mathbf{A}^{-1} + \mathbf{A}^{-1}\mathbf{B}(\mathbf{D} - \mathbf{C}\mathbf{A}^{-1}\mathbf{B})^{-1}\mathbf{C}\mathbf{A}^{-1} & -\mathbf{A}^{-1}\mathbf{B}(\mathbf{D}-\mathbf{C}\mathbf{A}^{-1}\mathbf{B})^{-1} \\ -(\mathbf{D} - \mathbf{C}\mathbf{A}^{-1}\mathbf{B})^{-1}\mathbf{C}\mathbf{A}^{-1} & (\mathbf{D} - \mathbf{C}\mathbf{A}^{-1}\mathbf{B})^{-1} \end{bmatrix},$$

$$(2.1)$$

where A, B, C, and D are block matrices with appropriate sizes. A must be square, so that it can be inverted. Furthermore, A and $D - CA^{-1}B$ must be nonsingular. By this method to compute the block matrix is particularly appealing if A is diagonal and $D - CA^{-1}B$ (the Schur complement of A) is a small matrix, since they are the only matrices requiring inversion.

The nullity theorem says that the nullity of A equals the nullity of the subblock in the lower right of the inverse matrix, whilst the nullity of B equals the nullity of the subblock in the upper right of the inverse matrix.

The inversion procedure leading to Eq. (2.1) is performed in the matrix blocks that is operated on C and D first. Instead, if A and B are operated on first, and provided that D and $A - BD^{-1}C$ are nonsingular, the result is

$$\begin{bmatrix} \mathbf{A} & \mathbf{B} \\ \mathbf{C} & \mathbf{D} \end{bmatrix}^{-1} = \begin{bmatrix} (\mathbf{A} - \mathbf{B}\mathbf{D}^{-1}\mathbf{C})^{-1} & -(\mathbf{A} - \mathbf{B}\mathbf{D}^{-1}\mathbf{C})^{-1}\mathbf{B}\mathbf{D}^{-1} \\ -\mathbf{D}^{-1}\mathbf{C}(\mathbf{A} - \mathbf{B}\mathbf{D}^{-1}\mathbf{C})^{-1} & \mathbf{D}^{-1} + \mathbf{D}^{-1}\mathbf{C}(\mathbf{A} - \mathbf{B}\mathbf{D}^{-1}\mathbf{C})^{-1}\mathbf{B}\mathbf{D}^{-1} \end{bmatrix}.$$

$$(2.2)$$

Equating Eqs. (2.1), (2.2) leads to

$$(\mathbf{A} - \mathbf{BD}^{-1}\mathbf{C})^{-1} = \mathbf{A}^{-1} + \mathbf{A}^{-1}\mathbf{B}(\mathbf{D} - \mathbf{CA}^{-1}\mathbf{B})^{-1}\mathbf{CA}^{-1}, \tag{2.3}$$

which is the Woodbury matrix identity, being equivalent to the binomial inverse theorem, and

$$(\mathbf{A} - \mathbf{BD}^{-1}\mathbf{C})^{-1}\mathbf{BD}^{-1} = \mathbf{A}^{-1}\mathbf{B}(\mathbf{D} - \mathbf{CA}^{-1}\mathbf{B})^{-1}$$
$$\mathbf{D}^{-1}\mathbf{C}(\mathbf{A} - \mathbf{BD}^{-1}\mathbf{C})^{-1} = (\mathbf{D} - \mathbf{CA}^{-1}\mathbf{B})^{-1}\mathbf{CA}^{-1}$$
$$\mathbf{D}^{-1} + \mathbf{D}^{-1}\mathbf{C}(\mathbf{A} - \mathbf{BD}^{-1}\mathbf{C})^{-1}\mathbf{BD}^{-1} = (\mathbf{D} - \mathbf{CA}^{-1}\mathbf{B})^{-1}.$$

Neumann series: If a matrix A has the property that

$$\lim_{n\to\infty} (\mathbf{I} - \mathbf{A})^n = 0,$$

then A is nonsingular and its inverse can be expressed by a Neumann series:

$$\mathbf{A}^{-1} = \sum_{n=0}^{\infty} (\mathbf{I} - \mathbf{A})^n.$$

Truncating the sum results in an approximate inverse that may be useful as a preconditioner. Note that a truncated series can be accelerated exponentially by noting that the Neumann series is a geometric sum. As such, it satisfies

$$\sum_{n=0}^{2^L-1} (\mathbf{I} - \mathbf{A})^n = \prod_{l=0}^{L-1} (\mathbf{I} + (\mathbf{I} - \mathbf{A})^{2^l}).$$

Therefore, only $2L - 2$ matrix multiplications are needed to compute 2^L terms of the summation. More generally, if A is close to the invertible matrix X in the sense that

$$\lim_{n\to\infty} (\mathbf{I} - \mathbf{X}^{-1}\mathbf{A})^n = 0 \quad \text{or} \quad \lim_{n\to\infty} (\mathbf{I} - \mathbf{AX}^{-1})^n = 0,$$

then A is nonsingular and its inverse is

$$\mathbf{A}^{-1} = \sum_{n=0}^{\infty} \left(\mathbf{X}^{-1}(\mathbf{X} - \mathbf{A})\right)^n \mathbf{X}^{-1}.$$

If it is also the case that $A - X$ has rank 1 then this is reduced to

$$\mathbf{A}^{-1} = \mathbf{X}^{-1} - \frac{\mathbf{X}^{-1}(\mathbf{A} - \mathbf{X})\mathbf{X}^{-1}}{1 + \text{tr}(\mathbf{X}^{-1}(\mathbf{A} - \mathbf{X}))}.$$

Derivative of the matrix inverse: Suppose that the invertible matrix A depends on a parameter t. Then the derivative of the inverse of A with respect to t is given by

$$\frac{d\mathbf{A}^{-1}(t)}{dt} = -\mathbf{A}^{-1}(t)\frac{d\mathbf{A}(t)}{dt}\mathbf{A}^{-1}(t).$$

Similarly, if ϵ is a small number then

$$(\mathbf{A} + \epsilon \mathbf{X})^{-1} = \mathbf{A}^{-1} - \epsilon \mathbf{A}^{-1} \mathbf{X} \mathbf{A}^{-1} + \mathcal{O}(\epsilon^2).$$

Schur decomposition: The Schur decomposition reads as follows: if A is a $n \times n$ square matrix with complex entries, then A can be expressed as

$$A = QUQ^{-1},$$

where Q is a unitary matrix (so that its inverse Q^{-1} is also the conjugate transpose $Q*$ of Q), and U is an upper triangular matrix, which is called a Schur form of A. Since U is similar to A, it has the same multiset of eigenvalues, and since it is triangular, those eigenvalues are the diagonal entries of U.

The Schur decomposition implies that there exists a nested sequence of A-invariant subspaces $\{0\} = V_0 \subset V_1 \subset \cdots \subset V_n = C^n$, and that there exists an ordered orthogonal basis (for the standard Hermitian form of C^n) such that the first i basis vectors span V_i for each i occurring in the nested sequence. In particular, a linear operator J on a complex finite-dimensional vector space stabilizes a complete flag $(V_1, \ldots, V_n)$.

Generalized Schur decomposition: Given square matrices A and B, the generalized Schur decomposition factorizes both matrices as $A = QSZ^*$ and $B = QTZ^*$, where Q and Z are unitary, and S and T are upper triangular. The generalized Schur decomposition is also sometimes called the QZ decomposition.

The generalized eigenvalues λ that solve the generalized eigenvalue problem $Ax = \lambda Bx$ (where x is an unknown nonzero vector) can be calculated as the ratio of the diagonal elements of S to those of T. That is, using subscripts to denote matrix elements, the ith generalized eigenvalue λ_i satisfies $\lambda_i = S_{ii}/T_{ii}$.

Properties of matrix inverse: Some important properties of matrix inverse are:

- If A and B are square matrices with the order n and their product produces an identity matrix, i.e., $A \times B = I_n = B \times A$, then we have $B = A^{-1}$.
- If a square matrix A has an inverse (nonsingular), then the inverse matrix is unique.
- A square matrix A has an inverse matrix if and only if the determinant is not zero: $|A| \neq 0$. Similarly, the matrix A is singular (has no inverse) if and only if its determinant is zero: $|A| = 0$.
- A square matrix A with order n has an inverse matrix if and only if the rank of this matrix is full, that is rank$(A) = n$.
- If a square matrix A has an inverse, the determinant of an inverse matrix is the reciprocal of the matrix determinant. That is $|A^{-1}| = \frac{1}{|A|}$.
- If a square matrix A has an inverse, for a scalar $k \neq 0$ then the inverse of a scalar multiple is equal to the product of their inverse. That is $(kA)^{-1} = \frac{1}{k}A^{-1}$.
- If a square matrix A has an inverse, the transpose of an inverse matrix is equal to the inverse of the transposed matrix. That is $(A^{-1})^T = (A^T)^{-1}$.
- If A and B are square nonsingular matrices both with the order n then the inverse of their product is equal to the product of their inverse in reverse order. That is $(AB)^{-1} = B^{-1}A^{-1}$.
- Let A and B are square matrices with the order n. If $A \times B = 0$ then either $A = 0$ or $B = 0$ or both A and B are singular matrices with no inverse.

2.4   Solving system of linear equation

2.4.1   Gauss method

A linear equation in variables $x_1, x_2, \ldots, x_n$ has the form

$$a_1 x_1 + a_2 x_2 + a_3 x_3 + \cdots + a_n x_n = d,$$

where the numbers $a_1, \ldots, a_n \in \mathbb{R}$ are the equation's coefficients and $d \in \mathbb{R}$ is a constant. An n-tuple $(s_1, s_2, \ldots, s_n) \in \mathbb{R}^n$ is a solution of that equation if substituting the numbers $s_1, s_2, \ldots, s_n$ for the variables gives a true statement:

$$a_1 s_1 + a_2 s_2 + a_3 s_3 + \cdots + a_n s_n = d.$$

A system of linear equations

$$a_{1,1} x_1 + a_{1,2} x_2 + \cdots + a_{1,n} x_n = d_1,$$
$$a_{2,1} x_1 + a_{2,2} x_2 + \cdots + a_{2,n} x_n = d_2,$$
$$\cdots$$
$$a_{n,1} x_1 + a_{n,2} x_2 + \cdots + a_{n,n} x_n = d_n$$

has the solution $(s_1, s_2, \ldots, s_n)$ if that n-tuple is a solution of all of the equations in the system.

Theorem 2.4.1 (Gauss method). *If a linear system is changed to another by one of the following operations:*

- *an equation is swapped with another,*
- *an equation has both sides to be multiplied by a nonzero constant,*
- *an equation is replaced by the sum of itself and a multiple of another,*

then the two systems have the same set of solutions.

Note that each of the above three operations has a restriction. Multiplying a row by 0 is not allowed because obviously that can change the solution set of the system. Similarly, adding a multiple of a row to itself is not allowed because adding -1 times the row to itself has the effect of multiplying the row by 0. Finally, swap ping a row with itself is disallowed.

Definition 2.4.1. The three operations from Gauss method are the elementary reduction operations, or row operations, or Gaussian operations. They are swapping, multiplying by a scalar or re-scaling, and pivoting.

2.4.2   A general scheme for solving system of linear equation

Linear equation of system can be written into

$$\begin{bmatrix} a_{11} & a_{12} & \cdots & a_{1n} \\ a_{21} & a_{22} & \cdots & a_{2n} \\ \vdots & \vdots & \ddots & \vdots \\ a_{m1} & a_{m2} & \cdots & a_{mn} \end{bmatrix} \begin{bmatrix} x_1 \\ x_2 \\ \vdots \\ x_n \end{bmatrix} = \begin{bmatrix} b_1 \\ b_2 \\ \vdots \\ b_n \end{bmatrix}.$$

The above linear system can be further simplified into a matrix product $Ax = b$. A solution of linear system is an order collection of n numbers that satisfies the m linear equations, which can be written in short as a vector solution x. A linear system $Ax = b$ is called a nonhomogeneous system when vector b is not a zero vector. A linear system $Ax = 0$ is called a homogeneous system when the vector b is a zero vector.

Rank of matrix A denoted by $R(A)$ is used to determine whether the linear system is consistent (has a solution), has many solutions or has a unique set of solutions, or inconsistent (has no solution) using matrix inverse. Diagram of Fig. 2.1 shows the solution of the system of linear equations based on rank of the coefficient matrix $R(A)$

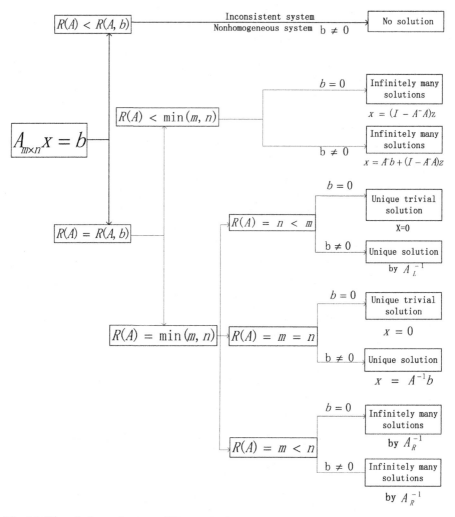

Fig. 2.1 The solutions of system of linear equation.
(modified from [43])

in comparison with the matrix size and rank of the augmented matrix coefficients A and the vector constants b: $R(A : b)$.

The equation $Ax = 0$ has infinitely many nontrivia solutions if and only if the matrix coefficient A is singular (i.e., it has no inverse, or $\det(A) = 0$), which happens when the number of equations is less than the unknowns ($m < n$). Otherwise, the homogeneous system only has the unique trivial solution of $x = 0$. General solution for homogeneous system is

$$x = (I - A^- A)z,$$

where z is an arbitrary nonzero vector and A^- is a generalized inverse ({1}-inverse) matrix of A satisfying $AA^- A = A$.

The linear system $Ax = b$ is called consistent if $AA^- b = b$. A consistent system can be solved using matrix inverse $x = A^{-1}b$, left inverse $x = A_L^{-1}b$ or right inverse $x = A_R^{-1}b$. A full rank nonhomogeneous system (happening when $R(A) = \min(m, n)$) has three possible options:

- When the number of the unknowns in a linear system is the same as the number of equations ($m = n$), the system is called uniquely determined system. There is only one possible solution to the system computed using matrix inverse $x = A^{-1}b$.
- When we have more equations than the unknown ($m > n$), the system is called overdetermined system. The system is usually inconsistent with no possible solution. It is still possible to find unique solution using left inverse $x = A_L^{-1}b$.
- When you have more unknowns than the equations ($m < n$), your system is called an undetermined system. The system usually has many possible solutions. The standard solution can be computed using right inverse $x = A_R^{-1}b$.

When a nonhomogeneous system $Ax = b$ is not full rank or when the rank of the matrix coefficients is less than the rank of the augmented coefficients matrix and the vector constants, that is $R(A) < R(A : b)$, then the system is usually inconsistent with no possible solution using matrix inverse. It is still possible to find the approximately least square solution that minimizes the norm of error. That is, using the generalized inverse of the matrix A and by

$$\min \|b - Ax\|,$$

we obtain

$$x = A^- b + (I - A^- A)z,$$

where z is an arbitrary nonzero vector.

2.5 Linear differential equation

2.5.1 Introduction

Linear differential equations are of the form

$$Ly = f,$$

where the differential operator L is a linear operator, y is the unknown function (such as a function of time $y(t)$), and the right hand side f is a given function of the same nature as y (called the source term). For a function dependent on time we may write the equation more expressly as

$$Ly(t) = f(t)$$

and, even more precisely by bracketing

$$L[y(t)] = f(t).$$

The linear operator L may be considered to be of the form

$$L_n(y) \equiv \frac{d^n y}{dt^n} + A_1(t)\frac{d^{n-1}y}{dt^{n-1}} + \cdots + A_{n-1}(t)\frac{dy}{dt} + A_n(t)y.$$

The linearity condition on L rules out operations such as taking the square of the derivative of y; but permits, for example, taking the second derivative of y. It is convenient to rewrite this equation in an operator form

$$L_n(y) \equiv [D^n + A_1(t)D^{n-1} + \cdots + A_{n-1}(t)D + A_n(t)]y,$$

where D is the differential operator d/dt (i.e., $Dy = y', D^2y = y'',\dots$), and $A_1(t),\dots,A_n(t)$ are given functions. Such an equation is said to have order n; the index of the highest derivative of y that is involved.

If y is assumed to be a function of only one variable, one terms it an ordinary differential equation, else the derivatives and their coefficients must be understood as (contracted) vectors, matrices or tensors of higher rank, and we have a (linear) partial differential equation.

The case where $f = 0$ is called a homogeneous equation and its solutions are called complementary functions. It is particularly important to the solution of the general case, since any complementary function can be added to a solution of the inhomogeneous equation to give another solution (by a method traditionally called *particular integral and complementary function*). When the components A_i are numbers, the equation is said to have constant coefficients.

2.5.2 Homogeneous equations with constant coefficients

The first method of solving linear homogeneous ordinary differential equations with constant coefficients is due to Euler, who worked out that the solutions have the form e^{zx}, for possibly complex values of z. The exponential function is one of the few functions to keep its shape after differentiation, allowing the sum of its multiple derivatives to cancel out the zeros, as required by the equation. Thus, for constant values $A_1,\dots,A_n$, to solve

$$y^n + A_1 y^{(n-1)} + \cdots + A_n y = 0$$

by setting $y = e^{zx}$ leads to

$$z^n e^{zx} + A_1 z^{n-1} e^{zx} + \cdots + A_n e^{zx} = 0.$$

Division by e^{zx} gives the nth-order polynomial

$$F(z) = z^n + A_1 z^{n-1} + \cdots + A_n = 0.$$

This algebraic equation $F(z) = 0$ is the characteristic equation.

Formally, the terms

$$y^{(k)} \quad (k = 1, 2, \ldots, n)$$

of the original differential equation are replaced by z^k. Solving the polynomial then gives n values of $z, z_1, \ldots, z_n$. Substitution of any of those values for z into e^{zx} gives a solution $e^{z_i x}$. Since homogeneous linear differential equations obey the superposition principle, any linear combination of these functions also satisfy the differential equation.

When the above roots are all distinct, we have n distinct solutions to the differential equation. It can be shown that these are linearly independent, by applying the Vandermonde determinant, and therefore they form a basis of the space of all solutions of the differential equation.

The procedure gives a solution for the case when all zeros are distinct, that is, each has multiplicity 1. For the general case, if z is a (possibly complex) zero (or root) of $F(z)$ having multiplicity m, then for $k \in \{0, 1, \ldots, m-1\}$, $y = x^k e^{zx}$ is a solution of the ordinary differential equation. Applying this to all roots gives a collection of n distinct and linearly independent functions, where n is the degree of $F(z)$. As before, these functions make up a basis of the solution space.

If the coefficients A_i of the differential equation are real, then real-valued solutions are generally preferable. Since nonreal roots z then come in conjugate pairs, so do their corresponding basis functions $x^k e^{zx}$, and the desired result is obtained by replacing each pair with their real-valued linear combinations $Re(y)$ and $Im(y)$, where y is one of the pair.

A case that involves complex roots can be solved with the aid of Euler's formula.

Solving of characteristic equation: A characteristic equation also called auxiliary equation of the form

$$r^2 + pr + q = 0.$$

Case 1: Two distinct roots, r_1 and r_2 .
Case 2: One real repeated root, r.
Case 3: Complex roots, $\alpha + \beta i$.

In case 1, the solution is given by

$$y = C_1 e^{r_1 x} + C_2 e^{r_2 x}.$$

In case 2, the solution is given by

$$y = (C_1 + C_2 x) e^{rx}.$$

In case 3, using Euler's equation the solution is given by

$$y = e^{\alpha x}(C_1 \cos \beta x + C_2 \sin \beta x).$$

2.5.3 Nonhomogeneous equation with constant coefficients

To obtain the solution to the nonhomogeneous equation (sometimes called inhomogeneous equation), find a particular integral $yP(x)$ by either the method of undetermined coefficients or the method of variation of parameters; the general solution to the linear differential equation is the sum of the general solution of the related homogeneous equation and the particular integral. Or, when the initial conditions are set, use Laplace transform to obtain the particular solution directly. Suppose we have

$$\frac{d^n y(x)}{dx^n} + A_1 \frac{d^{n-1} y(x)}{dx^{n-1}} + \cdots + A_n y(x) = f(x).$$

For later convenience, define the characteristic polynomial

$$P(v) = v^n + A_1 v^{n-1} + \cdots + A_n.$$

We find a solution basis $\{y_1(x), y_2(x), \ldots, y_n(x)\}$ for the homogeneous ($f(x) = 0$) case. We now seek a particular integral $y_p(x)$ by the variation of parameters method. Let the coefficients of the linear combination be functions of x, for which we formulate

$$y_p(x) = u_1(x)y_1(x) + u_2(x)y_2(x) + \cdots + u_n(x)y_n(x).$$

For ease of notation we will drop the dependency on x (i.e., the various (x)). Using the operator notation $D = d/dx$, the ODE in question is $P(D)y = f$; so

$$f = P(D)y_p = P(D)(u_1 y_1) + P(D)(u_2 y_2) + \cdots + P(D)(u_n y_n).$$

With the constraints

$$0 = u_1' y_1 + u_2' y_2 + \cdots + u_n' y_n$$
$$0 = u_1' y_1' + u_2' y_2' + \cdots + u_n' y_n'$$
$$\cdots$$
$$0 = u_1' y_1^{(n-2)} + u_2' y_2^{(n-2)} + \cdots + u_n' y_n^{(n-2)}$$

the parameters commute out,

$$f = u_1 P(D)y_1 + u_2 P(D)y_2 + \cdots + u_n P(D)y_n + u_1' y_1^{(n-1)} + u_2' y_2^{(n-1)} + \cdots + u_n' y_n^{(n-1)}.$$

But $P(D)y_j = 0$, therefore

$$f = u_1' y_1^{(n-1)} + u_2' y_2^{(n-1)} + \cdots + u_n' y_n^{(n-1)}.$$

This, with the constraints, gives a linear system in the u_j. Combining the Cramer's rule with the Wronskian, we have

$$u_j' = (-1)^{n+j} \frac{W(y_1, \ldots, y_{j-1}, y_{j+1} \ldots, y_n) \binom{0}{f}}{W(y_1, y_2, \ldots, y_n)}.$$

In terms of the nonstandard notation used above, one should take the i, n-minor of W and multiply it by f. That's why we get a minus-sign. Alternatively, forget about the minus sign and just compute the determinant of the matrix obtained by substituting the jth W column with $(0, 0, \ldots, f)$. The rest is a matter of integrating u_j. The particular integral is not unique; $y_p + c_1 y_1 + \cdots + c_n y_n$ also satisfies the ODE for any set of constants c_j.

Example 2.5.1. Suppose we have the following nonhomogeneous deferential equation with constant coefficients

$$y'' - 4y' + 5y = \sin x.$$

We take the solution basis found above

$$y_1(x) = e^{(2+i)x}, \quad y_2(x) = e^{(2-i)x}.$$

$$W = \begin{vmatrix} e^{(2+i)x} & e^{(2-i)x} \\ (2+i)e^{(2+i)x} & (2-i)e^{(2-i)x} \end{vmatrix} = e^{4x} \begin{vmatrix} 1 & 1 \\ 2+i & 2-i \end{vmatrix} = -2ie^{4x}$$

$$u_1' = \frac{1}{W} \begin{vmatrix} 0 & e^{(2-i)x} \\ \sin x & (2-i)e^{(2-i)x} \end{vmatrix} = -\tfrac{i}{2} \sin x e^{(-2-i)x}$$

$$u_2' = \frac{1}{W} \begin{vmatrix} e^{(2+i)x} & 0 \\ (2+i)e^{(2+i)x} & \sin x \end{vmatrix} = \tfrac{i}{2} \sin x e^{(-2+i)x}.$$

Using the list of integrals of exponential functions

$$u_1 = -\tfrac{i}{2} \int \sin x e^{(-2-i)x}\, dx = \frac{ie^{(-2-i)x}}{8(1+i)} ((2+i)\sin x + \cos x)$$

$$u_2 = \tfrac{i}{2} \int \sin x e^{(-2+i)x}\, dx = \frac{ie^{(i-2)x}}{8(1-i)} ((i-2)\sin x - \cos x).$$

And so

$$y_p = u_1(x)y_1(x) + u_2(x)y_2(x)$$
$$= \frac{i}{8(1+i)} ((2+i)\sin x + \cos x) + \frac{i}{8(1-i)} ((i-2)\sin x - \cos x),$$
$$= \frac{\sin x + \cos x}{8}.$$

(Notice that u_1 and u_2 had factors that canceled y_1 and y_2; that is typical.)

For interest's sake, this ODE has a physical interpretation as a driven damped harmonic oscillator; y_p represents the steady state, and $c_1 y_1 + c_2 y_2$ is the transient.

2.5.4 Equation with variable coefficients

A linear ODE of order n with variable coefficients has the general form

$$p_n(x)y^{(n)}(x) + p_{n-1}(x)y^{(n-1)}(x) + \cdots + p_0(x)y(x) = r(x).$$

First-order equation with variable coefficients: A linear ODE of order 1 with variable coefficients has the general form

$$Dy(x) + f(x)y(x) = g(x),$$

where D is the differential operator. Equations of this form can be solved by multiplying the integrating factor

$$e^{\int f(x)\,dx}$$

throughout to obtain

$$Dy(x)e^{\int f(x)\,dx} + f(x)y(x)e^{\int f(x)\,dx} = g(x)e^{\int f(x)\,dx},$$

which simplifies due to the product rule (applied backwards) to

$$D\left(y(x)e^{\int f(x)\,dx}\right) = g(x)e^{\int f(x)\,dx},$$

which, on integrating both sides and solving for $y(x)$ gives

$$y(x) = e^{-\int f(x)\,dx}\left(\int g(x)e^{\int f(x)\,dx}\,dx + \kappa\right).$$

In other words, the solution of a first-order linear ODE

$$y'(x) + f(x)y(x) = g(x)$$

with coefficients, that may or may not vary with x, is

$$y = e^{-a(x)}\left(\int g(x)e^{a(x)}\,dx + \kappa\right),$$

where κ is the constant of integration, and

$$a(x) = \int f(x)\,dx.$$

A compact form of the general solution based on a Green's function is

$$y(x) = \int_a^x [y(a)\delta(t-a) + g(t)]e^{-\int_t^x f(u)\,du}\,dt,$$

where $\delta(x)$ is the generalized Dirac delta function.

2.5.5 Systems of linear differential equations

An arbitrary linear ordinary differential equation or even a system of such equations can be converted into a first order system of linear differential equations by adding variables for all but the highest order derivatives. A linear system can be viewed as a single equation with a vector-valued variable. The general treatment is analogous to the treatment above of ordinary first order linear differential equations, but with complications stemming from noncommutativity of matrix multiplication. To solve

$$\begin{cases} \mathbf{y}'(x) & = A(x)\mathbf{y}(x) + \mathbf{b}(x), \\ \mathbf{y}(x_0) & = \mathbf{y}_0, \end{cases}$$

where $\mathbf{y}(x)$ is a vector or matrix, and $A(x)$ is a matrix, let $U(x)$ be the solution of $\mathbf{y}'(x) = A(x)\mathbf{y}(x)$ with $U(x_0) = I$ (the identity matrix). U is a fundamental matrix for the equation. The columns of U form a complete linearly independent set of solutions for the homogeneous equation. After substituting

$$\mathbf{y}(x) = U(x)\mathbf{z}(x)$$

the equation

$$\mathbf{y}'(x) = A(x)\mathbf{y}(x) + \mathbf{b}(x)$$

is reduced to

$$U(x)\mathbf{z}'(x) = \mathbf{b}(x).$$

Thus,

$$\mathbf{y}(x) = U(x)\mathbf{y}_0 + U(x) \int_{x_0}^{x} U^{-1}(t)\mathbf{b}(t)\, dt.$$

If $A(x_1)$ commutes with $A(x_2)$ for all x_1 and x_2, then

$$U(x) = e^{\int_{x_0}^{x} A(x)\, dx}$$

and thus

$$U^{-1}(x) = e^{-\int_{x_0}^{x} A(x)\, dx}.$$

However, in the general case there is no closed form solution, and an approximation method such as Magnus expansion may have to be used. Note that the exponentials are matrix exponentials.

2.6 Matrix differential equation

2.6.1 Introduction

A differential equation is a mathematical equation for an unknown function of one or several variables that relates the values of the function itself and of its derivatives of various orders. A matrix differential equation contains more than one function stacked into vector form with a matrix relating the functions to their derivatives. For example, a simple matrix ordinary differential equation is

$$\mathbf{x}'(t) = \mathbf{A}\mathbf{x}(t),$$

where $x(t)$ is an $n \times 1$ vector of functions of an underlying variable t, $x'(t)$ is the vector of first derivatives of these functions, and A is a matrix, of which all elements are constants.

In the case where A has n distinct eigenvalues, this differential equation has the following general solution

$$\mathbf{x}(t) = c_1 e^{\lambda_1 t}\mathbf{u}_1 + c_2 e^{\lambda_2 t}\mathbf{u}_2 + \cdots + c_n e^{\lambda_n t}\mathbf{u}_n,$$

where $1, 2, \ldots, n$ are the eigenvalues of A, $u_1, u_2, \ldots u_n$ are the respective eigenvectors of A; and $c_1, c_2, \ldots, c_n$ are constants.

By taking advantage of the Cayley-Hamilton theorem and Vandermonde-type matrices, this formal matrix exponential solution may be reduced to a simpler form. Below, this solution is displayed in terms of Putzer's algorithm.

2.6.2 Stability and steady state of the matrix system

The matrix equation

$$\mathbf{x}'(t) = \mathbf{A}\mathbf{x}(t) + \mathbf{b}$$

with $n \times 1$ parameter vector b is stable if and only if all eigenvalues of the matrix A have a negative real part. If the system is stable, the steady state x^* to which it converges, is found by setting

$$\mathbf{x}'(t) = \mathbf{0},$$

thus yielding

$$\mathbf{x}^* = -\mathbf{A}^{-1}\mathbf{b},$$

where we have assumed A is invertible. Therefore, the original equation can be written in homogeneous form in terms of deviations from the steady state

$$\mathbf{x}'(t) = \mathbf{A}[\mathbf{x}(t) - \mathbf{x}^*].$$

An equivalent way of expressing this is that x^* is a particular solution to the nonhomogeneous equation, while all solutions are in the form

$$\mathbf{x}_h + \mathbf{x}^*$$

with $\mathbf{x}_h$ a solution to the homogeneous equation ($b = 0$).

2.6.3 Solution in matrix form

The formal solution of $\mathbf{x}'(t) = \mathbf{A}[\mathbf{x}(t) - \mathbf{x}^*]$ is the celebrated matrix exponential,

$$\mathbf{x}(t) = \mathbf{x}^* + e^{\mathbf{A}t}[\mathbf{x}(0) - \mathbf{x}^*]$$

evaluated in a multitude of techniques.

2.6.4 Solving matrix ordinary differential equations

The process of solving the above equations and finding the required functions, of this particular order and form, consists of three main steps. Brief descriptions of each of these steps are listed below:

- finding the eigenvalues,
- finding the eigenvectors,
- finding the needed functions.

2.7 Laplace transform

2.7.1 Introduction

In mathematics the Laplace transform is an integral transform named after its discoverer Pierre-Simon Laplace. It takes a function of a positive real variable t (often time) to a function of a complex variable s (frequency).

The Laplace transform is very similar to the Fourier transform. While the Fourier transform of a function is a complex function of a real variable (frequency), the Laplace transform of a function is a complex function of a complex variable. Laplace transforms are usually restricted to functions of t with $t > 0$. A consequence of this restriction is that the Laplace transform of a function is a holomorphic function of the variable s. Unlike the Fourier transform, the Laplace transform of a distribution is generally a well-behaved function. Also techniques of complex variables can be used directly to study Laplace transforms. As a holomorphic function, the Laplace transform has a power series representation. This power series expresses a function as a linear superposition of moments of the function. This perspective has applications in probability theory.

The Laplace transform is invertible on a large class of functions. The inverse Laplace transform takes a function of a complex variable s (often frequency) and yields a function of a real variable t (time). Given a simple mathematical or functional description of an original or the resulted system, the Laplace transform provides an alternative functional description that often simplifies the process of analyzing the behavior of the system, or in synthesizing a new system based on a set of specifications. So, for example, Laplace transformation from the time domain to the frequency domain transforms differential equations into algebraic equations and convolution into multiplication. It has many applications in the sciences and technology.

2.7.2 Formal definition

The Laplace transform is a frequency-domain approach for continuous time signals irrespective of whether the system is stable or unstable. The Laplace transform of a function $f(t)$, defined for all real numbers $t \geq 0$, is the function $F(s)$, which is a unilateral transform defined by

$$F(s) = \int_0^{+\infty} e^{-st} f(t) dt,$$

where s is a complex variable given by $s = \sigma + i\omega$, with real numbers σ and ω. Other notations for the Laplace transform include Lf or alternatively $Lf(t)$ instead of F.

The meaning of the integral depends on types of functions of interest. A necessary condition for existence of the integral is that f must be locally integrable on $[0, \infty)$. For locally integrable functions that decay at infinity or are of exponential type, the integral can be understood to be a (proper) Lebesgue integral. However, for many applications it is necessary to regard it to be a conditionally convergent improper integral at ∞. Still more generally, the integral can be understood in a weak sense, and this is dealt with below.

One can define the Laplace transform of a finite Borel measure by the Lebesgue integral

$$\mathcal{L}\{\mu\}(s) = \int_{[0,\infty)} e^{-st} d\mu(t).$$

An important special case is where μ is a probability measure, for example, the Dirac delta function. In operational calculus, the Laplace transform of a measure is often treated as though the measure came from a probability density function f. In that case, to avoid potential confusion, one often writes

$$\mathcal{L}\{f\}(s) = \int_{0^-}^{\infty} e^{-st} f(t) \, dt,$$

where the lower limit of 0^- is the shorthand notation for

$$\lim_{\varepsilon \downarrow 0} \int_{-\varepsilon}^{\infty}.$$

This limit emphasizes that any point mass located at 0 is entirely captured by the Laplace transform. Although with the Lebesgue integral, it is not necessary to take such a limit, it does appear more naturally in connection with the Laplace-Stieltjes transform.

2.7.3 Region of convergence

If f is a locally integrable function (or more generally a Borel measure locally bounded variation), then the Laplace transform $F(s)$ of $f(t)$ converges, provided that the limit

$$\lim_{R \to \infty} \int_0^R f(t) e^{-st} \, dt$$

exists. The Laplace transform converges absolutely if the integral

$$\int_0^{\infty} \left| f(t) e^{-st} \right| \, dt$$

exists (as a proper Lebesgue integral). The Laplace transform is usually understood as conditionally convergent, meaning that it converges in the former instead of the latter sense.

Following the dominated convergence theorem, one notes that the set of values, for which $F(s)$ converges absolutely satisfies $Re(s) \geq a$, where a is an extended real constant. The constant a is known as the abscissa of absolute convergence, and depends on the growth behavior of $f(t)$. Analogously, the two-sided transform converges absolutely in a strip of the form $a \leq Re(s) \leq b$. The subset of values of s for which the Laplace transform converges absolutely is called the region of absolute convergence or the domain of absolute convergence. In the two-sided case, it is sometimes called the strip of absolute convergence. The Laplace transform is analytic in the region of absolute convergence.

Similarly, the set of values for which $F(s)$ converges (conditionally or absolutely) is known as the region of conditional convergence, or simply the region of convergence (ROC). If the Laplace transform converges (conditionally) at $s = s_0$, then it automatically converges for all s with $Re(s) > Re(s_0)$. Therefore, the region of convergence is a half-plane of the form $Re(s) > a$, possibly including some points of the boundary line $Re(s) = a$.

In the region of convergence $Re(s) > Re(s_0)$, the Laplace transform of f can be expressed by integrating by parts as the integral

$$F(s) = (s - s_0) \int_0^\infty e^{-(s-s_0)t} \beta(t)\, dt, \quad \beta(u) = \int_0^u e^{-s_0 t} f(t)\, dt.$$

That is, in the region of convergence $F(s)$ can effectively be expressed as the absolutely convergent Laplace transform of some other function.

There are several Paley-Wiener theorems concerning the relationship between the decay properties of f and the properties of the Laplace transform within the region of convergence.

In engineering applications, a function corresponding to a linear time-invariant (LTI) system is stable if every bounded original produces a bounded output. This is equivalent to the absolute convergence of the Laplace transform of the impulse response function in the region $Re(s) \geq 0$. As a result, LTI systems are stable provided the poles of the Laplace transform of the impulse response function have a negative real part. This ROC is used in determining the causality and stability of a system.

2.7.4 Laplace transform pair table

It is convenient to display the Laplace transforms of standard signals in one table. Table 2.1 displays the time Signal $x(t)$ and its corresponding Laplace transform and region of absolute convergence.

2.7.5 Properties and theorems

We have the initial value theorem:

$$f(0+) = \lim_{s \to \infty} sF(s),$$

Table 2.1 **Laplace transform pairs**

Time domain $f(t) = \mathcal{L}^{-1}\{F(s)\}$	Laplace s-domain $F(s) = \mathcal{L}\{f(t)\}$	Region of convergence
$u(t)$	$\frac{1}{s}$	$Re(s) > 0$
$\delta(t)$	1	all s
$e^{-at}u(t)$	$\frac{1}{s+a}$	$Re(s) > -Re(a)$
$t^k e^{-at}u(-t)$	$\frac{k!}{(s+a)^{k+1}}$	$Re(s) > -Re(a)$
$e^{-at}u(-t)$	$\frac{1}{(s+a)}$	$Re(s) < -Re(a)$
$(-t)^k e^{-at}u(-t)$	$\frac{k!}{(s+a)^{k+1}}$	$Re(s) < -Re(a)$
$\frac{d^k \delta(t)}{dt^k}$	s^k	all s
$t^k u(t)$	$\frac{k!}{s^{k+1}}$	$Re(s) > 0$
$\sin \omega_0 t u(t)$	$\frac{\omega_0}{s^2+\omega_0^2}$	$Re(s) > 0$
$\cos \omega_0 t u(t)$	$\frac{s}{s^2+\omega_0^2}$	$Re(s) > 0$
$e^{-at} \sin \omega_0 t u(t)$	$\frac{\omega_0}{(s+a)^2+\omega_0^2}$	$Re(s) > -Re(a)$
$e^{-at} \cos \omega_0 t u(t)$	$\frac{s+a}{(s+a)^2+\omega_0^2}$	$Re(s) > -Re(a)$

and the final value theorem:

$$f(\infty) = \lim_{s \to 0} sF(s),$$

if all poles of $sF(s)$ are in the left half-plane. The final value theorem is useful because it gives the long-term behavior without having to perform partial fraction decompositions or other difficult algebra. If $F(s)$ has a pole in the right-hand plane or poles on the imaginary axis, e.g., if $f(t) = e^t$ or $f(t) = \sin(t)$, the behavior of this formula is undefined.

The Laplace transform has a number of properties that make it useful for analyzing linear dynamical systems. The most significant advantage is that differentiation and integration become multiplication and division, respectively, by s (similarly to logarithms changing multiplication of numbers to addition of their logarithms).

Because of this property, the Laplace variable s is also known as the operator variable in the s domain: either derivative operator or (for s^{-1}) integration operator. The transform turns integral equations and differential equations to polynomial equations, which are much easier to solve. Once solved, use of the inverse Laplace transform reverts to the time domain.

Given the functions $f(t)$ and $g(t)$, and their respective Laplace transforms $F(s)$ and $G(s)$,

$$f(t) = \mathcal{L}^{-1}\{F(s)\},$$

$$g(t) = \mathcal{L}^{-1}\{G(s)\}.$$

Table 2.2 is a list of properties of unilateral Laplace transform.

Table 2.2 **Properties of the unilateral Laplace transform**

Time domain	s domain
$af(t) + bg(t)$	$aF(s) + bG(s)$
$tf(t)$	$-F'(s)$
$t^n f(t)$	$(-1)^n F^{(n)}(s)$
$f''(t)$	$s^2 F(s) - sf(0) - f'(0)$
$f^{(n)}(t)$	$s^n F(s) - \sum_{k=1}^{n} s^{n-k} f^{(k-1)}(0)$
$\frac{1}{t} f(t)$	$\int_s^\infty F(\sigma)\, d\sigma$
$f(at)$	$\frac{1}{a} F\left(\frac{s}{a}\right)$
$(f * g)(t) = \int_0^t f(\tau) g(t - \tau)\, d\tau$	$F(s) \cdot G(s)$
$f(t)g(t)$	$\frac{1}{2\pi i} \lim_{T \to \infty} \int_{c-iT}^{c+iT} F(\sigma) G(s - \sigma)\, d\sigma$
$e^{at} f(t)$	$F(s - a)$
$\int_0^t f(\tau)\, d\tau = (u * f)(t)$	$\frac{1}{s} F(s)$

2.7.6 Inverse Laplace transform

Two integrable functions have the same Laplace transform only if they differ on a set of Lebesgue measure zero. This means that, on the range of the transform, there is an inverse transform. In fact, besides integrable functions, the Laplace transform is a one-to-one mapping from one function space into another and in many other function spaces as well, although there is usually no easy characterization of the range. Typical function spaces in which this is true include the spaces of bounded continuous functions, the space $L^\infty(0, \infty)$, or more generally tempered functions (i.e., functions of at worst polynomial growth) on $(0, \infty)$. The Laplace transform is also defined and injective for suitable spaces of tempered distributions.

In these cases, the image of the Laplace transform lives in a space of analytic functions in the region of convergence. The inverse Laplace transform is given by the following complex integral, which is known by various names (the Bromwich integral, the Fourier-Mellin integral, and Mellin's inverse formula):

$$f(t) = \mathcal{L}^{-1}\{F\}(t) = \frac{1}{2\pi i} \lim_{T \to \infty} \int_{\gamma - iT}^{\gamma + iT} e^{st} F(s)\, ds,$$

where γ is a real number so that the contour path of integration is in the region of convergence of F(s). An alternative formula for the inverse Laplace transform is given by Post's inversion formula. In practice, it is typically more convenient to decompose a Laplace transform into the known transforms of functions obtained from a table, and construct the inverse by inspection.

Generalized inverse of matrix and solution of linear system equation

3

In this chapter, we briefly present some background for generalized inverse of matrix and its relation to solution of linear system equation, that is needed in the sequel. This chapter can be skimmed very quickly and used mainly as a quick reference. There have been a huge number of reference resources for this topic, to mention a few [43, 44, 51–54].

3.1 The generalized inverse of matrix

Recall how we defined the matrix inverse. A matrix inverse A^{-1} is defined as a matrix that produces identity matrix when we multiply with the original matrix A, that is, we define $AA^{-1} = I = A^{-1}A$. Matrix inverse exists only for square matrices.

Real world data are not always square. Furthermore, real world data are not always consistent and might contain many repetitions. To deal with real world data, generalized inverse for rectangular matrix is needed.

Generalized inverse matrix is defined as $AA^-A = A$. Notice that the usual matrix inverse is covered by this definition because $AA^{-1}A = A$. We use the term "generalized" inverse for a general rectangular matrix and to distinguish it from the inverse matrix that is for a square matrix. Generalized inverse is also called the pseudo-inverse.

Unfortunately there are many types of generalized inverse. Most generalized inverses are not unique. Some generalized inverses are reflexive satisfying $(A^-)^- = A$ and some are not reflexive. In this tutorial, we will only discuss a few of them that are often used in practical applications.

- A reflexive generalized inverse is defined as

$$AA^-A = A,$$
$$A^-AA^- = A^-.$$

- A minimum norm generalized inverse, such that $x = A^-b$ will minimize $\|x\|$ is defined as

$$AA^-A = A,$$
$$A^-AA^- = A^-,$$
$$(A^-A)^- = A^-A.$$

A Generalized Framework of Linear Multivariable Control. http://dx.doi.org/10.1016/B978-0-08-101946-7.00003-2

- A generalized inverse that produces the least square solution that will minimize the residual or error $\min_x \|b - Ax\|$, is defined as

$$AA^-A = A,$$
$$(AA^-)^T = AA^-.$$

3.1.1 The left inverse and right inverse

The usual matrix inverse is defined as a two-side inverse, i.e., $AA^{-1} = I = A^{-1}A$ because we can multiply the inverse matrix from the left or from the right of matrix A and we still get the identity matrix. This property is only true for a square matrix A.

For a rectangular matrix A, we may have a generalized left inverse or left inverse for short when we multiply the inverse from the left to get identity matrix $A_{\text{left}}^{-1}A = I$. Similarly, we may have a generalized right inverse, or right inverse for short, when we multiply the inverse from the right to get the identity matrix $AA_{\text{right}} = I$.

In general, the left inverse is not equal to the right inverse. The generalized inverse of a rectangular matrix is related to the solving of system linear equations $Ax = b$. The solution to a normal equation is $x = (A^TA)^{-1}A^Tb$, which is equal to $x = A^-b$. The term

$$A^- = A_{\text{left}}^{-1} = (A^TA)^{-1}A^T$$

is often called as generalized left inverse. Yet another pseudo-inverse can also be obtained by multiplying the transpose matrix from the right and this is called a generalized right inverse

$$A^- = A_{\text{right}}^{-1} = A^T(AA^T)^{-1}.$$

3.1.2 Moore-Penrose inverse

It is possible to obtain a unique generalized matrix. To distinguish the unique generalized inverse from other nonunique generalized inverses A^-, we use the symbol A^+. The unique generalized inverse is called the Moore-Penrose inverse. It is defined using the following four conditions:

(1) $AA^+A = A$,
(2) $A^+AA^+ = A^+$,
(3) $(AA^+)^T = AA^+$,
(4) $(A^+A)^T = A^+A$.

The first condition $AA^+A = A$ is the definition of a generalized inverse. Together with the first condition, the second condition indicates the generalized inverse is reflexive $(A^-)^- = A$. Together with the first condition, the third condition indicates that the generalized inverse is the least square solution that will minimize the norm of error $\min_x \|b - Ax\|$. The fourth condition above demonstrates the unique generalized inverse.

Properties of generalized inverse of matrix: Some important properties of generalized inverse of matrix are:

1. The transpose of the left inverse of A is the right inverse $A_{\text{right}}^{-1} = (A_{\text{left}}^{-1})^T$. Similarly, the transpose of the right inverse of A is the left inverse $A_{\text{left}}^{-1} = (A_{\text{right}}^{-1})^T$.

2. A matrix $A_{m \times n}$ has a left inverse A_{left}^{-1} if and only if its rank equals its number of columns and the number of rows is more than the number of columns $\rho(A) = n < m$. In this case $A^+A = A_{\text{left}}^{-1}A = I$.

3. A matrix $A_{m \times n}$ has a right inverse A_{right}^{-1} if and only if its rank equals its number of rows and the number of rows is less than the number of columns $\rho(A) = m < n$. In this case $A^+A = AA_{\text{right}}^{-1} = I$.

4. The Moore-Penrose inverse is equal to left inverse $A^+ = A_{\text{left}}^{-1}$, when $\rho(A) = n < m$ and equals the right inverse $A^+ = A_{\text{right}}^{-1}$, when $\rho(A) = m < n$. The Moore-Penrose inverse is equal to the matrix inverse $A^+ = A^{-1}$, when $\rho(A) = m = n$.

3.1.3 The minimization approach to solve an algebraic matrix equation

The generalized inverse of the matrix has been used extensively in the areas of modern control, least square estimation and aircraft structural analysis. It is the purpose of this note to extend the results by presenting a unified framework that provides geometric insight and highlights certain optimal features imbedded in the generalized inverse.

Consider the algebraic matrix equation

$$y = Ax, \tag{3.1}$$

where A is an $n \times m$ constant matrix, y is a given n vector, and x is an m vector to be determined. For the trivial case where $n = m$ and A is a nonsingular matrix, i.e., rank$(A) = n$, a unique solution to Eq. (3.1) exists and is given by

$$x = A^{-1}y, \tag{3.2}$$

where A^{-1} denotes the inverse of A.

For the case $n \neq m$, the expression of x in terms of y involves the generalized inverse of A, denoted A^+, and thus

$$x = A^+y. \tag{3.3}$$

In the following cases it will be shown that A^+, for either $n > m$ or $n < m$, may be viewed as a solution to a certain minimization problem.

(1) Case A: $n > m$

With no loss of generality it can be assumed that A is of full rank, i.e.,

$$\text{rank}(A) = m \tag{3.4}$$

however, rank $(A) < n$. It is possible to delete the dependent columns of A, set the respective unknown components of x equal to zero, and reduce the problem to the case in which Eq. (3.4) is satisfied. Since the m columns of A do not span the n dimensional space R_n, an exact solution to Eq. (3.1) cannot be obtained if y is not contained in the subspace spanned by the columns of A. Thus, one is motivated to seek approximate solutions, the best of which is the one that minimizes the Euclidian norm of the error. Let the error e be given by

$$e = y - Ax. \tag{3.5}$$

Then let z be given by

$$z = \|e\|^2 = e^T e = (y - Ax)^T (y - Ax), \tag{3.6}$$

where the superscript "T" denotes the transpose. Evaluation of the gradient of z with respect to x yields

$$\frac{\partial z}{\partial x} = -2A^T y + 2A^T Ax = 0. \tag{3.7}$$

The Hessian matrix is

$$\frac{\partial^2 z}{\partial x^2} = 2A^T A. \tag{3.8}$$

From Eq. (3.7), x is given by

$$x = (A^T A)^{-1} A^T y. \tag{3.9}$$

By virtue of A having full rank, $(A^T A)$ is a positive definite matrix. Thus $(A^T A)^{-1}$ exists and the Hessian matrix is positive definite, implying that a minimum was obtained. In this case $(n > m)$ the generalized inverse of A is given by

$$A^+ = (A^T A)^{-1} A^T. \tag{3.10}$$

It is interesting to note that if y is contained in the subspace spanned by the columns of A, Eq. (3.9) yields an exact solution to Eq. (3.1), i.e., $\|e\| = 0$. The optimal feature of Eq. (3.9) has found extensive applications in data processing for least square approximations. In closing it should be noted that Eq. (3.9) can be obtained by invoking the Orthogonal Projection Lemma, thus providing a geometric interpretation to the optimal feature of Eq. (3.10).

(2) Case B: $n < m$

Again, without loss of generality, it can be assumed that A is of full rank, i.e.,

$$\text{rank}(A) = n. \tag{3.11}$$

If, however, rank(A) $< n$, it implies that some of the equations are merely a linear combination of the others and therefore may be deleted without loss of information, thereby reducing the case rank(A) $< n$ to the case rank(A) $= n$. Moreover, if A is a square singular matrix, it can be reduced to case B after proper deletion of the dependent rows of A.

Eqs. (3.1), (3.11) with $n < m$ yield an infinite number of solutions, the "optimal" of which is the one having the smallest norm. Therefore one is confronted with a constrained minimization problem, where the minimization of $\|x\|$ (or equivalently $\frac{\|x\|^2}{2}$) is to be accomplished subject to Eq. (3.1). Adjoining the constraint, via a vector of Lagrange multipliers (λ), the objective function to be minimized is

$$H = \frac{1}{2}\|x\|^2 + \lambda^T(y - Ax) = \frac{1}{2}x^Tx + \lambda^T(y - Ax). \tag{3.12}$$

By evaluating the respective gradients

$$\frac{\partial H}{\partial x} = x - A^T\lambda = 0, \tag{3.13}$$

$$\frac{\partial H}{\partial \lambda} = y - Ax = 0. \tag{3.14}$$

From Eq. (3.13), one has

$$x = A^T\lambda. \tag{3.15}$$

Substitution of Eq. (3.15) into Eq. (3.14) yields

$$y = AA^T\lambda. \tag{3.16}$$

That is

$$\lambda = (AA^T)^{-1}y. \tag{3.17}$$

The existence of $(AA^T)^{-1}$ is guaranteed by virtue of Eq. (3.11). Substitution of Eq. (3.17) into Eq. (3.15) yields

$$x = A^T(AA^T)^{-1}y. \tag{3.18}$$

In this case ($n < m$) the generalized inverse of A is given by

$$A^+ = A^T(AA^T)^{-1}. \tag{3.19}$$

For the sake of completeness it should be noted that the norm minimization of $\|e\|$ and $\|x\|$ of Eqs. (3.6), (3.12), respectively, can be performed by the extended vector norms, where a norm of a vector w is defined as w^TQw and Q is a compatible positive definite weighting matrix. By doing so, one can choose the relative emphasis of the magnitudes of the vector components that is minimized.

For case A, Eq. (3.6) becomes

$$z = e^T Qe = (y - Ax)^T Q(y - Ax) \tag{3.20}$$

with the solution

$$x = (A^T QA)^{-1} A^T Qy. \tag{3.21}$$

From case B, Eq. (3.12) becomes

$$H = \frac{1}{2} x^T Qx + \lambda^T (y - Ax) \tag{3.22}$$

with the solution

$$x = Q^{-1} A^T (AQ^{-1} A^T)^{-1}. \tag{3.23}$$

In summary, a unified framework has been presented showing that the generalized inverse of a matrix can be viewed as a result of a minimization problem leading to a practical interpretation.

3.2 The full rank decomposition theorem

Concerning the situations of whether a matrix being full row rank or full column rank, we have the following obvious statement.

Theorem 3.2.1. *When $A \in C^{m \times n}$, if*

(1) rank $A = m$, A is of full row rank,
(2) rank $A = n$, A is of full column rank.

It is obvious that if A is of full row rank, then $m \leq n$ and if A is of full column rank, then $m \geq n$. A sufficient and necessary condition for a matrix to be of full rank is that it is both of full row rank and of full column rank.

Theorem 3.2.2. *Assuming $A \in C^{m \times n}$, if there is a full column rank matrix F and full row rank matrix G making that*

$$A = FG,$$

then A has a full rank decomposition.

Theorem 3.2.3. *When $A \in C^{m \times n}$, rank $A = r > 0$, there must be a full rank factorization for it.*

Proof. For the matrix A, if rank $A = r > 0$, one is able to choose r linear independent columns from the columns of A: $A_{i_1}, A_{i_2}, \ldots, A_{i_r}$. The remaining columns can been expressed by these columns. That is, we have

$$F = [A_{i_1} A_{i_2} \ \ldots \ A_{i_r}].$$

It is seen that, the matrix F is a full column rank matrix. Therefore, there exists permutation matrix Q and S such that

$$AQ = [A_{i_1} \ \ldots \ A_{i_2} \ \widetilde{A_1} \ \ldots \ \widetilde{A_{n-1}}] = [F \ \ FS] = F[I_r \ \ S].$$

Letting $G = [I_r \ S]Q^{-1}$, G is a full row rank matrix. So A has a full rank decomposition

$$A = FG. \qquad \qquad \qquad \qquad \qquad \square$$

Theorem 3.2.4. *Some important properties of full column (row) rank matrix: if $F \in C^{m \times r}$ is a full column rank matrix, then one has*

- *the eigenvalue of $F^H F$ is larger than zero;*
- *$F^H F$ is positive definite Hermite matrix;*
- *$F^H F$ is an r order invertible matrix.*

Proof. If λ is any eigenvalue of $F^H F$, the vector x is the corresponding nonzero feature vector, one has

$$F^H Fx = \lambda x.$$

In this case, we have

$$\lambda \|x\|_2^2 = \lambda x^H x = x^H (F^H Fx) = (Fx)^H (Fx) = \|Fx\|_2^2 \geq 0.$$

Considering F is a full column rank matrix, and $Fx \neq 0$, so $\|Fx\|_2^2 > 0$. We therefore draw the conclusion: $\lambda > 0$. $\qquad \square$

Analogously, if $G \in C^{r \times n}$ is a full row rank matrix, then GG^H has the above properties. When $A \in C^{m \times n}, B \in C^{n \times m}$, if $AB = I_m$, then B is the right inverse of A denoted by $B = A_{\text{right}}^{-1}$ and A is the left inverse of B denoted by $A = B_{\text{left}}^{-1}$.

3.3 The least square solution to an algebraic matrix equation

The method of least squares is a standard approach in regression analysis to the approximate solution of the over determined systems, in which among the set of equations there are more equations than unknowns. The term "least squares" refers to this situation, the overall solution minimizes the summation of the squares of the errors, which are brought by the results of every single equation.

3.3.1 The solution to the compatible linear equations

We have the following general solution of a given linear matrix equation, which is formulated in terms of the Moore-Penrose inverse.

Theorem 3.3.1. *If $A \in C^{m \times n}, b \in C^m, x \in C^n$, and if the system of linear equations $Ax = b$ is compatible, the general solution is given by*

$$x = A^+ b + (I_n - A^+ A)t \quad (\forall t \in C^n).$$

Proof. Because $Ax = b$ is compatible, there is a x_0 making $Ax_0 = b$. So we have on one hand

$$
\begin{aligned}
Ax &= AA^+ b + A(I_n - A^+ A)t \\
&= AA^+ Ax_0 + (A - AA^+ A)t \\
&= Ax_0 + Ot \\
&= b.
\end{aligned}
$$

On the other hand, if x_0 is any solution of $Ax = b$, such that $t = x_0$. Therefore,

$$
\begin{aligned}
A^+ b + (I_n - A^+ A)t &= A^+ b + (I_n - A^+ A)x_0 \\
&= A^+ b + x_0 - A^+ Ax_0 \\
&= A^+ b + x_0 - A^+ b \\
&= x_0.
\end{aligned}
$$

So far, we have proved that

$$x = A^+ b + (I_n - A^+ A)t \quad (\forall t \in C^n)$$

is the general solution of the equation $Ax = b$, which concludes the proof. $\square$

In particular, when $b = 0$, we obtain the general solution of the homogeneous linear equations $Ax = 0$ given by

$$x = (I_n - A^+ A)t \quad (\forall t \in C^n).$$

3.3.2 The least square solution of incompatible equation

Considering in many applications, such as data processing, multivariate analysis, optimization theory, modern control theory, and networking theory, the mathematical model of linear matrix equation is often an incompatible one. We hereafter use the generalized inverse matrix to represent the following general solution of incompatible linear equations.

Theorem 3.3.2. *When $A \in C^{m \times n}, b \in C^m$, if there is $x^* \in C^n$ such that*

$$\|Ax^* - b\|_2 = \min_{x \in C^n}\{\|Ax - b\|_2\},$$

then the component x^* is termed as the least square solutions to the system of equations $Ax = b$.

Theorem 3.3.3. When $A \in C^{m \times n}$, all the least square solutions of system of equation $Ax = b$ are given by

$$x^* = A^+ b + (I_n - A^+ A)t \quad (t \in C^n).$$

Proof. $\forall x \in C^n$, we define

$$\varphi(x) = \|Ax - b\|_2^2 = x^H A^H Ax - 2x^H A^H b + b^H b$$

and denote $x = \mu + iv$ $(\mu, v \in R^n)$, we then have

$$\frac{\partial \varphi}{\partial x} \triangleq \frac{\partial \varphi}{\partial \mu} + i \frac{\partial \varphi}{\partial v} = 2A^H Ax - 2A^H b = 0.$$

Therefore we have $A^H Ax = A^H b$. Because

$$\text{rank } A^H A \leq \text{rank}[A^H A \vdots A^H b] = \text{rank } A^H[A \vdots b] \leq \text{rank } A = \text{rank } A^H A,$$

due to the rank of the augmented matrix is equal to that of the coefficient matrix, so $A^H Ax = A^H b$ is compatibility equations. Assuming $\varphi(x)$ is nonnegative real function, according to multivariate function extreme value theory, the solution of normal equations is the minimum point of $\varphi(x)$. Subsequently, the total least square solution of equations $Ax = b$ is

$$x^* = (A^H A)^+ A^H b + (I_n - (A^H A)^+ A^H A)t$$
$$= A^+ b + (I_n - A^+ A)t \quad (\forall t \in C^n).$$

$\square$

3.3.3 The minimum norm least squares solution for the equations

From the above discussions, whether the equations are compatible or not, the least-square solutions can be expressed as

$$x^* = A^+ b + (I_n - A^+ A)t \quad (t \in C^n).$$

Because the least squares solution is not unique, one is able to find a least square solution of minimum norm, which is given by

$$\|x^{**}\|_2 = \min\{\|x^*\|_2 : \ x^* = A^+ b + (I_n - A^+ A)t, \ t \in C^n\},$$

where x^{**} is referred to as the very minimal norm least square solution to the equations $Ax = b$. We have the following theorem:

Theorem 3.3.4. $x^{**} = A^+ b$ is the minimal norm least squares solution for the equations

Proof. By the nature of the A^+, we have

$$
\begin{aligned}
\langle A^+b,\ (I_n - A^+A)t\rangle &= b^H(A^+)^H(I_n - A^+A)t \\
&= b^H[(A^+)^H - (A^+)^H A^+A]t \\
&= b^H[(A^+)^H - (A^+)^H(A^+A)^H]t \\
&= b^H[(A^+)^H - (A^+AA^+)^H]t \\
&= b^H[(A^+)^H - (A^+)^H]t \\
&= 0.
\end{aligned}
$$

The least-square solutions of equations $Ax = b$ is composed of two orthogonal vectors, so

$$
\|x^*\|_2^2 = \|A^+b\|_2^2 + \|(I_n - A^+A)t\|_2^2 \geq \|A^+b\|_2^2 \quad (t \in C^n).
$$

This suggests that $x^{**} = A^+b$ is the minimal norm least square solution for the equation $Ax = b$. $\qquad\square$

3.4 The singular value decomposition

The Moore-Penrose inverse can be obtained through singular value decomposition (SVD): $A = UDV^T$, such that $A^+ = VD^{-1}U^T$. We have the following definition

Definition 3.4.1. When $A \in C^{m \times n}$, denote the following positive characteristic values for Hermite matrix $A^H A$

$$
\lambda_1,\ \lambda_2, \ldots, \lambda_n.
$$

Then

$$
\sigma_1 = \sqrt{\lambda_1},\ \sigma_2 = \sqrt{\lambda_2}, \ldots, \sigma_n = \sqrt{\lambda_n}
$$

are the single values of A.

If $A \in C^{m \times n}$, rank $A = r$, then $A^H A$ has r positive eigenvalues and the remaining eigenvalues are zeros. Further, we assume that they can be ordered in the following manner

$$
\sigma_1 \geq \sigma_2 \geq \cdots \sigma_r > \sigma_{r+1} = \sigma_{r+2} = \cdots = \sigma_n = 0.
$$

Theorem 3.4.1. When $A \in C^{m \times n}$, rank $A = r > 0$, the so-called singular value decomposition (SVD) of A is

$$
U^H AV = \begin{bmatrix} \Sigma & 0 \\ 0 & 0 \end{bmatrix},
$$

where the matrix U is the m order unitary matrix, the matrix V is the n order unitary matrix, and the matrix Σ is the following diagonal matrix

$$\Sigma = \mathrm{diag}(\sigma_1, \sigma_2, \ldots, \sigma_r), \ \sigma_1 \geq \sigma_2 \geq \cdots \geq \sigma_r > 0.$$

Note in the above the component r is the positive singular value of A.

Proof. If $\sigma_1^2, \sigma_2^2, \ldots, \sigma_r^2, 0, \ldots, 0$ are all the eigenvalues of A, and $\sigma_1 \geq \sigma_2 \geq \cdots \geq \sigma_r > 0$. Because $A^H A$ is nonnegative definite Hermite matrix, there is a unitary matrix V with the order n, satisfying the following associations

$$U^H(A^H A)V = \begin{bmatrix} \sigma_1^2 & & & & & \\ & \ddots & & & & \\ & & \sigma_r^2 & & & \\ & & & 0 & & \\ & & & & \ddots & \\ & & & & & 0 \end{bmatrix} \triangleq \begin{bmatrix} \Sigma^2 & 0 \\ 0 & 0 \end{bmatrix}.$$

Among them $\Sigma = \mathrm{diag}(\sigma_1, \sigma_2, \ldots, \sigma_r)$. By partitioning the matrix V into some subblocks

$$V = [V_1, V_2], \quad V_1 \in C^{n \times r}, \quad V_2 \in C^{n \times (n \times r)}$$

then we will have

$$\begin{bmatrix} V_1^H \\ V_2^H \end{bmatrix} (A^H A)[V_1 \ V_2] = \begin{bmatrix} V_1^H A^H A V_1 & V_1^H A^H A V_2 \\ V_2^H A^H A V_1 & V_2^H A^H A V_2 \end{bmatrix}$$

$$= \begin{bmatrix} \Sigma^2 & 0 \\ 0 & 0 \end{bmatrix}.$$

Then

$$V_1^H A^H A V_1 = \Sigma^2, \quad V_1^H A^H A V_2 = O, \quad V_2^H A^H A V_1 = O, \quad V_2^H A^H A V_2 = O.$$

So

$$(A V_1 \Sigma^{-1})^H (A V_1 \Sigma^{-1}) = I_r, \quad (A V_2)^H (A V_2) = O, \quad A V_2 = O.$$

By letting

$$U_1 = A V_1 \Sigma^{-1},$$

where U_1 is part of the unitary matrix and extending it into an m order unitary matrix $U = [U_1 \ U_2]$, U_2 is part of the column unitary matrix, we thus have

$$
\begin{aligned}
U^H A V &= \begin{bmatrix} U_1^H \\ U_2^H \end{bmatrix} A \ [V_1 \ V_2] \\
&= \begin{bmatrix} U_1^H A V_1 & U_1^H A V_2 \\ U_2^H A V_1 & U_2^H A V_2 \end{bmatrix} \\
&= \begin{bmatrix} U_1^H (U_1 \Sigma) & O \\ U_2^H (U_1 \Sigma) & O \end{bmatrix} \\
&= \begin{bmatrix} \Sigma & 0 \\ 0 & 0 \end{bmatrix},
\end{aligned}
$$

which concludes the proof.   □

Singular value decomposition is a factorization of a rectangular matrix A into three matrices U, D, and V. The two matrices U and V are orthogonal matrices ($U^T = U^{-1}, VV^T = I$) while D is a diagonal matrix. The factorization means that we can multiply the three matrices to get back the original matrix $A = UDV^T$. The transpose matrix is obtained through $A^T = VDU^T$. Since both orthogonal matrix and diagonal matrix have many nice properties, *SVD* is one of the most powerful matrix decomposition and is used in many applications, such as the least square solution (regression), feature selection (PCA, MDS), spectral clustering, image restoration and three-dimensional computer vision (fundamental matrix estimation), equilibrium of Markov Chain, and many others.

Matrix U and V are not unique, their columns come from the concatenation of eigenvectors of symmetric matrices AA^T and $A^T A$. Since eigenvectors of symmetric matrix are orthogonal (and linearly independent), they can be used as basis vectors (coordinate system) to span a multidimensional space. The absolute value of the determinant of orthogonal matrix is one, thus the matrix always has inverse. Furthermore, each column (and each row) of orthogonal matrix has unit norm. The diagonal matrix D contains the square of eigenvalues of symmetric matrix $A^T A$. The diagonal elements are nonnegative numbers and they are called singular values. Because they come from a symmetric matrix, the eigenvalues (and the eigenvectors) are all real numbers (no complex numbers).

Numerical computation of *SVD* is stable in terms of round off error. When some of the singular values are nearly zero, we can truncate them as zero and it yields numerical stability. If the SVD factor matrix $A = UDV^T$, then the diagonal matrix can also be obtained from $D = U^T A V$. The eigenvectors represent many solutions of the homogeneous equation. They are not unique and correct up to a scalar multiple.

Properties: SVD can reveal many things:

1. Singular value gives valuable information as to whether a square matrix A is singular. A square matrix A is nonsingular (i.e., have inverse) if and only if all its singular values are different from zero.

2. If the square matrix A is nonsingular, the inverse matrix can be obtained by

$$A^{-1} = VD^{-1}U^T.$$

3. The number of nonzero singular values is equal to the rank of any rectangular matrix. In fact, SVD is a robust technique to compute matrix rank against ill-conditioned matrices.

4. The ratio between the largest and the smallest singular value is called the condition number, which measures the degree of singularity and reveals ill-conditioned matrix.

5. SVD can produce one of matrix norms, which is called the Frobenius norm, by taking the sum of square of singular values

$$\|A\|_F = \sum_i \sigma_i^2 = \sum_i \sum_j a_{ij}^2.$$

The Frobenius norm is computed by taking the sums of the square elements in the matrix.

6. SVD can also produce a generalized inverse (pseudo-inverse) for any rectangular matrix. In fact, the generalized inverse is also a Moore-Penrose inverse by setting

$$A^+ = VD_0^{-1}U^T,$$

where the matrix D_0^{-1} is equal to D^{-1} but all nearly zero values are set to zero.

7. SVD also approximates the solution of the nonhomogenous linear system $AX = b$ such that the norm is minimum min $\|A - bx\|$. This is the basic of least square, orthogonal projection and regression analysis.

8. SVD also solves the homogenous linear system by taking the column of V^T, which represents the eigenvector corresponding to the zero eigenvalue of symmetric matrix $A^T A$.

Polynomial fraction description

4

This chapter is dedicated to giving a full introduction to polynomial fraction description, which is important to the main theme in the sequel. The complete description of polynomial fraction description can be found in [32, 55]. For polynomial fraction descriptions of time-varying linear systems, one is referred to [56]. For an introduction to coprime factorization, one finds details in [40]. For the expositions of the Smith form for a polynomial matrix and Smith-McMillan form for a rational matrix, one finds the discussions, for example, in [57]. For the polynomial matrix description and structural controllability of the composite system connected by some structural controllable and observable linear time-invariant subsystems in the representations of the irreducible matrix fraction description, recent developments can be found in [58]. For expositions on system poles and zeros, see [59–65]. A recent paper [66] introduces a block decoupling control technique, which is based on the spectral factors of the denominator of the right matrix fraction description in control design and low order controller. Applications of polynomial matrix factorization into resource allocation of optical communication systems are seen from [67]. The paper [68] proposes an alternative option to model a dynamical power system process, which is based on a matrix polynomial form. The model can be obtained naturally writing the high order differential equations of some components (e.g., transmission line models modeled by sequences of RLC PI circuits in power systems) or some of the subsystems that compose the whole system. For other relevant discussions, see also [13, 66, 69–75].

4.1 Introduction

The polynomial fraction description is a mathematically efficient and useful representation of a matrix of rational functions. With application to the transfer function of a multiinput, multioutput linear state equation, polynomial fraction description can inherit the structural features that, for example, permit natural generalization of minimal realization having established single-input, single-output state equations. The main aim of this chapter is to lay out a basic introduction to the polynomial fraction description, which is a key component in the current exposition of the generalized linear multivariable control theory.

We assume throughout this chapter a continuous-time setting, with $G(s)$ a $p \times m$ matrix of strictly proper rational functions of s. Then $G(s)$ is realizable by a time-invariant linear state equation. Reinterpretation for discrete time is straightforward by replacement of the Laplace-transform s by a z-transform z.

A Generalized Framework of Linear Multivariable Control. http://dx.doi.org/10.1016/B978-0-08-101946-7.00004-4

4.2   Right polynomial fractions

Matrices of real-coefficient polynomials in s, equivalently polynomials in s with coefficients that are real matrices, provide the mathematical foundation for the new transfer function representation.

Definition 4.2.1. A $p \times r$ polynomial matrix $P(s)$ is a matrix with entries that are real-coefficient polynomials in s. A square $(p = r)$ polynomial matrix $P(s)$ is called nonsingular if det $P(s)$ is a nonzero polynomial, and unimodular if det $P(s)$ is a nonzero real number.

The determinant of a square polynomial is a polynomial (a sum of products of the polynomial entries). Thus an alternative characterization is that a square polynomial matrix $P(s)$ is nonsingular if and only if det $P(s_0) \neq 0$ for all but a finite number of complex numbers s_0. And $P(s)$ is unimodular if and only if det $P(s_0) \neq 0$ for all complex numbers s_0.

The adjugate-overdeterminant formula shows that if $P(s)$ is square and nonsingular, then $P^{-1}(s)$ exits and (each entry) is a rational function of s. Also $P^{-1}(s)$ is a polynomial matrix if $P(s)$ is unimodular. Sometimes a polynomial is viewed as a rational function with unity denominator. From the reciprocal-determinant relationship between a matrix and its inverse, $P^{-1}(s)$ is unimodular if $P(s)$ is unimodular. Conversely if $P(s)$ and $P^{-1}(s)$ both are polynomial matrices, then both are unimodular.

Definition 4.2.2. A right polynomial fraction description for the $p \times m$ strictly proper rational transfer function $G(s)$ is an expression of the form

$$G(s) = N(s)D^{-1}(s), \tag{4.1}$$

where $N(s)$ is a $p \times m$ polynomial matrix and $D(s)$ is an $m \times m$ nonsingular polynomial matrix. A left polynomial fraction description for $G(s)$ is an expression

$$G(s) = D_L^{-1}(s)N_L(s), \tag{4.2}$$

where $N_L(s)$ is a $p \times m$ polynomial matrix and $D_L(s)$ is a $p \times p$ nonsingular polynomial matrix. The degree of a right polynomial fraction description is the degree of the polynomial det $D(s)$. Similarly the degree of a left polynomial fraction is the degree of det $D_L(s)$.

Of course this definition is familiar if $m = p = 1$. In the multiinput, multioutput case, a simple device can be used to exhibit so-called elementary polynomial fractions for $G(s)$. Suppose $d(s)$ is a least common multiple of the denominator polynomials of entries of $G(s)$. (In fact, any common multiple of the denominators can be used.) Then

$$N_d(s) = d(s)G(s)$$

is a $p \times m$ polynomial matrix, and we can write either a right or left polynomial fraction description

$$G(s) = N_d(s)[d(s)I_m]^{-1} = [d(s)I_p]^{-1}N_d(s). \tag{4.3}$$

The degrees of the two descriptions are different in general, and it should not be surprising that lower-degree polynomial fraction descriptions typically can be found if some effort is invested.

In the single-input, single-output case, the issue of common factors in the scalar numerator and denominator polynomials of $G(s)$ arises at this point. The utility of the polynomial fraction representation begins to emerge from the corresponding concept in the matrix case.

Definition 4.2.3. An $r \times r$ polynomial matrix $R(s)$ is called a right divisor of the $p \times r$ polynomial matrix $P(s)$ if there exists a $p \times r$ polynomial matrix $\widetilde{P}(s)$ such that

$$P(s) = \widetilde{P}(s)R(s).$$

If a right divisor $R(s)$ is nonsingular, then $P(s)R^{-1}(s)$ is a $p \times r$ polynomial matrix. Also if $P(s)$ is square and nonsingular, then every right divisor of $P(s)$ is nonsingular.

To become accustomed to these notions, it helps to reflect on the case of scalar polynomials. There a right divisor is simply a factor of the polynomial. For polynomial matrices the situation is roughly similar.

Next we consider a matrix-polynomial extension of the concept of a common factor of two scalar polynomials. Since one of the polynomial matrices is always square in our application to transfer function representation, attention is restricted to that situation.

Definition 4.2.4. Suppose $P(s)$ is a $p \times r$ polynomial matrix and $Q(s)$ is an $r \times r$ polynomial matrix. If the $r \times r$ polynomial matrix $R(s)$ is a right divisor of both, then $R(s)$ is called a common right divisor of $P(s)$ and $Q(s)$. We call $R(s)$ a greatest common right divisor of $P(s)$ and $Q(s)$ if it is a common right divisor, and if any other common right divisor of $P(s)$ and $Q(s)$ is a right divisor of $R(s)$. If all common right divisors of $P(s)$ and $Q(s)$ are unimodular, then $P(s)$ and $Q(s)$ are called right coprime.

For polynomial fraction descriptions of a transfer function, one of the polynomial matrices is always nonsingular, so only nonsingular common right divisors occur. Suppose $G(s)$ is given by the right polynomial fraction description

$$G(s) = N(s)D^{-1}(s)$$

and that $R(s)$ is a common right divisor of $N(s)$ and $D(s)$. Then

$$\widetilde{N}(s) = N(s)R^{-1}(s), \quad \widetilde{D}(s) = D(s)R^{-1}(s) \tag{4.4}$$

are polynomial matrices, and they provide another right polynomial fraction description for $G(s)$ since

$$\widetilde{N}(s)\widetilde{D}^{-1}(s) = N(s)R^{-1}(s)R(s)D^{-1}(s) = G(s).$$

The degree of this new polynomial fraction description is no greater than the degree of the original since

$$\deg\,[\det\,D(s)] = \deg\,[\det\,\widetilde{D}(s)] + \deg\,[\det\,R(s)].$$

Of course the largest degree reduction occurs if $R(s)$ is a greatest common right divisor, and no reduction occurs if $N(s)$ and $D(s)$ are right coprime. This discussion indicates that exacting common right divisors of a right polynomial fraction is a generalization of the process of canceling common factors in a scalar rational function.

Computation of the greatest common right divisors can be based on the capabilities of elementary row operations on a polynomial matrix-operation similar to the elementary row operations on a matrix of real numbers. To set up this approach we present a preliminary result.

Theorem 4.2.1. *Suppose $P(s)$ is a $p \times r$ polynomial matrix and $Q(s)$ is an $r \times r$ polynomial matrix. If a unimodular $(p + r) \times (p + r)$ polynomial matrix $U(s)$ and an $r \times r$ polynomial matrix $R(s)$ are such that*

$$U(s)\begin{bmatrix} Q(s) \\ P(s) \end{bmatrix} = \begin{bmatrix} R(s) \\ 0 \end{bmatrix} \tag{4.5}$$

then $R(s)$ is a greatest common right divisor of $P(s)$ and $Q(s)$.

Proof. Partition $U(s)$ in the form

$$U(s) = \begin{bmatrix} U_{11}(s) & U_{12}(s) \\ U_{21}(s) & U_{22}(s) \end{bmatrix}, \tag{4.6}$$

where $U_{11}(s)$ is $r \times r$, and $U_{22}(s)$ is $p \times p$. Then the polynomial matrix $U^{-1}(s)$ can be partitioned similarly as

$$U^{-1}(s) = \begin{bmatrix} U_{11}^{-}(s) & U_{12}^{-}(s) \\ U_{21}^{-}(s) & U_{22}^{-}(s) \end{bmatrix}.$$

Using this notation to rewrite Eq. (4.5) gives

$$\begin{bmatrix} Q(s) \\ P(s) \end{bmatrix} = \begin{bmatrix} U_{11}^{-}(s) & U_{12}^{-}(s) \\ U_{21}^{-}(s) & U_{22}^{-}(s) \end{bmatrix}\begin{bmatrix} R(s) \\ 0 \end{bmatrix}.$$

That is,

$$Q(s) = U_{11}^{-}(s)R(s), \quad P(s) = U_{21}^{-}(s)R(s).$$

Therefore $R(s)$ is a common right divisor of $P(s)$ and $Q(s)$. But, from Eqs. (4.5), (4.6),

$$R(s) = U_{11}(s)Q(s) + U_{12}(s)P(s), \tag{4.7}$$

so that if $R_a(s)$ is another common right divisor of $P(s)$ and $Q(s)$, say

$$Q(s) = \widetilde{Q}_a(s)R_a(s), \quad P(s) = \widetilde{P}_a(s)R_a(s),$$

then Eq. (4.7) gives

$$R(s) = [U_{11}(s)\widetilde{Q}_a(s) + U_{12}(s)\widetilde{P}_a(s)]R_a(s).$$

This shows $R_a(s)$ also is a right divisor of $R(s)$, and thus $R(s)$ is a greatest common right divisor of $P(s)$ and $Q(s)$.   □

To calculate the greatest common right divisors using Theorem 4.2.1, we consider three types of elementary row operations on a polynomial matrix. First is the interchange of two rows, and second is the multiplication of a row by a nonzero real number. The third is to add to any row a polynomial multiple of another row. Each of these elementary row operations can be represented by premultiplication by a unimodular matrix, as is easily seen by filling in the following argument.

Interchange of rows i and $j \neq i$ correspond to premultiplying by a matrix E_a that has a very simple form. The diagonal entries are unity, except that $[E_a]_{ii} = [E_a]_{jj} = 0$, and the off-diagonal entries are zero, except that $[E_a]_{ij} = [E_a]_{ji} = 1$. Multiplication of the ith-row by a real number $\alpha \neq 0$ corresponds to premultiplication by a matrix E_b that is diagonal with all diagonal entries unity, except $[E_b]_{ii} = \alpha$. Finally, adding to row i a polynomial $p(s)$ times row $j, j \neq i$, corresponds to premultiplication by a matrix $E_c(s)$ that has unity diagonal entries, with off-diagonal entries zero, except $[E_c(s)]_{ij} = p(s)$.

It is straightforward to show that the determinants of matrices of the form E_a, E_b, and $E_c(s)$ described above are nonzero real numbers. That is, these matrices are unimodular. Also it is easy to show that the inverse of any of these matrices corresponds to another elementary row operation. The diligent might prove that multiplication of a row by a polynomial is not an elementary row operation in the sense of multiplication by a unimodular matrix, thereby burying a frequent misconception.

It should be clear that a sequence of elementary row operations can be represented as premultiplication by a sequence of these elementary unimodular matrices, and thus as a single unimodular premultiplication. We also want to show the converse—that premultiplication by any unimodular matrix can be represented by a sequence of elementary row operations. Then Theorem 4.2.1 provides a method based on elementary row operations for computing a greatest common right divisor $R(s)$ via Eq. (4.5).

That any unimodular matrix can be written as a product of matrices of the form E_a, E_b, and $E_c(s)$ derives easily from a special form for polynomial matrices. We present this special form for the particular case where the polynomial matrix contains a full-dimension nonsingular partition. This suffices for our application to polynomial fraction descriptions, and also avoids some fussy but trivial issues, such as how to handle identical columns, or all-zero columns. Recall the terminology; that a scalar polynomial is called monic if the coefficient of the highest power of s is unity, that the degree of a polynomial is the highest power of s with nonzero coefficient, and that the degree of the zero polynomial is, by convention, $-\infty$.

Theorem 4.2.2. *Suppose $P(s)$ is a $p \times r$ polynomial matrix and $Q(s)$ is an $r \times r$, nonsingular polynomial matrix. Then elementary row operations can be used to transform*

$$M(s) = \begin{bmatrix} Q(s) \\ P(s) \end{bmatrix} \tag{4.8}$$

into row Hermite form described as follows. For $k = 1, \ldots, r$, all entries of the kth-column below the k, k-entry are zero, and the k, k-entry is nonzero and monic with higher degree than every entry above it in column k. (If the k, k-entry is unity, then all entries above it are zero.)

Proof. Row Hermite form can be computed by an algorithm that is similar to the row reduction process for constant matrices.

Step (i): In the first column of $M(s)$ use row interchange to bring to the first row a lowest-degree entry among nonzero first-column entries. (By nonsingularity of $Q(s)$, there is a nonzero first-column entry.)

Step (ii): Multiply the first row by a real number so that the first column entry is monic.

Step (iii): For each entry $m_{i1}(s)$ below the first row in the first column, use polynomial division to write

$$m_{i1}(s) = q_i(s)m_{11}(s) + r_{i1}(s), \quad i = 2, \ldots, p + r, \tag{4.9}$$

where each remainder is such that deg $r_{i1}(s) <$ deg $m_{11}(s)$. (If $m_{i1}(s) = 0$, that is deg $m_{i1}(s) = -\infty$, we set $q_i(s) = r_{i1}(s) = 0$. If deg $m_{i1}(s) = 0$, then by Step (i) deg $m_{11}(s) = 0$. Therefore deg $q_i(s) = 0$ and deg $r_{i1} = -\infty$, that is, $r_{i1}(s) = 0$.)

Step (iv): For $i = 2, \ldots, p + r$, add to the ith-row the product of $-q_i(s)$ and the first row. The resulting entries in the first column, below the first row, are $r_{21}(s), \ldots, r_{p+r,1}(s)$, all of which have degrees less than deg $m_{11}(s)$.

Step (v): Repeat Steps (i) through (iv) until all entries of the first column are zero except the first entry. Since the degrees of the entries below the first entry are lowered by at least one in each iteration, a finite number of operations is required.

□

Proceed to the second column of $M(s)$ and repeat the above steps while ignoring the first row. This results in a monic, nonzero entry $m_{22}(s)$, with all entries below it zero. If $m_{12}(s)$ does not have lower degree than $m_{22}(s)$, then polynomial division of $m_{12}(s)$ by $m_{22}(s)$ as in Step (iii) and an elementary row operation as in Step (iv) replaces $m_{12}(s)$ by a polynomial of degree less than deg$m_{22}(s)$. Next repeat the process for the third column of $M(s)$, while ignoring the first two rows. Continuing yields the claimed form on exhausting the columns of $M(s)$.

To complete the connection between unimodular matrices and elementary row operations, suppose in Theorem 4.2.2 that $p = 0$, and $Q(s)$ is unimodular. Of course the resulting row Hermite form is upper triangular. The diagonal entries must be unity, for a diagonal entry of positive degree would yield a determinant of positive degree, contradicting unimodularity. But then entries above the diagonal must have degree $-\infty$. Thus row Hermite form for a unimodular matrix is the identity matrix. In other words for a unimodular polynomial matrix $U(s)$ there is a sequence of elementary row operations, say $E_a, E_b, E_c(s), \ldots, E_b$, such that

$$[E_a E_b E_c(s) \cdots E_b]U(s) = I. \tag{4.10}$$

This obviously gives $U(s)$ as the sequence of elementary row operations on the identity specified by

$$U(s) = [E_b^{-1} \cdots E_c^{-1}(s)E_b^{-1}E_a^{-1}]I$$

and premultiplication of a matrix by $U(s)$ thus corresponds to application of a sequence of elementary row operations. Therefore Theorem 4.2.1 can be restated, for the case of nonsingular $Q(s)$, in terms of elementary row operations rather than premultiplication by a unimodular $U(s)$. If reduction to row Hermite form is used in implementing Eq. (4.5), then the greatest common right divisor $R(s)$ will be an upper-triangular polynomial matrix. Furthermore, if $P(s)$ and $Q(s)$ are right coprime, then Theorem 4.2.2 shows that there is a unimodular $U(s)$ such that Eq. (4.5) is satisfied for $R(s) = I_r$.

Two different characterizations of right coprimeness are used in the sequel. One is in the form of a polynomial matrix equation, while the other involves rank properties of a complex matrix obtained by evaluation of a polynomial matrix at complex values of s.

Theorem 4.2.3. *For a $p \times r$ polynomial matrix $P(s)$ and a nonsingular $r \times r$ polynomial matrix $Q(s)$, the following statements are equivalent:*

(i) *The polynomial matrices $P(s)$ and $Q(s)$ are right coprime.*

(ii) *There exist an $r \times p$ polynomial matrix $X(s)$ and an $r \times r$ polynomial matrix $Y(s)$ satisfying the so-called Bezout identity*

$$X(s)P(s) + Y(s)Q(s) = I_r. \tag{4.11}$$

(iii) *For every complex number s_0,*

$$\text{rank} \begin{bmatrix} Q(s_0) \\ P(s_0) \end{bmatrix} = r. \tag{4.12}$$

Proof. Beginning a demonstration that each claim implies the next, first we show that (i) implies (ii). If $P(s)$ and $Q(s)$ are right coprime, then reduction to row Hermite form as in Eq. (4.5) yields polynomial matrices $U_{11}(s)$ and $U_{12}(s)$ such that

$$U_{11}(s)Q(s) + U_{12}(s)P(s) = I_r$$

and this has the form of Eq. (4.11).

To prove that (ii) implies (iii), write the condition (4.11) in the matrix form

$$[Y(s) \quad X(s)] \begin{bmatrix} Q(s) \\ P(s) \end{bmatrix} = I_r.$$

If s_0 is a complex number for which

$$\text{rank} \begin{bmatrix} Q(s_0) \\ P(s_0) \end{bmatrix} < r,$$

then we have a rank contradiction.   □

To show (iii) implies (i), suppose that Eq. (4.12) holds and $R(s)$ is a common right divisor of $P(s)$ and $Q(s)$. Then for some $p \times r$ polynomial matrix $\tilde{P}(s)$ and some $r \times r$ polynomial matrix $\tilde{Q}(s)$

$$\begin{bmatrix} Q(s) \\ P(s) \end{bmatrix} = \begin{bmatrix} \tilde{Q}(s) \\ \tilde{P}(s) \end{bmatrix} R(s). \tag{4.13}$$

If $\det R(s)$ is a polynomial of degree at least one and s_0 is a root of this polynomial, then $R(s_0)$ is a complex matrix of less than full rank. Thus we obtain the contradiction

$$\text{rank} \begin{bmatrix} Q(s_0) \\ P(s_0) \end{bmatrix} \le \text{rank } R(s_0) < r.$$

Therefore $\det R(s)$ is a nonzero constant, that is, $R(s)$ is unimodular. This proves that $P(s)$ and $Q(s)$ are right coprime.

A right polynomial fraction description with $N(s)$ and $D(s)$ right coprime is called simply a coprime right polynomial fraction description. The next result shows that in an important sense all coprime right polynomial fraction descriptions of a given transfer function are equivalent. In particular they all have the same degree.

Theorem 4.2.4. *For any two coprime right polynomial fraction descriptions of a strictly proper rational transfer function*

$$G(s) = N(s)D^{-1}(s) = N_a(s)D_a^{-1}(s),$$

there exists a unimodular polynomial matrix $U(s)$ such that

$$N(s) = N_a(s)U(s), \quad D(s) = D_a(s)U(s).$$

Proof. By Theorem 4.2.3 there exist polynomial matrices $X(s), Y(s), A(s)$, and $B(s)$ such that

$$X(s)N_a(s) + Y(s)D_a(s) = I_m \tag{4.14}$$

and

$$A(s)N(s) + B(s)D(s) = I_m. \tag{4.15}$$

Since $N(s)D^{-1}(s) = N_a(s)D_a^{-1}(s)$, we have $N_a(s) = N(s)D^{-1}(s)D_a(s)$. Substituting this into Eq. (4.14) gives

$$X(s)N(s)D^{-1}(s)D_a(s) + Y(s)D_a(s) = I_m$$

or

$$X(s)N(s) + Y(s)D(s) = D_a^{-1}(s)D(s).$$

A similar calculation using $N(s) = N_a(s)D_a^{-1}(s)D(s)$ in Eq. (4.15) gives

$$A(s)N_a(s) + B(s)D_a(s) = D^{-1}(s)D_a(s).$$

Therefore both $D_a^{-1}(s)D(s)$ and $D^{-1}(s)D_a(s)$ are polynomial matrices, and since they are inverses of each other both must be unimodular. Let

$$U(s) = D_a^{-1}(s)D(s).$$

Then

$$N(s) = N_a(s)U(s), \quad D(s) = D_a(s)D_a^{-1}(s)D(s) = D_a(s)U(s)$$

and the proof is complete. $\qquad\qquad\qquad\qquad\qquad\qquad\qquad\qquad\qquad\quad\square$

4.3 Left polynomial fraction

Before going further we pause to consider left polynomial fraction descriptions and their relation to right polynomial fraction descriptions of the same transfer function. This means repeating much of the right-handed development, and proofs of the results are left as unlisted exercises.

Definition 4.3.1. A $q \times q$ polynomial matrix $L(s)$ is called a left divisor of the $q \times p$ polynomial matrix $P(s)$ if there exists a $q \times p$ polynomial matrix $\widetilde{P}(s)$ such that

$$P(s) = L(s)\widetilde{P}(s). \tag{4.16}$$

Definition 4.3.2. If $P(s)$ is a $q \times p$ polynomial matrix and $Q(s)$ is a $q \times q$ polynomial matrix, then a $q \times q$ polynomial matrix $L(s)$ is called a common left divisor of $P(s)$ and $Q(s)$ if $L(s)$ is a left divisor of both $P(s)$ and $Q(s)$. We call $L(s)$ a greatest common left divisor of $P(s)$ and $Q(s)$ if it is a common left divisor, and if any other common left divisor of $P(s)$ and $Q(s)$ is a left divisor of $L(s)$. If all common left divisors of $P(s)$ and $Q(s)$ are unimodular, then $P(s)$ and $Q(s)$ are called left coprime.

Theorem 4.3.1. *Suppose $P(s)$ is a $q \times p$ polynomial matrix and $Q(s)$ is a $q \times q$ polynomial matrix. If a $(q + p) \times (q + p)$ unimodular polynomial matrix $U(s)$ and a $q \times q$ polynomial matrix $L(s)$ are such that*

$$[Q(s) \quad P(s)]U(s) = [L(s) \quad 0], \tag{4.17}$$

then $L(s)$ is a greatest common left divisor of $P(s)$ and $Q(s)$.

Three types of elementary column operations can be represented by postmultiplication by a unimodular matrix. The first is interchange of two columns, and the second is multiplication of any column by a nonzero real number. The third elementary column operation is addition to any column of a polynomial multiple of another column. It is easy to check that a sequence of these elementary column operations can be represented by postmultiplication by a unimodular matrix. That postmultiplication by any unimodular matrix can be represented by an appropriate sequence of elementary column operations is a consequence of another special form, introduced below for the class of polynomial matrices of interest.

Theorem 4.3.2. *Suppose $P(s)$ is a $q \times p$ polynomial matrix and $Q(s)$ is a $q \times q$ nonsingular polynomial matrix. Then elementary column operations can be used to transform*

$$M(s) = [Q(s) \quad P(s)]$$

into a column Hermite form described as follows. For $k = 1, \ldots, q$, all entries of the kth-row to the right of the k, k-entry are zero, and the k, k-entry is monic with higher degree than any entry to its left. (If the k, k-entry is unity, all entries to its left are zero.)

Theorems 4.3.1 and 4.3.2 together provide a method for computing greatest common left divisors using elementary column operations to obtain column Hermite form. The polynomial matrix $L(s)$ in Eq. (4.17) will be lower-triangular.

Theorem 4.3.3. *For a $q \times p$ polynomial matrix $P(s)$ and a nonsingular $q \times q$ polynomial matrix $Q(s)$, the following statements are equivalent:*

(i) The polynomial matrices $P(s)$ and $Q(s)$ are left coprime.

(ii) There exist a $p \times q$ polynomial matrix $X(s)$ and a $q \times q$ polynomial matrix $Y(s)$ such that

$$P(s)X(s) + Q(s)Y(s) = I_q. \tag{4.18}$$

(iii) For every complex number s_0,

$$\text{rank } [Q(s_0) \quad P(s_0)] = q. \tag{4.19}$$

Naturally a left polynomial fraction description composed of left coprime polynomial matrices is called a coprime left polynomial fraction description.

Theorem 4.3.4. *For any two coprime left polynomial fraction descriptions of a strictly proper rational transfer function*

$$G(s) = D^{-1}(s)N(s) = D_a^{-1}(s)N_a(s),$$

there exists a unimodular polynomial matrix $U(s)$ such that

$$N(s) = U(s)N_a(s), \quad D(s) = U(s)D_a(s).$$

Suppose that we begin with the elementary right polynomial fraction description and the elementary left polynomial fraction description in Eq. (4.3) for a given strictly proper rational transfer function $G(s)$. Then appropriate greatest common divisors can be extracted to obtain a coprime right polynomial fraction description, and a coprime left polynomial fraction description for $G(s)$. We now show that these two coprime polynomial fraction descriptions have the same degree. An economical demonstration relies on a particular polynomial-matrix inversion formula.

Lemma 4.3.1. *Suppose that $V_{11}(s)$ is a $m \times m$ nonsingular polynomial matrix and*

$$V(s) = \begin{bmatrix} V_{11}(s) & V_{12}(s) \\ V_{21}(s) & V_{22}(s) \end{bmatrix}, \tag{4.20}$$

is an $(m + p) \times (m + p)$ nonsingular polynomial matrix. Then defining the matrix of rational functions $V_a(s) = V_{22}(s) - V_{21}(s)V_{11}^{-1}(s)V_{12}(s)$

(i) $\det V(s) = \det[V_{11}(s)] \cdot \det[V_a(s)]$,
(ii) $\det V_a(s)$ *is a nonzero rational function,*
(iii) *the inverse of $V(s)$ is*

$$V^{-1}(s) = \begin{bmatrix} V_{11}^{-1}(s) + V_{11}^{-1}(s)V_{12}(s)V_a^{-1}(s)V_{21}(s)V_{11}^{-1}(s) & -V_{11}^{-1}(s)V_{12}(s)V_a^{-1}(s) \\ -V_a^{-1}(s)V_{21}(s)V_{11}^{-1}(s) & V_a^{-1}(s). \end{bmatrix}$$

Proof. A partitioned calculation verifies

$$\begin{bmatrix} I_m & 0_{m \times p} \\ -V_{21}(s)V_{11}^{-1}(s) & I_p \end{bmatrix} V(s) = \begin{bmatrix} V_{11}(s) & V_{12}(s) \\ 0 & V_a(s) \end{bmatrix}. \tag{4.21}$$

Using the obvious determinant identity for block-triangular matrices, in particular

$$\det \begin{bmatrix} I_m & 0_{m \times p} \\ -V_{21}(s)V_{11}^{-1}(s) & I_p \end{bmatrix} = 1$$

gives

$$\det V(s) = \det[V_{11}(s)] \cdot \det[V_a(s)].$$

Since $V(s)$ and $V_{11}(s)$ are nonsingular polynomial matrices, this proves that $\det V_a(s)$ is a nonzero rational function, that is, $V_a^{-1}(s)$ exists. To establish (iii), multiply Eq. (4.21) on the left by

$$\begin{bmatrix} V_{11}^{-1}(s) & 0 \\ 0 & V_a^{-1}(s) \end{bmatrix} \begin{bmatrix} I_m & -V_{12}(s)V_a^{-1}(s) \\ 0 & I_p \end{bmatrix}$$

to obtain

$$\begin{bmatrix} V_{11}^{-1}(s) + V_{11}^{-1}(s)V_{12}(s)V_a^{-1}(s)V_{21}(s)V_{11}^{-1}(s) & -V_{11}^{-1}(s)V_{12}(s)V_a^{-1}(s) \\ -V_a^{-1}(s)V_{21}(s)V_{11}^{-1}(s) & V_a^{-1}(s) \end{bmatrix}$$

$$V(s) = \begin{bmatrix} I_m & 0 \\ 0 & I_p \end{bmatrix}$$

and the proof is complete.   □

Theorem 4.3.5. *Suppose that a strictly proper rational transfer function is represented by a coprime right polynomial fraction and a coprime left polynomial fraction*

$$G(s) = N(s)D^{-1}(s) = D_L^{-1}(s)N_L(s). \tag{4.22}$$

Then there exists a nonzero constant α such that $\det D(s) = \alpha \det D_L(s)$.

Proof. By right-coprimeness of $N(s)$ and $D(s)$ there exists an $(m + p) \times (m + p)$ unimodular polynomial matrix

$$U(s) = \begin{bmatrix} U_{11}(s) & U_{12}(s) \\ U_{21}(s) & U_{22}(s) \end{bmatrix}$$

such that

$$\begin{bmatrix} U_{11}(s) & U_{12}(s) \\ U_{21}(s) & U_{22}(s) \end{bmatrix} \begin{bmatrix} D(s) \\ N(s) \end{bmatrix} = \begin{bmatrix} I_m \\ 0 \end{bmatrix}. \tag{4.23}$$

For notational convenience let

$$\begin{bmatrix} U_{11}(s) & U_{12}(s) \\ U_{21}(s) & U_{22}(s) \end{bmatrix}^{-1} = \begin{bmatrix} V_{11}(s) & V_{12}(s) \\ V_{21}(s) & V_{22}(s) \end{bmatrix}.$$

Each $V_{ij}(s)$ is a polynomial matrix, and in particular Eq. (4.23) gives

$$V_{11}(s) = D(s), \quad V_{21}(s) = N(s).$$

Therefore $V_{11}(s)$ is nonsingular, and calling on Lemma 4.3.1 we have that

$$U_{22}(s) = [V_{22}(s) - V_{21}(s)V_{11}^{-1}(s)V_{12}(s)]^{-1},$$

which of course is a polynomial matrix, is nonsingular. Furthermore writing

$$\begin{bmatrix} U_{11}(s) & U_{12}(s) \\ U_{21}(s) & U_{22}(s) \end{bmatrix} \begin{bmatrix} V_{11}(s) & V_{12}(s) \\ V_{21}(s) & V_{22}(s) \end{bmatrix} = \begin{bmatrix} I_m & 0 \\ 0 & I_p \end{bmatrix} \tag{4.24}$$

gives, in the 2,2-block,

$$U_{21}(s)V_{12}(s) + U_{22}(s)V_{22}(s) = I_p.$$

By Theorem 4.3.3 this implies that $U_{21}(s)$ and $U_{22}(s)$ are left coprime. Also, from the 2,1-block,

$$U_{21}(s)V_{11}(s) + U_{22}(s)V_{21}(s) = U_{21}(s)D(s) + U_{22}(s)N(s) = 0. \tag{4.25}$$

Thus we can write, from Eq. (4.25)

$$G(s) = N(s)D^{-1}(s) = -U_{22}^{-1}(s)U_{21}(s). \tag{4.26}$$

This is a coprime left polynomial fraction description for $G(s)$. Again using Lemma 4.3.1, and the unimodularity of $V(s)$, there exists a nonzero constant α such that

$$\det \begin{bmatrix} V_{11}(s) & V_{12}(s) \\ V_{21}(s) & V_{22}(s) \end{bmatrix} = \det [V_{11}(s)] \cdot \det [V_{22}(s) - V_{21}(s)V_{11}^{-1}(s)V_{12}(s)]$$

$$= \det [D(s)] \cdot \det [U_{22}^{-1}(s)] \tag{4.27}$$

$$= \frac{\det D(s)}{\det U_{22}(s)} = \frac{1}{\alpha}.$$

□

Therefore, for the coprime left polynomial fraction description in Eq. (4.26), we have $\det U_{22}(s) = \alpha \det D(s)$. Finally, using the unimodular relation between coprime left polynomial fractions in Theorem 4.3.4, such a determinant formula, with possibly a different nonzero constant, must hold for any coprime left polynomial fraction description for $G(s)$.

4.4 Column and row degrees

There is an additional technical consideration that complicates the representation of a strictly proper rational transfer function by polynomial fraction descriptions. First, we introduce terminology for matrix polynomials that is related to the notion of the degree of a scalar polynomial. Recall again conventions that the degree of a nonzero constant is zero, and the degree of the polynomial 0 is $-\infty$.

Definition 4.4.1. For a $p \times r$ polynomial matrix $P(s)$, the degree of the highest-degree polynomial in the jth-column of $P(s)$, written $c_j[P]$, is called the jth-column degree of $P(s)$. The column degree coefficient matrix for $P(s)$, written P_{hc}, is the real $p \times r$ matrix with i, j-entry given by the coefficient of $s^{c_j[P]}$ in the i, j-entry of $P(s)$. If $P(s)$ is square and nonsingular, then it is called column reduced if

$$\deg [\det P(s)] = c_1[P] + \cdots + c_p[P]. \tag{4.28}$$

If $P(s)$ is square, then the Laplace expansion of the determinant about columns shows that the degree of $\det P(s)$ cannot be greater than $c_1[P] + \cdots + c_p[P]$. But it can be less.

The issue that requires attention involves the column degrees of $D(s)$ in a right polynomial fraction description for a strictly proper rational transfer function. It is clear in the $m = p = 1$ case that this column degree plays an important role in realization considerations, for example. The same is true in the multiinput, multioutput case, and the complication is that column degrees of $D(s)$ can be artificially high, and they can change in the process of postmultiplication by a unimodular matrix. Therefore two coprime right polynomial fraction descriptions for $G(s)$, as in Theorem 4.2.4, can be such that $D(s)$ and $D_a(s)$ have different column degrees, even though the degrees of the polynomials det $D(s)$ and det $D_a(s)$ are the same.

The first step in investigating this situation is to characterize column-reduced polynomial matrices in a way that does not involve computing a determinant. Using Definition 4.4.1 it is convenient to write a $p \times p$ polynomial matrix $P(s)$ in the form

$$P(s) = P_{hc} \begin{bmatrix} s^{c_1[P]} & 0 & \cdots & 0 \\ 0 & s^{c_2[P]} & \cdots & 0 \\ \vdots & \vdots & \vdots & \vdots \\ 0 & 0 & \cdots & s^{c_p[P]} \end{bmatrix} + P_l(s), \tag{4.29}$$

where $P_l(s)$ is a $p \times p$ polynomial matrix in which each entry of the jth-column has degree strictly less than $c_j[P]$. (We use this notation only when $P(s)$ is nonsingular, so that $c_1[P], \ldots, c_p[P] \geq 0$.)

Theorem 4.4.1. *If $P(s)$ is a $p \times p$ nonsingular polynomial matrix, then $P(s)$ is column reduced if and only if P_{hc} is invertible.*

Proof. We can write, using the representation (4.29)

$$s^{-c_1[P]-\cdots-c_p[P]} \cdot \det P(s) = \det [P(s) \cdot diagonal \{s^{-c_1[P]}, \ldots, s^{-c_p[P]}\}]$$

$$= \det [P_{hc} + P_l(s) \cdot diagonal \{s^{-c_1[P]}, \ldots, s^{-c_p[P]}\}] \tag{4.30}$$

$$= \det [P_{hc} + \widetilde{P}(s^{-1})],$$

where $\widetilde{P}(s^{-1})$ is a matrix with entries that are polynomials in s^{-1} that have no constant terms, that is, no s^0 terms. A key fact in the remaining argument is that, viewing s as real and positive, letting $s \to \infty$ yields $\widetilde{P}(s^{-1}) \to 0$. Also the determinant of a matrix is a continuous function of the matrix entries, so limit and determinant can be interchanged. In particular we can write

$$\lim_{s \to \infty} [s^{-c_1[P]-\cdots-c_p[P]} \cdot \det P(s)] = \lim_{s \to \infty} \det [P_{hc} + \widetilde{P}(s^{-1})]$$

$$= \det \{\lim_{s \to \infty} [P_{hc} + \widetilde{P}(s^{-1})]\} \tag{4.31}$$

$$= \det P_{hc}.$$

Using Eq. (4.28) the left side of Eq. (4.31) is a nonzero constant if and only if $P(s)$ is column reduced, and thus the proof is complete.   □

Consider a coprime right polynomial fraction description $N(s)D^{-1}(s)$, where $D(s)$ is not column reduced. We next show that elementary column operations on $D(s)$ (postmultiplication by a unimodular matrix $U(s)$) can be used to reduce individual column degrees, and thus compute a new coprime right polynomial fraction description

$$\widetilde{N}(s) = N(s)U(s), \quad \widetilde{D}(s) = D(s)U(s), \tag{4.32}$$

where $\widetilde{D}(s)$ is column reduced. Of course $U(s)$ need not be constructed, explicitly simply perform the same sequence of elementary column operations on $N(s)$ as on $D(s)$ to obtain $\widetilde{N}(s)$ along with $\widetilde{D}(s)$.

To describe the required calculations, suppose the column degrees of the $m \times m$ polynomial matrix $D(s)$ satisfy $c_1[D] \geq c_2[D], \ldots, c_m[D]$, as can achieved by column interchanges. Using the notation

$$D(s) = D_{hc}\Delta(s) + D_l(s), \tag{4.33}$$

there exists a nonzero $m \times 1$ vector z such that $D_{hc}z = 0$, since $D(s)$ is not column reduced. Suppose that the first nonzero entry in z is z_k, and define a corresponding polynomial vector by

$$z = \begin{bmatrix} 0 \\ \vdots \\ 0 \\ z_k \\ z_{k+1} \\ \vdots \\ z_m \end{bmatrix} \quad \rightarrow \quad z(s) = \begin{bmatrix} 0 \\ \vdots \\ 0 \\ z_k \\ z_{k+1}s^{c_k[D]-c_{k+1}[D]} \\ \vdots \\ z_m s^{c_k[D]-c_m[D]} \end{bmatrix}. \tag{4.34}$$

Then

$$\begin{aligned} D(s)z(s) &= D_{hc}\Delta(s)z(s) + D_l(s)z(s) \\ &= D_{hc}z s^{c_k[D]} + D_l(s)z(s) \\ &= D_l(s)z(s) \end{aligned} \tag{4.35}$$

and all entries of $D_l(s)z(s)$ have degree no greater than $c_k[D] - 1$. Choosing the unimodular matrix

$$U(s) = [e_1 \cdots e_{k-1}\ z(s)\ e_{k+1} \cdots e_m],$$

where e_i denotes the ith-column of I_m, it follows that $\widetilde{D}(s) = D(s)U(s)$ has column degrees satisfying

$$c_k[\widetilde{D}] < c_k[D]; \quad c_j[\widetilde{D}] = c_j[D], \quad j = 1, \ldots, k-1, k+1, \ldots, m.$$

If $\widetilde{D}(s)$ is not column reduced, then the process is repeated, beginning with the reordering of columns to obtain nonincreasing column degrees. A finite number of such repetitions builds a unimodular $U(s)$ such that $\widetilde{D}(s)$ in Eq. (4.32) is column reduced.

Another aspect of the column degree issue involves determining when a given $N(s)$ and $D(s)$ are such that $N(s)D^{-1}(s)$ is a strictly proper rational transfer function. The relative column degrees of $N(s)$ and $D(s)$ play important roles, but not as simply as the single-input, single-output case suggests.

Theorem 4.4.2. *If the polynomial fraction description $N(s)D^{-1}(s)$ is a strictly proper rational function, then $c_j[N] < c_j[D]$, $j = 1, \ldots, m$. If $D(s)$ is column reduced and $c_j[N] < c_j[D]$, $j = 1, \ldots, m$, then $N(s)D^{-1}(s)$ is a strictly proper rational function.*

Proof. Suppose $G(s) = N(s)D^{-1}(s)$ is strictly proper. Then $N(s) = G(s)D(s)$, and in particular

$$N_{ij}(s) = \sum_{k=1}^{m} G_{ik}(s)D_{kj}(s), \quad i = 1, \ldots, p; \quad j = 1, \ldots, m. \tag{4.36}$$

Then for any fixed value of j,

$$N_{ij}(s)s^{-c_j[D]} = \sum_{k=1}^{m} G_{ik}(s)D_{kj}(s)s^{-c_j[D]}, \quad i = 1, \ldots, p. \tag{4.37}$$

As we let (real) $s \to \infty$ each strictly proper rational function $G_{ik}(s)$ approaches 0, and each $D_{kj}(s)s^{-c_j[D]}$ approaches a finite constant, possibly zero. In any case this gives

$$\lim_{s \to \infty} N_{ij}(s)s^{-c_j[D]} = 0, \quad i = 1, \ldots, p.$$

Therefore deg $N_{ij}(s) < c_j[D], i = 1, \ldots, p$, which implies $c_j[N] < c_j[D]$.

Now suppose that $D(s)$ is column reduced, and $c_j[N] < c_j[D], j = 1, \ldots, m$. We can write

$$N(s)D^{-1}(s) = [N(s) \cdot diagonal \ \{s^{-c_1[D]}, \ldots, s^{-c_m[D]}\}]$$
$$\cdot [D(s) \cdot diagonal \ \{s^{-c_1[D]}, \ldots, s^{-c_m[D]}\}]^{-1} \tag{4.38}$$

and since $c_j[N] < c_j[D], j = 1, \ldots, m$,

$$\lim_{s \to \infty} [N(s) \cdot diagonal \ \{s^{-c_1[D]}, \ldots, s^{-c_m[D]}\}] = 0.$$

The adjugate-overdeterminant formula implies that each entry in the inverse of a matrix is a continuous function of the entries of the matrix. Thus limit can be interchanged with matrix inversion,

$$\lim_{s \to \infty} [D(s) \cdot diagonal \ \{s^{-c_1[D]}, \ldots, s^{-c_m[D]}\}]^{-1}$$
$$= [\lim_{s \to \infty} (D(s) \cdot diagonal \ \{s^{-c_1[D]}, \ldots, s^{-c_m[D]}\})]^{-1}. \tag{4.39}$$

Writing $D(s)$ in the form (4.29), the limit yields D_{hc}^{-1}. Then, from Eq. (4.38),

$$\lim_{s \to \infty} N(s)D^{-1}(s) = 0 \cdot D_{hc}^{-1} = 0,$$

which implies strict properness. $\qquad\square$

It remains to give the corresponding development for left polynomial fraction descriptions, though details are omitted.

Definition 4.4.2. For a $q \times p$ polynomial matrix $P(s)$, the degree of the highest-degree polynomial in the ith-row of $P(s)$, written $r_i[P]$, is called the ith-row degree of $P(s)$. The row degree coefficient matrix of $P(s)$, written P_{hr}, is the real $q \times p$ matrix with i, j-entry given by the coefficient of $s^{r_i[P]}$ in $P_{ij}(s)$. If $P(s)$ is square and nonsingular, then it is called row reduced if

$$\deg \ [\det \ P(s)] = r_1[P] + \cdots + r_q[P]. \tag{4.40}$$

Theorem 4.4.3. *If $p(s)$ is a $p \times p$ nonsingular polynomial matrix, then $P(s)$ is row reduced if and only if P_{hr} is invertible.*

Theorem 4.4.4. *If the polynomial fraction description $D^{-1}(s)N(s)$ is a strictly proper rational function, then $r_i[N] < r_i[D]$, $i = 1, \ldots, p$. If $D(s)$ is row reduced and $r_i[N] < r_i[D]$, $i = 1, \ldots, p$, then $D^{-1}(s)N(s)$ is a strictly proper rational function.*

Finally, if $G(s) = D^{-1}(s)N(s)$ is a polynomial fraction description and $D(s)$ is not row reduced, then a unimodular matrix $U(s)$ can be computed such that $D_b(s) = U(s)D(s)$ is row reduced. Letting $N_b(s) = U(s)N(s)$, the left polynomial fraction description

$$D_b^{-1}(s)N_b(s) = [U(s)D(s)]^{-1}U(s)N(s) = G(s) \tag{4.41}$$

has the same degree as the original.

Because of machinery developed in this chapter, a polynomial fraction description for a strictly proper rational transfer function $G(s)$ can be assumed as either a coprime right polynomial fraction description with column-reduced $D(s)$, or a coprime left polynomial fraction with row-reduced $D_L(s)$. In either case the degree of the polynomial fraction description is the same, and is given by the sum of the column degrees or, respectively, the sum of the row degrees.

4.5   Minimal realization

We assume that a $p \times m$ strictly proper rational transfer function is specified by a coprime right polynomial fraction description

$$G(s) = N(s)D^{-1}(s) \tag{4.42}$$

with $D(s)$ column reduced. Then the column degrees of $N(s)$ and $D(s)$ satisfy $c_j[N] < c_j[D]$, $j = 1, \ldots, m$. Some simplification occurs if one uninteresting case is ruled out. If $c_j[D] = 0$ for some j, then by Theorem 4.4.2 $G(s)$ is strictly proper if and only if all entries of the jth-column of $N(s)$ are zero, that is, $c_j[N] = -\infty$. Therefore a standing assumption in this chapter is that $c_1[D], \ldots, c_m[D] \geq 1$, which turns out to be compatible with assuming rank $B = m$ for a linear state equation. Recall that the degree of the polynomial fraction description (4.42) is $c_1[D] + \cdots + c_m[D]$, since $D(s)$ is column reduced.

We know there exists a minimal realization for $G(s)$,

$$\begin{aligned} \dot{x}(t) &= Ax(t) + Bu(t), \\ y(t) &= Cx(t). \end{aligned} \tag{4.43}$$

In exploring the connection between a transfer function and its minimal realizations, an additional bit of terminology is convenient.

Definition 4.5.1. Suppose $N(s)D^{-1}(s)$ is a coprime right polynomial fraction description for the $p \times m$, strictly proper, rational transfer function $G(s)$. Then the degree of this polynomial fraction description is called the McMillan degree of $G(s)$.

The first objective is to show that the McMillan degree of $G(s)$ is precisely the dimension of minimal realizations of $G(s)$. Our roundabout strategy is to prove that minimal realizations cannot have dimension less than the McMillan degree, and then compute a realization of dimension equal to the McMillan degree. This forces the conclusion that the computed realization is a minimal realization.

Lemma 4.5.1. *The dimension of any realization of a strictly proper rational transfer function $G(s)$ is at least the McMillan degree of $G(s)$.*

Proof. Suppose that the linear state equation (4.43) is a dimension-n minimal realization for the $p \times m$ transfer function $G(s)$. Then Eq. (4.43) is both controllable and observable, and

$$G(s) = C(sI - A)^{-1}B.$$

Define a $n \times m$ strictly proper transfer function $H(s)$ by the left polynomial fraction description

$$H(s) = D_L^{-1}(s)N_L(s) = (sI - A)^{-1}B. \tag{4.44}$$

Clearly this left polynomial fraction description has degree n. Since the state equation (4.43) is controllable, then

$$\text{rank } [D_L(s_0) \quad N_L(s_0)] = \text{rank } [(s_0 I - A) \quad B]$$
$$= n \tag{4.45}$$

for every complex s_0. Thus by Theorem 4.3.3 the left polynomial fraction description (4.44) is coprime. Now suppose $N_a(s)D_a^{-1}(s)$ is a coprime right polynomial fraction description for $H(s)$. Then this right polynomial fraction description also has degree n, and

$$G(s) = [CN_a(s)]D_a^{-1}(s) \tag{4.46}$$

is a degree-n right polynomial fraction description for $G(s)$, though not necessarily coprime. Therefore the McMillan degree of $G(s)$ is no greater than n, the dimension of a minimal realization of $G(s)$. $\qquad\square$

For notational assistance in the construction of a minimal realization, recall the integrator coefficient matrices corresponding to a set of k positive integers, $\alpha_1, \ldots, \alpha_k$, with $\alpha_1 + \cdots + \alpha_k = n$. These matrices are

$$A_0 = block\ diagonal \left\{ \begin{bmatrix} 0 & 1 & \cdots & 0 \\ 0 & 0 & \cdots & 0 \\ \vdots & \vdots & \vdots & \vdots \\ 0 & 0 & \cdots & 1 \\ 0 & 0 & \cdots & 0 \end{bmatrix}_{(\alpha_i \times \alpha_i)} , \quad i = 1, \ldots, k \right\}, \tag{4.47}$$

$$B_0 = block\ diagonal \left\{ \begin{bmatrix} 0 \\ \vdots \\ 0 \\ 1 \end{bmatrix}_{(\alpha_i \times 1)} , \quad i = 1, \ldots, k \right\}. \tag{4.48}$$

Define the corresponding integrator polynomial matrices by

$$\psi(s) = block\ diagonal \left\{ \begin{bmatrix} 1 \\ s \\ \vdots \\ s^{\alpha_i-1} \end{bmatrix}, \quad i = 1, \ldots, k \right\}, \tag{4.49}$$

$$\Delta(s) = diagonal\ \{s^{\alpha_1}, \ldots, s^{\alpha_k}\}. \tag{4.50}$$

The terminology couldn't be more appropriate, as we now demonstrate.

Lemma 4.5.2. *The integrator polynomial matrices provide a right polynomial fraction description for the corresponding integrator state equation. That is,*

$$(sI - A_0)^{-1}B_0 = \psi(s)\Delta^{-1}(s). \tag{4.51}$$

Proof. To verify Eq. (4.51), first multiply on the left by $(sI - A_0)$ and on the right by $\Delta(s)$ to obtain

$$B_0\Delta(s) = s\psi(s) - A_0\psi(s). \tag{4.52}$$

This expression is easy to check in a column-by-column fashion using the structure of the various matrices. For example the first column of Eq. (4.52) is the obvious

$$\begin{bmatrix} 0 \\ \vdots \\ 0 \\ s^{\alpha_1} \\ 0 \\ \vdots \\ 0 \end{bmatrix} = \begin{bmatrix} s \\ \vdots \\ s^{\alpha_1-1} \\ s^{\alpha_1} \\ 0 \\ \vdots \\ 0 \end{bmatrix} - \begin{bmatrix} s \\ \vdots \\ s^{\alpha_1-1} \\ 0 \\ 0 \\ \vdots \\ 0 \end{bmatrix}. \tag{4.53}$$

Proceeding similarly through the remaining columns in Eq. (4.52) yields the proof. $\square$

Completing our minimal realization strategy now reduces to comparing a special representation for the polynomial fraction description and a special structure for a dimension-n state equation.

Theorem 4.5.1. *Suppose that a strictly proper rational transfer function is described by a coprime right polynomial fraction description (4.42), where $D(s)$ is column reduced with column degrees $c_1[D], \ldots, c_m[D] \geq 1$. Then the McMillan degree of $G(s)$ is given by $n = c_1[D] + \cdots + c_m[D]$, and minimal realizations of $G(s)$ have dimension n. Furthermore, writing*

$$\begin{aligned} N(s) &= N_l\psi(s) \\ D(s) &= D_{hc}\Delta(s) + D_l\psi(s), \end{aligned} \tag{4.54}$$

where $\psi(s)$ and $\Delta(s)$ are the integrator polynomial matrices corresponding to $c_1[D], \ldots, c_m[D]$, a minimal realization for $G(s)$ is

$$\begin{aligned} \dot{x}(t) &= (A_0 - B_0D_{hc}^{-1}D_l)x(t) + B_0D_{hc}^{-1}u(t) \\ y(t) &= N_lx(t), \end{aligned} \tag{4.55}$$

where A_0 and B_0 are the integrator coefficient matrices corresponding to $c_1[D], \ldots, c_m[D]$.

Proof. First we verify that Eq. (4.55) is a realization for $G(s)$. It is straightforward to write down the representation in Eq. (4.54), where N_l and D_l are constant matrices that select for appropriate polynomial entries of $N(s)$, and $D_l(s)$. Then solving for $\triangle(s)$ in Eq. (4.54) and substituting into Eq. (4.52) gives

$$
\begin{aligned}
B_0 D_{hc}^{-1} D(s) &= s\psi(s) - A_0\psi(s) + B_0 D_{hc}^{-1} D_l\psi(s) \\
&= (sI - A_0 + B_0 D_{hc}^{-1} D_l)\psi(s).
\end{aligned} \tag{4.56}
$$

This implies

$$
(sI - A_0 + B_0 D_{hc}^{-1} D_l)^{-1} B_0 D_{hc}^{-1} = \psi(s)D^{-1}(s) \tag{4.57}
$$

from which the transfer function for Eq. (4.55) is

$$
N_l(sI - A_0 + B_0 D_{hc}^{-1} D_l)^{-1} B_0 D_{hc}^{-1} = N(s)D^{-1}(s). \tag{4.58}
$$

Thus Eq. (4.55) is a realization of $G(s)$ with dimension $c_1[D] + \cdots + c_m[D]$, which is the McMillan degree of $G(s)$. Then by invoking Lemma 4.5.1 we conclude that the McMillan degree of $G(s)$ is the dimension of minimal realizations of $G(s)$. $\square$

In the minimal realization (4.55), note that if D_{hc} is upper triangular with unity diagonal entries, then the realization is in the controller form. (Upper triangular structure for D_{hc} can be obtained by elementary column operations on the original polynomial fraction description.) If Eq. (4.55) is in controller form, then the controllability indices are precisely $\rho_1 = c_1[D], \ldots, \rho_m = c_m[D]$. We see that all minimal realizations of $N(s)D^{-1}(s)$ have the same controllability indices up to reordering. All minimal realizations of a strictly proper rational transfer function $G(s)$ have the same controllability indices up to reordering.

Calculations similar to those in the proof of Theorem 4.5.1 can be used to display a right polynomial fraction description for a given linear state equation.

Theorem 4.5.2. *Suppose the linear state equation (4.43) is controllable with controllability indices $\rho_1, \ldots, \rho_m \geq 1$. Then the transfer function for Eq. (4.43) is given by the right polynomial fraction description*

$$
C(sI - A)^{-1}B = N(s)D^{-1}(s), \tag{4.59}
$$

where

$$
\begin{aligned}
N(s) &= CP^{-1}\psi(s) \\
D(s) &= R^{-1}\triangle(s) - R^{-1}UP^{-1}\psi(s)
\end{aligned} \tag{4.60}
$$

and $D(s)$ is column reduced. Here $\psi(s)$ and $\triangle(s)$ are the integrator polynomial matrices corresponding to $\rho_1, \ldots, \rho_m, P$ is the controller-form variable change, and U and R are the coefficient matrices. If the state equation (4.43) also is observable, then $N(s)D^{-1}(s)$ is coprime with degree n.

Proof. We can write

$$
PAP^{-1} = A_0 + B_0 UP^{-1}, \quad PB = B_0 R, \tag{4.61}
$$

where A_0 and B_0 are the integrator coefficient matrices corresponding to $\rho_1, \ldots, \rho_m$. Let $\triangle(s)$ and $\psi(s)$ be the corresponding integrator polynomial matrices. Using Eq. (4.60) to substitute for $\triangle(s)$ in Eq. (4.52) gives

$$B_0 R D(s) + B_0 U P^{-1} \psi(s) = s\psi(s) - A_0 \psi(s). \tag{4.62}$$

Rearranging this expression yields

$$\psi(s)D^{-1}(s) = (sI - A_0 - B_0 U P^{-1})^{-1} B_0 R \tag{4.63}$$

and therefore

$$\begin{aligned} N(s)D^{-1}(s) &= CP^{-1}(sI - A_0 - B_0 U P^{-1})^{-1} B_0 R \\ &= CP^{-1}(sI - PAP^{-1})^{-1} PB \\ &= C(sI - A)^{-1}B. \end{aligned} \tag{4.64}$$

This calculation verifies that the polynomial fraction description defined by Eq. (4.60) represents the transfer function of the linear state equation (4.43). Also, $D(s)$ in Eq. (4.60) is column reduced because $D_{hc} = R^{-1}$. Since the degree of the polynomial fraction description is n, if the state equation also is observable, hence a minimal realization of its transfer function, then n is the McMillan degree of the polynomial fraction description (4.60).   □

For left polynomial fraction descriptions, the strategy for right fraction descriptions applies since the McMillan degree of $G(s)$ also is the degree of any coprime left polynomial fraction description for $G(s)$. The only details that remain in proving a left-handed version of Theorem 4.5.1 involve construction of a minimal realization. But this construction is not difficult to deduce from a summary statement.

Theorem 4.5.3. *Suppose that a strictly proper rational transfer function is described by a coprime left polynomial fraction description $D^{-1}(s)N(s)$, where $D(s)$ is row reduced with row degrees $r_1[D], \ldots, r_p[D] \geq 1$. Then the McMillan degree of $G(s)$ is given by $n = r_1[D] + \cdots + r_p[D]$, and minimal realizations of $G(s)$ have dimension n. Furthermore, writing*

$$\begin{aligned} N(s) &= \psi^T(s)N_l \\ D(s) &= \triangle(s)D_{hr} + \psi^T(s)D_l, \end{aligned} \tag{4.65}$$

where $\psi(s)$ and $\triangle(s)$ are the integrator polynomial matrices corresponding to $r_1[D], \ldots, r_p[D]$, a minimal realization for $G(s)$ is

$$\begin{aligned} \dot{x}(t) &= (A_0^T - D_l D_{hr}^{-1} B_0^T)x(t) + N_l u(t) \\ y(t) &= D_{hr}^{-1} B_0^T x(t), \end{aligned} \tag{4.66}$$

where A_0 and B_0 are the integrator coefficient matrices corresponding to $r_1[D], \ldots, r_p[D]$.

Analogous to the discussion following Theorem 4.5.3, in the setting of Theorem 4.5.3 the observability indices of minimal realizations of $D^{-1}(s)N(s)$ are the same, up to reordering, as the row degrees of $D(s)$.

Theorem 4.5.4. *Suppose the linear state equation (4.43) is observable with observability indices $\eta_1, \ldots, \eta_p \geq 1$. Then the transfer function for Eq. (4.43) is given by the left polynomial fraction description*

$$C(sI - A)^{-1}B = D^{-1}(s)N(s),$$

where

$$N(s) = \psi^T(s)Q^{-1}B,$$
$$D(s) = \triangle(s)S^{-1} - \psi^T(s)Q^{-1}VS^{-1} \tag{4.67}$$

and $D(s)$ is row reduced. Here $\psi(s)$ and $\triangle(s)$ are the integrator polynomial matrices corresponding to $\eta_1, \ldots, \eta_p$, Q is the observer-form variable change, and V and S are the coefficient matrices. If the state equation (4.43) also is controllable, then $D^{-1}(s)N(s)$ is coprime with degree n.

4.6 Poles and zeros

The connections between a coprime polynomial fraction description for a strictly proper rational transfer function $G(s)$ and minimal realizations of $G(s)$ can be used to define notions of poles and zeros of $G(s)$ that generalize the familiar notions for scalar transfer functions. In addition we characterize these concepts in terms of response properties of a minimal realization of $G(s)$. For the peer results for discrete time, some translation and modification of these results are required.

Given coprime polynomial fraction descriptions

$$G(s) = N(s)D^{-1}(s) = D_L^{-1}(s)N_L(s), \tag{4.68}$$

it follows from Theorem 4.3.5 that the polynomials det $D(s)$ and det $D_L(s)$ have the same roots. Furthermore from Theorem 4.2.4 it is clear that these roots are the same for every coprime polynomial description. This permits the introduction of terminology in terms of either a right or left polynomial fraction description, though we adhere to a societal bias and use right.

Definition 4.6.1. Suppose $G(s)$ is a strictly proper rational transfer function. A complex number s_0 is called a pole of $G(s)$ if det $D(s_0) = 0$, where $N(s)D^{-1}(s)$ is a coprime right polynomial fraction description for $G(s)$. The multiplicity of a pole s_0 is the multiplicity of s_0 as a root of the polynomial det $D(s)$.

This terminology is compatible with customary usage in the $m = p = 1$ case. Specifically if s_0 is a pole of $G(s)$, then some entry $G_{ij}(s)$ is such that $|G_{ij}(s_0)| = \infty$. Conversely if some entry of $G(s)$ has infinite magnitude when evaluated at the complex number s_0, then s_0 is a pole of $G(s)$. A linear state equation with transfer function $G(s)$ is uniformly bounded-input, bounded-output stable if and only if all poles of $G(s)$ have negative real parts, that is, all roots of det $D(s)$ have negative real parts.

The relation between eigenvalues of A in the linear state equation (4.43) and poles of the corresponding transfer function

$$G(s) = C(sI - A)^{-1}B$$

is a crucial feature in some of our arguments. Writing $G(s)$ in terms of a coprime right polynomial fraction description gives

$$\frac{N(s) \cdot adj\ D(s)}{det\ D(s)} = \frac{C[adj\ (sI - A)]B}{det\ (sI - A)}. \tag{4.69}$$

Using Lemma 4.5.1, Eq. (4.69) reveals that if s_0 is a pole of $G(s)$ with multiplicity σ_0, then s_0 is an eigenvalue of A with multiplicity at least σ_0. But simple single-input, single-output examples confirm that multiplicities can be different, and in particular an eigenvalue of A might not be a pole of $G(s)$. The remedy for this displeasing situation is to assume 4.43 is controllable and observable. Then Eq. (4.69) shows that, since the denominator polynomials are identical up to a constant multiplier, the set of poles of $G(s)$ is identical to the set of eigenvalues of a minimal realization of $G(s)$.

This discussion leads to an interpretation of a pole of a transfer function in terms of zero-input response properties of a minimal realization of the transfer function.

Theorem 4.6.1. *Suppose the linear state equation (4.43) is controllable and observable. Then the complex number s_0 is a pole of*

$$G(s) = C(sI - A)^{-1}B$$

if and only if there exists a complex $n \times 1$ vector x_0 and a complex $p \times 1$ vector $y_0 \neq 0$ such that

$$Ce^{At}x_0 = y_0 e^{s_0 t}, \quad t \geq 0. \tag{4.70}$$

Proof. If s_0 is a pole of $G(s)$, then s_0 is an eigenvalue of A. With x_0 an eigenvector of A corresponding to the eigenvalue s_0, we have

$$e^{At}x_0 = e^{s_0 t}x_0.$$

This easily gives Eq. (4.70), where $y_0 = Cx_0$ is nonzero by the observability of Eq. (4.43).

On the other hand if Eq. (4.70) holds, then taking Laplace transforms gives

$$C(sI - A)^{-1}x_0 = y_0(s - s_0)^{-1}$$

or,

$$(s - s_0)C[adj\ (sI - A)]x_0 = y_0 \cdot det\ (sI - A). \tag{4.71}$$

Evaluating this at $s = s_0$ shows that, since $y_0 \neq 0$, $det\ (s_0 I - A) = 0$. Therefore s_0 is an eigenvalue of A and, by minimality of the state equation, a pole of $G(s)$. $\qquad\square$

Of course if s_0 is a real pole of $G(s)$, then Eq. (4.70) directly gives a corresponding zero-input response property of minimal realizations of $G(s)$. If s_0 is complex, then the real initial state $x_0 + \bar{x}_0$ gives an easily computed real response that can be written as a product of an exponential with exponent $(Re\ [s_0])t$ and a sinusoid with frequency $Im\ [s_0]$.

The concept of a zero of a transfer function is more delicate. For a scalar function $G(s)$ with coprime numerator and denominator polynomials, a zero is a complex number s_0 such that $G(s_0) = 0$. Evaluations of a scalar $G(s)$ at particular complex numbers can result in a zero or nonzero complex value, or can be undefined (at a pole). These possibilities multiply for multiinput, multioutput systems, where a corresponding notion of a zero is a complex s_0 where the matrix $G(s_0)$ "loses rank."

To carefully define the concept of a zero, the underlying assumption we make is that rank $G(s) = \min[m, p]$ for almost all complex values of s. (By "almost all" we mean "all but a finite number.") In particular, at poles of $G(s)$ at least one entry of $G(s)$ is ill-defined, and so poles are among those values of s ignored when checking rank. (Another phrasing of this assumption is that $G(s)$ is assumed to have rank $\min[m, p]$ over the field of rational functions, a more sophisticated terminology that we do not further employ.) Now consider coprime polynomial fraction descriptions

$$G(s) = N(s)D^{-1}(s) = D_L^{-1}(s)N_L(s) \tag{4.72}$$

for $G(s)$. Since both $D(s)$ and $D_L(s)$ are nonsingular polynomial matrices, assuming rank $G(s) = \min[m, p]$ for almost all complex values of s is equivalent to assuming rank $N(s) = \min[m, p]$ for almost all complex values of s, and also equivalent to assuming rank $N_L(s) = \min[m, p]$ for almost all complex values of s. The agreeable feature of polynomial fraction descriptions is that $N(s)$ and $N_L(s)$ are well-defined for all values of s. Either right or left polynomial fractions can be adopted as the basis for defining transfer-function zeros.

Definition 4.6.2. Suppose $G(s)$ is a strictly proper rational transfer function with rank $G(s) = \min[m, p]$ for almost all complex numbers s. A complex number s_0 is called a transmission zero of $G(s)$ if rank $N(s_0) < \min[m, p]$, where $N(s)D^{-1}(s)$ is any coprime right polynomial fraction description for $G(s)$.

This reduces to the customary definition in the single-input, single-output ease. But a look at multiinput, multioutput examples reveals subtleties in the concept of transmission zero.

Another complication arises as we develop a characterization of transmission zeros in terms of identically zero response of a minimal realization of $G(s)$ to a particular initial state and particular input signal. Namely with $m \geq 2$ there can exist a nonzero $m \times 1$ vector $U(s)$ of strictly proper rational functions such that $G(s)U(s) = 0$. In this situation multiplying all the denominators in $U(s)$ by the same nonzero polynomial in s generates whole families of inputs for which the zero-state response is identically zero. This inconvenience always occurs when $m > p$. Here we add an assumption that forces $m \leq p$.

The basic idea is to devise an input $U(s)$ such that the zero-state response component contains exponential terms due solely to poles of the transfer function, and such that these exponential terms can be canceled by terms in the zero-input response component.

Theorem 4.6.2. *Suppose the linear state equation (4.43) is controllable and observable, and*

$$G(s) = C(sI - A)^{-1}B \tag{4.73}$$

has rank m for almost all complex numbers s. If the complex number s_0 is not a pole of $G(s)$, then it is a transmission zero of $G(s)$ if and only if there is a nonzero, complex $m \times 1$ vector u_0 and a complex $n \times 1$ vector x_0 such that

$$Ce^{At}x_0 + \int_0^t Ce^{A(t-\sigma)}Bu_0e^{s_0\sigma}\,d\sigma = 0, \quad t \geq 0. \tag{4.74}$$

Proof. Suppose $N(s)D^{-1}(s)$ is a coprime right polynomial fraction description for Eq. (4.73). If s_0 is not a pole of $G(s)$, then $D(s_0)$ is invertible and s_0 is not an eigenvalue of A. If x_0 and $u_0 \neq 0$ are such that Eq. (4.74) holds, then the Laplace of Eq. (4.74) gives

$$C(sI - A)^{-1}x_0 + N(s)D^{-1}(s)u_0(s - s_0)^{-1} = 0$$

or

$$(s - s_0)C(sI - A)^{-1}x_0 + N(s)D^{-1}(s)u_0 = 0.$$

Evaluating this expression at $s = s_0$ yields

$$N(s_0)D^{-1}(s_0)u_0 = 0$$

and this implies that rank $N(s_0) < m$. That is, s_0 is a transmission zero of $G(s)$. $\square$

On the other hand suppose s_0 is not a pole of $G(s)$. Using the easily verified identity

$$(s_0I - A)^{-1}(s - s_0)^{-1} = (sI - A)^{-1}(s_0I - A)^{-1} + (sI - A)^{-1}(s - s_0)^{-1}, \tag{4.75}$$

we can write, for any $m \times 1$ complex vector u_0 and corresponding $n \times 1$ complex vector $x_0 = (s_0I - A)^{-1}Bu_0$, the Laplace transform expression

$$\begin{aligned}
\mathcal{L}\{Ce^{At}x_0 &+ \int_0^t Ce^{A(t-\sigma)}Bu_0e^{s_0\sigma}\,d\sigma\} \\
&= C(sI - A)^{-1}x_0 + C(sI - A)^{-1}Bu_0(s - s_0)^{-1} \\
&= C[(sI - A)^{-1}(s_0I - A)^{-1} + (sI - A)^{-1}(s - s_0)^{-1}]Bu_0 \\
&= G(s_0)u_0(s - s_0)^{-1} \\
&= N(s_0)D^{-1}(s_0)u_0(s - s_0)^{-1}.
\end{aligned} \tag{4.76}$$

Taking the inverse Laplace transform gives, for the particular choice of x_0 above,

$$Ce^{At}x_0 + \int_0^t Ce^{A(t-\sigma)}Bu_0e^{s_0\sigma}\,d\sigma = N(s_0)D^{-1}(s_0)u_0e^{s_0t}, \quad t \geq 0. \tag{4.77}$$

Clearly the $m \times 1$ vector u_0 can be chosen so that this expression is zero for $t \geq 0$ if rank $N(s_0) < m$, that is, if s_0 is a transmission zero of $G(s)$.

Of course if a transmission zero s_0 is real and not a pole, then we can take u_0 real, and the corresponding $x_0 = (s_0 I - A)^{-1} B u_0$ is real. Then Eq. (4.74) shows that the complete response for $x(0) = x_0$ and $u(t) = u_0 e^{s_0 t}$ is identically zero. If s_0 is a complex transmission zero, then specification of a real input and real initial state that provides identically zero response is left as a mild exercise.

4.7 State feedback

Properties of linear state feedback

$$u(t) = Kx(t) + Mr(t)$$

applied to a linear state equation (4.43). As noted, a direct approach to relating the closed-loop and plant transfer functions is unpromising in the case of state feedback. However, polynomial fraction descriptions and an adroit formulation lead to a way around the difficulty.

We assume that a strictly proper rational transfer function for the plant is given as a coprime right polynomial fraction $G(s) = N(s)D^{-1}(s)$ with $D(s)$ column reduced. To represent linear state feedback, it is convenient to write the input-output description

$$Y(s) = N(s)D^{-1}(s)U(s) \tag{4.78}$$

as a pair of equations with polynomial matrix coefficients

$$D(s)\xi(s) = U(s),$$
$$Y(s) = N(s)\xi(s). \tag{4.79}$$

The $m \times 1$ vector $\xi(s)$ is called the pseudo-state of the plant. This terminology can be motivated by considering a minimal realization of the form (4.55) for $G(s)$. From Eq. (4.57) we write

$$\psi(s)\xi(s) = \psi(s)D^{-1}(s)U(s)$$
$$= (sI - A_0 + B_0 D_{hc}^{-1} D_l)^{-1} B_0 D_{hc}^{-1} U(s) \tag{4.80}$$

or

$$s\psi(s)\xi(s) = (A_0 - B_0 D_{hc}^{-1} D_l)\psi(s)\xi(s) + B_0 D_{hc}^{-1} U(s). \tag{4.81}$$

Defining the $n \times 1$ vector $x(t)$ as the inverse Laplace transform

$$x(t) = \mathcal{L}^{-1}[\psi(s)\xi(s)],$$

we see that Eq. (4.81) is the Laplace transform representation of the linear state equation (4.55) with zero initial state. Beyond motivation for terminology, this

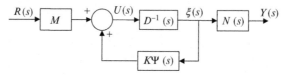

Fig. 4.1 Transfer function diagram for state feedback.

development shows that linear state feedback for a linear state equation corresponds to feedback of $\psi(s)\xi(s)$ in the associated pseudo-state representation (4.79).

Now, as illustrated in Fig. 4.1, consider linear state feedback for Eq. (4.79) represented by

$$U(s) = K\psi(s)\xi(s) + MR(s), \tag{4.82}$$

where K and M are real matrices of dimensions $m \times n$ and $m \times m$, respectively. We assume that M is invertible. To develop a polynomial fraction description for the resulting closed-loop transfer function, substitute Eq. (4.82) into Eq. (4.79) to obtain

$$[D(s) - K\psi(s)]\xi(s) = MR(s)$$
$$Y(s) = N(s)\xi(s). \tag{4.83}$$

Nonsingularity of the polynomial matrix $D(s) - K\psi(s)$ is assured, since its column degree coefficient matrix is the same as the assumed-invertible column degree coefficient matrix for $D(s)$. Therefore we can write

$$\xi(s) = [D(s) - K\psi(s)]^{-1}MR(s)$$
$$Y(s) = N(s)\xi(s). \tag{4.84}$$

Since M is invertible Eq. (4.84) gives a right polynomial fraction description for the closed-loop transfer function

$$N(s)\hat{D}^{-1}(s) = N(s)[M^{-1}D(s) - M^{-1}K\psi(s)]^{-1}. \tag{4.85}$$

This description is not necessarily coprime, though $\hat{D}(s)$ is column reduced.

Reflection on Eq. (4.85) reveals that choices of K and invertible M provide complete freedom to specify the coefficients of $\hat{D}(s)$. In detail, suppose

$$D(s) = D_{hc}\Delta(s) + D_l\psi(s)$$

and suppose the desired $\hat{D}(s)$ is

$$\hat{D}(s) = \hat{D}_{hc}\Delta(s) + \hat{D}_l\psi(s).$$

Then the feedback gains

$$M = D_{hc}\hat{D}_{hc}^{-1}, \quad K = -M\hat{D}_l + D_l$$

accomplish the task. Although the choices of K and M do not directly affect $N(s)$, there is an indirect effect in that Eq. (4.85) might not be coprime. This occurs in a more obvious fashion in the single-input, single-output case when linear state feedback places a root of the denominator polynomial coincident with a root of the numerator polynomial.

Stability

There have been a large number of references on stability theory for ordinary differential equations, to mention a few [70, 76–86]. On the basis of the above references, this chapter serves a tutorial summary on stability of linear system, which covers the topics of internal stability, Lyapunov stability, and input-output stability.

5.1 Internal stability

Internal stability deals with boundedness properties and asymptotic behavior (as $t \to \infty$) of solutions of the zero-input linear state equation

$$\dot{x}(t) = A(t)x(t), \quad x(t_0) = x_0. \tag{5.1}$$

While bounds on solutions might be of interest for fixed t_0 and x_0, or for various initial states at a fixed t_0, we focus on boundedness properties that hole regardless of the choice of t_0 or x_0. In a similar fashion the concept we adopt relative to asymptotically zero solutions is independent of the choice of initial time. The reason is that these "uniform in t_0" concepts are most appropriate in relation to input-output stability properties of linear state equations.

It is natural to begin by characterizing stability of linear state equation (5.1) in terms of bounds on the transition matrix $\Phi(t, \pi)$ for $A(t)$. This leads to a well-known eigenvalue condition when $A(t)$ is constant, but does not provide a generally useful stability test for time-varying examples because of the difficulty of computing $\Phi(t, \pi)$.

The first stability notion involves boundedness of solutions of Eq. (5.1). Because solutions are linear in the initial state, it is convenient to express the bound as a linear function of the norm of the initial state.

Definition 5.1.1. The linear state equation (5.1) is called uniformly stable if there exists a finite positive constant γ such that for any t_0 and x_0 the corresponding solution satisfies

$$\|x(t)\| \leq \gamma \|x_0\|, \quad t \geq t_0. \tag{5.2}$$

Evaluation of Eq. (5.2) at $t = t_0$ shows that the constant γ must satisfy $\gamma \geq 1$. The adjective uniform in the definition refers precisely to the fact that γ must not depend on the choice of initial time, as illustrated in Fig. 5.1. A "nonuniform" stability concept can be defined by permitting γ to depend on the initial time, but this is not considered here except to show that there is a difference via a standard example.

A Generalized Framework of Linear Multivariable Control. http://dx.doi.org/10.1016/B978-0-08-101946-7.00005-6

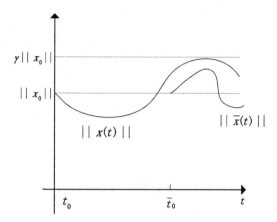

Fig. 5.1 Uniform stability implies the γ-bound is independent of t_0.

Theorem 5.1.1. *The linear state equation (5.1) is uniformly stable if and only if there exists a finite positive constant γ such that*

$$\|\Phi(t, \tau)\| \le \gamma \tag{5.3}$$

for all t, τ such that $t \ge \tau$.

Proof. First suppose that such a γ exists. Then for any t_0 and x_0 the solution of Eq. (5.1) satisfies

$$\|x(t)\| = \|\Phi(t, t_0)x_0\| \le \|\Phi(t, t_0)\| \|x_0\| \le \gamma \|x_0\|, \quad t \ge t_0$$

and uniform stability is established.

For the reverse implication suppose that the state equation (5.1) is uniformly stable. Then there is a finite γ such that, for any t_0 and x_0, solutions satisfy

$$\|x(t)\| \le \gamma \|x_0\|, \quad t \ge t_0.$$

Given any t_0 and $t_a \ge t_0$, let x_a be such that

$$\|x_a\| = 1, \quad \|\Phi(t_a, t_0)x_a\| = \|\Phi(t_a, t_0)\|$$

(Such an x_a exists by definition of the induced norm.) Then the initial state $x(t_0) = x_a$ yields a solution of Eq. (5.1) that at time t_a satisfies

$$\|x(t_a)\| = \|\Phi(t_a, t_0)x_a\| = \|\Phi(t_a, t_0)\| \|x_a\| \le \gamma \|x_a\|. \tag{5.4}$$

Since $\|x_a\| = 1$, this shows that $\|\Phi(t_a, t_0)\| \le \gamma$. Because such an x_a can be selected for any t_0 and $t_a \ge t_0$, the proof is complete. $\qquad \square$

5.1.1 Uniform exponential stability

Next we consider a stability property for Eq. (5.1) that addresses both boundedness and asymptotic behavior of solution. It implies uniform stability, and imposes an additional requirement that all solutions approach zero exponentially as $t \rightarrow \infty$.

Definition 5.1.2. The linear state equation (5.1) is called uniformly exponentially stable if there exist finite positive constants γ, λ such that for any t_0 and x_0 the corresponding solution satisfies

$$\|x(t)\| \leq \gamma e^{-\lambda(t-t_0)} \|x_0\|, \quad t \geq t_0. \tag{5.5}$$

Again γ is no less than unity, and the adjective uniform refers to the fact that γ and λ are independent of t_0. This is illustrated in Fig. 5.2. The property of uniform exponential stability can be expressed in terms of an exponential bound on the transition matrix. The proof is similar to that of Theorem 5.1.1.

Theorem 5.1.2. *The linear state equation (5.1) is uniformly exponentially stable if and only if there exist finite positive constants γ and λ such that*

$$\|\Phi(t, \tau)\| \leq \gamma e^{-\lambda(t-\tau)} \tag{5.6}$$

for all t, τ such that $t \geq \tau$.

Uniform stability and uniform exponential stability are the only internal stability concepts used in the sequel. Uniform exponential stability is the most important of the two, and another theoretical characterization of uniform exponentially stability for the bounded-coefficient case will prove useful.

Theorem 5.1.3. *Suppose there exists a finite positive constant α such $\|A(t)\| \leq \alpha$ for all t. Then the linear state equation (5.1) is uniformly exponentially stable if and only if there exists a finite positive constant β such that*

$$\int_{\tau}^{t} \|\Phi(t, \sigma)\| d\sigma \leq \beta \tag{5.7}$$

for all t, τ such that $t \geq \tau$.

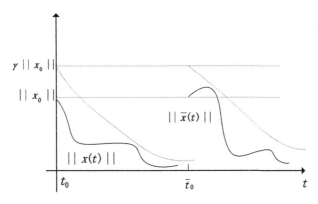

Fig. 5.2 A decaying-exponential bound independent of t_0.

Proof. If the state equation is uniformly exponentially stable, then by Theorem 5.1.1 there exist finite $\gamma, \lambda > 0$ such that

$$\|\Phi(t,\sigma)\| \le \gamma e^{-\lambda(t-\sigma)}$$

for all t, σ such that $t \ge \sigma$. Then

$$
\begin{aligned}
\int_\tau^t \|\Phi(t,\sigma)\| d\sigma &\le \int_\tau^t \gamma e^{-\lambda(t-\sigma)} d\sigma \\
&= \frac{\gamma(1 - e^{-\lambda(t-\tau)})}{\lambda} \\
&\le \frac{\gamma}{\lambda}
\end{aligned}
$$

for all t, τ such that $t \ge \tau$. Thus Eq. (5.7) is established with $\beta = \frac{\gamma}{\lambda}$. $\square$

Conversely suppose Eq. (5.7) holds. Basic calculus permit the representation

$$
\begin{aligned}
\Phi(t,\tau) &= I - \int_\tau^t \frac{\partial}{\partial\sigma} \Phi(t,\sigma) d\sigma \\
&= I + \int_\tau^t \Phi(t,\sigma) A(\sigma) d\sigma
\end{aligned}
$$

and thus

$$
\begin{aligned}
\|\Phi(t,\tau)\| &\le 1 + \alpha \int_\tau^t \|\Phi(t,\sigma)\| d\sigma \\
&\le 1 + \alpha\beta
\end{aligned}
\tag{5.8}
$$

for all t, τ such that $t \ge \tau$. In completing this proof the composition property of the transition matrix is crucial. So long as $t \ge \tau$ we can write, cleverly,

$$
\begin{aligned}
\|\Phi(t,\tau)\|(t-\tau) &= \int_\tau^t \|\Phi(t,\tau)\| d\sigma \\
&\le \int_\tau^t \|\Phi(t,\sigma)\| \|\Phi(\sigma,\tau)\| d\sigma \\
&\le \beta(1 + \alpha\beta).
\end{aligned}
$$

Therefore letting $T = 2\beta(1 + \alpha\beta)$ and $t = \tau + T$ gives

$$\|\Phi(\tau + T, \tau)\| \le \frac{1}{2} \tag{5.9}$$

for all τ. Applying Eqs. (5.8), (5.9), the following inequalities on time intervals of the form $[\tau + \kappa T, \tau + (\kappa + 1)T)$, where τ is arbitrary, are transparent:

$$\|\Phi(t,\tau)\| \le 1 + \alpha\beta, \quad t \in [\tau, \tau + T]$$

$$\begin{aligned}
\|\Phi(t,\tau)\| &= \|\Phi(t,\tau+T)\Phi(\tau+T,\tau)\| \\
&\le \|\Phi(t,\tau+T)\|\|\Phi(\tau+T,\tau)\| \\
&\le \frac{1+\alpha\beta}{2}, \quad t \in [\tau+T, \tau+2T],
\end{aligned}$$

$$\begin{aligned}
\|\Phi(t,\tau)\| &= \|\Phi(t,\tau+2T)\Phi(\tau+2T,\tau+T)\Phi(\tau+T,\tau)\| \\
&\le \|\Phi(t,\tau+2T)\|\|\Phi(\tau+2T,\tau+T)\|\|\Phi(\tau+T,\tau)\| \\
&\le \frac{1+\alpha\beta}{2^2}, \quad t \in [\tau+2T, \tau+3T].
\end{aligned}$$

Continuing in this fashion shows that for any value of τ

$$\|\Phi(t,\tau)\| \le \frac{1+\alpha\beta}{2^k}, \quad t \in [\tau + kT, \tau + (k+1)T]. \tag{5.10}$$

Finally, choose $\lambda = (-\frac{1}{T})\ln\left(\frac{1}{2}\right)$ and $\gamma = 2(1 + \alpha\beta)$. Fig. 5.3 presents a plot of the corresponding decaying exponential and the bound (5.10), from which it is clear that

$$\|\Phi(t,\tau)\| \le \gamma e^{-\lambda(t-\tau)}$$

for all t, τ such that $t \ge \tau$. Uniform exponential stability thus is a consequence of Theorem 5.1.1.

An alternate form for the uniform exponential stability condition in Theorem 5.1.3 is

$$\int_{-\infty}^{t} \|\Phi(t,\sigma)\| d\sigma \le \beta$$

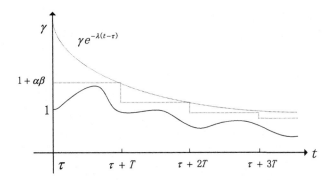

Fig. 5.3 Bounds constructed in the proof of 5.1.3.

for all t. For time-invariant linear state equations, where $\Phi(t, \sigma) = e^{A(t-\sigma)}$, an integration-variable change, in either form of the condition, shows that uniform exponential stability is equivalent to finiteness of

$$\int_0^\infty \|e^{At}\| dt. \tag{5.11}$$

The adjective "uniform" is superfluous in the time-invariant case, and we will drop it in clear contexts. Though exponential stability usually is called asymptotic stability when discussing time-invariant linear state equations, we retain the term exponential stability.

Combining an explicit representation for e^{At} with the finiteness condition on Eq. (5.11) yields a better-known characterization of exponential stability.

Theorem 5.1.4. *A linear state equation (5.1) with constant $A(t) = A$ is exponentially stable if and only if all eigenvalues of A have negative real parts.*

Proof. Suppose the eigenvalue condition holds. Then writing e^{At} in the explicit form, where $\lambda_1, \ldots \ldots, \lambda_m$ are the distinct eigenvalues of A, gives

$$
\begin{aligned}
\int_0^\infty \|e^{At}\| dt &= \int_0^\infty \| \sum_{k=1}^m \sum_{j=1}^{\sigma_k} W_{kj} \frac{t^{j-1}}{(j-1)!} e^{\lambda_k t} \| dt \\
&\leq \sum_{k=1}^m \sum_{j=1}^{\sigma_k} \|W_{kj}\| \int_0^\infty \frac{t^{j-1}}{(j-1)!} |e^{\lambda_k t}| dt.
\end{aligned}
\tag{5.12}
$$

Since $|e^{\lambda_k t}| = e^{Re[\lambda_k]t}$, the bound shows that the right side is finite, and exponential stability follows.

If the negative-real-part eigenvalue condition on A fails, then appropriate selection of an eigenvalue of A as an initial state can be used to show that the linear state equation is not exponentially stable. Suppose first that a real eigenvalue λ is nonnegative, and let p be an associated eigenvector. Then the power series representation for the matrix exponential easily shows that

$$e^{At} p = e^{\lambda t} p.$$

For the initial state $x_0 = p$, it is clear that the corresponding solution of Eq. (5.1), $x(t) = e^{\lambda t} p$, does not go to zero as $t \to \infty$. Thus the state equation is not exponentially stable.

Now suppose that $\lambda = \sigma + i\omega$ is a complex eigenvalue of A with $\sigma > 0$. Again let p be an eigenvector associate with λ, written

$$p = Re[p] + i\, Im[p].$$

Then

$$\|e^{At} p\| = |e^{\lambda t}|\|p\| = e^{\sigma t}\|p\| \geq \|p\|, \quad t \geq 0$$

and thus

$$e^{At}p = e^{At}Re[p] + i\, e^{At}Im[p]$$

does not approach zero as $t \to \infty$. Therefore at least one of the real initial states $x_0 = Re[p]$ *or* $x_0 = Im[p]$ yields a solution that does not approach zero as $t \to \infty$.

This proof, with a bit of elaboration, shows also that $\lim_{t \to \infty} e^{At} = 0$ is a necessary and sufficient condition for uniform exponential stability in the time-invariant case. The corresponding statement is not true for time-invariant linear state equations. □

5.1.2 Uniform asymptotic stability

Definition 5.1.3. The linear state equation (5.1) is called uniformly asymptotically stable if it is uniformly stable, and if given any positive constant δ there exists a positive T such that for any t_0 and x_0 the corresponding solution satisfies

$$\|x(t)\| \le \delta \|x_0\|, \quad t \ge t_0 + T. \tag{5.13}$$

Note that the elapsed time T, until the solution satisfies the bound (5.13), must be independence of the initial time. Some of the same tools used in proving Theorem 5.1.3 can be used to show that this "elapsed-time uniformity" is the key to uniform exponential stability.

Theorem 5.1.5. *The linear state equation (5.1) is uniformly asymptotically stable if and only if it is uniformly exponentially stable.*

Proof. Suppose that the state equation is uniformly exponentially stable, that is, there exist finite, positive γ and λ such that $\|\Phi(t, \tau)\| \le \gamma e^{-\lambda(t-\tau)}$ whenever $t \ge \tau$. Then the state equation clearly is uniformly stable. To show it is uniformly asymptotically stable, for a given $\sigma > 0$ pick T such that $e^{-\lambda T} \le \frac{\delta}{\gamma}$. Then for any t_0 and x_0, and $t \ge t_0 + T$,

$$\begin{aligned}
\|x(t)\| &= \|\Phi(t, t_0)x_0\| \\
&\le \|\Phi(t, t_0)\| \|x_0\| \\
&\le \gamma e^{-\lambda(t-t_0)} \|x_0\| \\
&\le \gamma e^{-\lambda T} \|x_0\| \\
&\le \sigma \|x_0\|, \quad t \ge t_0 + T.
\end{aligned}$$

This demonstrates uniform asymptotic stability. □

Conversely suppose the state equation is uniformly asymptotically stable. Uniform stability is implied by definition, so there exists a positive γ such that

$$\|\Phi(t, \tau)\| \le \gamma \tag{5.14}$$

for all t, τ such that $t \ge \tau$. Select $\delta = \frac{1}{2}$, and by Definition 5.1.3 let T be such that Eq. (5.13) is satisfied. Then given a t_0, let x_a be such that $\|x_a\| = 1$, and

$$\|\Phi(t_0 + T, t_0)x_a\| = \|\Phi(t_0 + T, t_0)\|.$$

With the initial state $x(t_0) = x_a$, the solution of Eq. (5.1) satisfies

$$\|x(t_0 + T)\| = \|\Phi(t_0 + T, t_0)x_a\|$$
$$= \|\Phi(t_0 + T, t_0)\|\|x_a\|$$
$$\leq \frac{1}{2}\|x_a\|$$

from which

$$\|\Phi(t_0 + T, t_0)\| \leq \frac{1}{2}. \tag{5.15}$$

Of course such an x_a exists for any given t_0, so the argument compels Eq. (5.15) for any t_0. Now uniform exponential stability is implied by Eqs. (5.14), (5.15), exactly as in the proof of Theorem 5.1.3.

5.1.3 Lyapunov transformation

The stability concepts under discussion are properties of particular linear state equation that presumably represent a system of interest in terms of physically meaningful variables. A basic question involves preservation of stability properties under a state variable change. Since time-varying variable changes are permitted, simple scalar examples can be generated to show that, for example, uniform stability can be created or destroyed by variable change. To circumvent this difficulty we must limit attention to a particular class of state variable changes.

Definition 5.1.4. An $n \times n$ matrix $P(t)$ that is continuously differentiable and invertible at each t is called a Lyapunov transformation if there exist finite positive constants ρ and η such that for all t

$$\|P(t)\| \leq \rho, \quad |\det P(t)| \geq \eta. \tag{5.16}$$

A condition equivalent to Eq. (5.16) is existence of a finite positive constant ρ such that for all t

$$\|P(t)\| \leq \rho, \quad \|P^{-1}(t)\| \leq \rho.$$

The lower bound on $|detP(t)|$ implies an upper bound on $\|P^{-1}(t)\|$.

Reflecting on the effect of a state variable change on the transition matrix, a detailed proof that Lyapunov transformations preserve stability properties is perhaps belaboring the evident.

Theorem 5.1.6. *Suppose the $n \times n$ matrix $P(t)$ is Lyapunov transformation. Then the linear state equation (5.1) is uniformly stable (respectively, uniformly exponentially stable) if and only if the state equation*

$$\dot{z}(t) = [P^{-1}(t)A(t)P(t) - P^{-1}\dot{P}(t)]z(t) \tag{5.17}$$

is uniformly stable (respectively, uniformly exponentially stable).

Proof. The linear state equations (5.1) and (5.17) are related by the variable change $z(t) = P^{-1}(t)x(t)$, and we note that the properties required of a Lyapunov transformation subsume those required of a variable change. Thus the relation between the two transition matrices is

$$\Phi_z(t, \tau) = P^{-1}(t)\Phi_x(t, \tau)P(\tau).$$

Now suppose Eq. (5.1) is uniformly stable. Then there exists γ such that $\|\Phi_x(t, \tau)\| \leq \gamma$ for all t, τ such that $t \geq \tau$, and, from Eq. (5.16)

$$\begin{aligned}
\|\Phi_z(t, \tau)\| &= \|p^{-1}(t)\Phi_x(t, \tau)P(\tau)\| \\
&\leq \|P^{-1}(t)\|\|\Phi_x(t, \tau)\|\|P(\tau)\| \qquad\qquad (5.18) \\
&\leq \frac{\gamma\rho^n}{\eta}
\end{aligned}$$

for all t, τ such that $t \geq \tau$. This shows that Eq. (5.17) is uniformly stable. An obviously similar argument applied to

$$\Phi_x(t, \tau) = P(t)\Phi_z(t, \tau)P^{-1}(\tau)$$

shows that if Eq. (5.17) is uniformly stable, then Eq. (5.1) is uniformly stable. The corresponding demonstrations for uniform exponential stability are similar.   □

The Floquet decomposition for T-periodic state equations provides a general illustration. Since $P(t)$ is the product of a transition matrix and a matrix exponential, it is continuously differentiable with respect to t. Since $P(t)$ is invertible, by continuity arguments there exist $\rho, \eta > 0$ such that Eq. (5.16) holds for all t in any $P(t)$ is Lyapunov transformation. It is easy to verify that $z(t) = P^{-1}(t)x(t)$ yields the time-invariant linear state equation

$$\dot{z}(t) = Rz(t).$$

By this connection stability properties of the original T-periodic state equation are equivalent to stability properties of a time-invariant linear state equation (though, it must be noted, the time-invariant state equation in general is complex).

5.2 Lyapunov stability

The origin of Lyapunov's so-called direct method for stability assessment is the notion that total energy of an unforced, dissipative mechanical system decreases as the state of the system evolves in time. Therefore the state vector approaches a constant value corresponding to zero energy as time increases. Phrased more generally, stability properties involve the growth properties of solutions of the state equation, and these properties can be measured by suitable (energy-like) scalar function of the state vector. The problem is finding a suitable scalar function.

5.2.1 Introduction

To illustrate the basic idea we consider conditions that imply all solutions of the linear
state equation

$$\dot{x}(t) = A(t)x(t), \quad x(t_0) = x_0 \tag{5.19}$$

are such that $\|x(t)\|^2$ monotonically decreases as $t \to \infty$. For any solution $x(t)$ of
Eq. (5.19), the derivative of the scalar function

$$\|x(t)\|^2 = x^T(t)x(t) \tag{5.20}$$

with respect to t can be written as

$$
\begin{aligned}
\frac{d}{dt}\|x(t)\|^2 &= \dot{x}^T(t)x(t) + x^T(t)\dot{x}(t) \\
&= x^T(t)[A^T(t) + A(t)]x(t).
\end{aligned}
\tag{5.21}
$$

In this computation $\dot{x}(t)$ is replaced by $A(t)x(t)$ precisely because $x(t)$ is a solution
of Eq. (5.19). Suppose that the quadratic form on the right side of Eq. (5.21) is negative
definite, that is, suppose the matrix $A^T(t) + A(t)$ is negative definite at each t. Then,
as shown in Fig. 5.4, $\|x(t)\|^2$ decreases as t increases. Further we can show that if this
negative definiteness does not asymptotically vanish, that is, if there is a constant $\nu > 0$
such that $A^T(t) + A(t) \le -\nu I$ for all t, then $\|x(t)\|^2$ goes to zero as $t \to \infty$. Notice that
the transition matrix for $A(t)$ is not needed in this calculation, and growth properties of
the scalar function (5.20) depend on sign-definiteness properties of the quadratic form
in Eq. (5.21). Admittedly this calculation results in a restrictive sufficient condition—
negative definiteness of $A^T(t)+A(t)$—for a type of asymptotic stability. However, more
general scalar functions than Eq. (5.20) can be considered.

Formalization of the above discussion involves somewhat intricate definitions of
time-dependent quadratic forms that are useful as scalar functions of the state vector of
Eq. (5.19) for stability purpose. Such quadratic forms are called quadratic Lyapunov

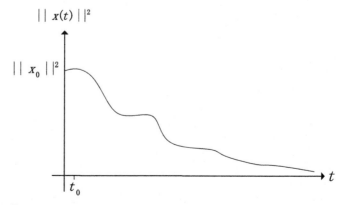

Fig. 5.4 If $A^T(t) + A(t) < 0$ at each t, the solution norm decreases for $t \ge t_0$.

functions. They can be written as $x^T Q(t)x$, where $Q(t)$ is assumed to be symmetric and continuously differentiable for all t. If $x(t)$ is any solution of Eq. (5.19) for $t \geq t_0$, then we are interested in the behavior of the real quantity $x^T(t)Q(t)x(t)$ for $t \geq t_0$. This behavior can be assessed by computing the time derivative using the product rule, and replacing $\dot{x}(t)$ by $A(t)x(t)$ to obtain

$$\frac{d}{dt}[x^T(t)Q(t)x(t)] = x^T(t)[A^T(t)Q(t) + Q(t)A(t) + \dot{Q}(t)]x(t). \tag{5.22}$$

To analyze stability properties, various bounds are required on quadratic Lyapunov functions and on the quadratic forms (5.22) that arise as their derivatives along solutions of Eq. (5.19). These bounds can be expressed in alternative ways. For example, the condition that there exists a positive constant η such that

$$Q(t) \geq \eta I$$

for all t is equivalent by definition to existence of a positive η such that

$$x^T Q(t)x \geq \eta \|x\|^2$$

for all t and all $n \times 1$ vectors x. Yet another way to write this is to require the existence of a symmetric, positive-definite constant matrix M such that

$$x^T Q(t)x \geq x^T M x$$

for all t and all $n \times 1$ vectors x. The choice is largely a matter of taste, and the most economical form is adopted here.

5.2.2 Uniform stability

We begin with a sufficient condition for uniform stability. The presentation style throughout is to list the requirement on $Q(t)$ so that the corresponding quadratic form can be used to prove the desired stability property.

Theorem 5.2.1. *The linear state equation (5.19) is uniformly stable if there exists an $n \times n$ matrix $Q(t)$ that for all t is symmetric, continuously differentiable, and such that*

$$\eta I \leq Q(t) \leq \rho I, \tag{5.23}$$

$$A^T(t)Q(t) + Q(t)A(t) + \dot{Q}(t) \leq 0, \tag{5.24}$$

where η and ρ are finite positive constants.

Proof. Given any t_0 and x_0, the corresponding solution $x(t)$ of Eq. (5.19) is such that, from Eqs. (5.22), (5.24),

$$x^T(t)Q(t)x(t) - x_0^T Q(t_0)x_0 = \int_{t_0}^{t} \frac{d}{d\sigma}[x^T(\sigma)Q(\sigma)x(\sigma)]d\sigma$$
$$\leq 0, \quad t \geq t_0.$$

Using the inequalities in Eq. (5.23) we obtain

$$x^T(t)Q(t)x(t) \leq x_0^T Q(t_0)x_0 \leq \rho\|x_0\|^2, \quad t \geq t_0$$

and then

$$\eta\|x(t)\|^2 \leq \rho\|x_0\|^2, \quad t \geq t_0.$$

Therefore

$$\|x(t)\| \leq \sqrt{\frac{\rho}{\eta}}\|x_0\|, \quad t \geq t_0. \tag{5.25}$$

Since Eq. (5.25) holds for any x_0 and t_0, the state equation (5.19) is uniformly stable by definition. $\qquad\square$

Typically it is profitable to use a quadratic Lyapunov function to obtain stability conditions for linear state equations, rather than a particular instance.

5.2.3 Uniform exponential stability

Theorem 5.2.2. *The linear state equation (5.19) is uniformly exponentially stable if there exists an $n \times n$ matrix function $Q(t)$ that for all t is symmetric, continuously differentiable, and such that*

$$\eta I \leq Q(t) \leq \rho I, \tag{5.26}$$

$$A^T Q(t) + Q(t)A(t) + \dot{Q}(t) \leq -\nu I, \tag{5.27}$$

when $\eta, \rho,$ and ν are finite positive constants.

Proof. For any t_0, x_0, and corresponding solution $x(t)$ of the state equation, the inequality (5.27) gives

$$\frac{d}{dt}[x^T(t)Q(t)x(t)] \leq -\nu\|x(t)\|^2, \quad t \geq t_0.$$

Also from Eq. (5.26)

$$x^T(t)Q(t)x(t) \leq \rho\|x(t)\|^2, \quad t \geq t_0,$$

$$-\|x(t)\|^2 \leq -\frac{1}{\rho}x^T(t)Q(t)x(t), \quad t \geq t_0.$$

Therefore

$$\frac{d}{dt}[x^T(t)Q(t)x(t)] \leq -\frac{\nu}{\rho}x^T(t)Q(t)x(t), \quad t \geq t_0 \tag{5.28}$$

and this implies, after multiplication by the appropriate exponential integrating factor, and integrating from t_0 to t,

$$x^T(t)Q(t)x(t) \leq e^{-\frac{v}{\rho}(t-t_0)} x_0^T Q(t_0)x_0, \quad t \geq t_0.$$

Summoning Eq. (5.26) again,

$$\|x(t)\|^2 \leq \frac{1}{\eta} x^T(t)Q(t)x(t)$$

$$\leq \frac{1}{\eta} e^{-\frac{v}{\rho}(t-t_0)} x_0^T Q(t_0)x_0, \quad t \geq t_0,$$

which in turn gives

$$\|x(t)\|^2 \leq \frac{\rho}{\eta} e^{-\frac{v}{\rho}(t-t_0)} \|x_0\|^2, \quad t \geq t_0. \tag{5.29}$$

Noting that Eq. (5.29) holds for any x_0 and t_0, and taking the positive square root of both sides, uniform exponential stability is established.   □

For $n = 2$ and constant $Q(t) = Q$, Theorem 5.22 admits a simple pictorial representation. The condition (5.26) implies that Q is positive definite, and therefore the level curves of the real-valued function $x^T Q x$ are ellipses in the (x_1, x_2)-plane. The condition (5.27) implies that for any solution $x(t)$ of the state equation the value of $x^T(t)Qx(t)$ is decreasing as t increases. Thus a plot of the solution $x(t)$ on the (x_1, x_2)-plane crosses smaller-value level curves as t increases, as shown in Fig. 5.5. Under the same assumptions, a similar pictorial interpretation can be given for Theorem 5.2.1. Note that if $Q(t)$ is not constant, the level curves vary with t and the picture is much less informative.

Just in case it appears that stability of linear state equations is reasonably intuitive. A first guess is that the state equation is uniformly exponentially stable if $a(t)$ is continuous and positive for all t, though suspicions might arise if $a(t) \to 0$ as $t \to \infty$. These suspicious would be well founded, but what is more surprising is that there are other obstructions to uniform exponential stability.

The stability criteria provided by the preceding theorems are sufficient conditions that depend on skill in selecting an appropriate $Q(t)$. It is comforting to show that there indeed exists a suitable $Q(t)$ for a large class of uniformly exponentially stable linear state equations. The dark side is that it can be roughly as hard to compute $Q(t)$ as it is to compute the transition matrix for $A(t)$.

Theorem 5.2.3. *Suppose that the linear state equation (5.19) is uniformly exponentially stable, and there exists a finite constant α such that $\|A(t)\| \leq \alpha$ for all t. Then*

$$Q(t) = \int_t^\infty \Phi^T(\sigma, t)\Phi(\sigma, t)d\sigma \tag{5.30}$$

satisfies all the hypotheses of Theorem 5.2.2.

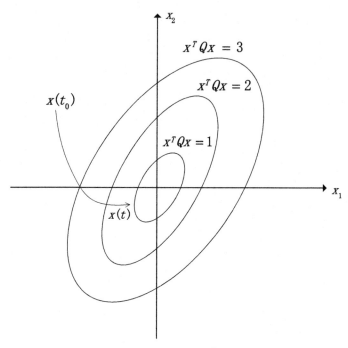

Fig. 5.5 A solution $x(t)$ in relation to level curves for $x^T Q x$.

Proof. First we show that the integral converges for each t, so that $Q(t)$ is well defined. Since the state equation is uniformly exponentially stable, there exists positive γ and λ such that

$$\|\Phi(t, t_0)\| \le \gamma e^{-\lambda(t-t_0)}$$

for all t, t_0 such that $t \ge t_0$. Thus

$$\left\| \int_t^\infty \Phi^T(\sigma, t)\Phi(\sigma, t)d\sigma \right\| \le \int_t^\infty \|\Phi^T(\sigma, t)\| \cdot \|\Phi(\sigma, t)\| d\sigma$$

$$\le \int_t^\infty \gamma^2 e^{-2\lambda(\sigma - t)} d\sigma$$

$$= \frac{\gamma^2}{2\lambda}$$

for all t. This calculation also defines ρ in Eq. (5.26). Since $Q(t)$ clearly is symmetric and continuously differentiable at each t, it remains only to show that there exist η, $\nu > 0$ as needed in Eqs. (5.26), (5.27). To obtain ν, differentiation of Eq. (5.30) gives

$$\dot{Q}(t) = -I + \int_t^\infty [-A^T(t)\Phi^T(\sigma, t)\Phi(\sigma, t) - \Phi^T(\sigma, t)\Phi(\sigma, t)A(t)]d\sigma$$

$$= -I - A^T(t)Q(t) - Q(t)A(t).$$

$$(5.31)$$

That is

$$A^T(t)Q(t) + Q(t)A(t) + \dot{Q}(t) = -I$$

and clearly a valid choice for ν in Eq. (5.27) is $\nu = 1$. Finally it must be shown that there exists a positive η such that $Q(t) \geq \eta I$ for all t, and for this we set up an adroit maneuver. For any x and t

$$\frac{d}{d\sigma}[x^T \Phi^T(\sigma, t)\Phi(\sigma, t)x] = x^T \Phi^T(\sigma, t)[A^T(\sigma) + A(\sigma)]\Phi(\sigma, t)x$$

$$\geq -\|A^T(\sigma) + A(\sigma)\| x^T \Phi^T(\sigma, t)\Phi(\sigma, t)x$$

$$\geq -2\alpha x^T \Phi^T(\sigma, t)\Phi(\sigma, t)x.$$

Using the fact that $\Phi(\sigma, t)$ approaches zero exponentially as $\sigma \to \infty$, we integrate both sides to obtain

$$\int_t^\infty \frac{d}{d\sigma}[x^T \Phi^T(\sigma, t)\Phi(\sigma, t)x]d\sigma \geq -2\alpha \int_t^\infty x^T \Phi^T(\sigma, t)\Phi(\sigma, t)x d\sigma \qquad (5.32)$$

$$= -2\alpha x^T Q(t)x.$$

Evaluating the integral gives

$$-x^T x \geq -2\alpha x^T Q(t)x$$

or

$$Q(t) \geq \frac{1}{2\alpha}I$$

for all t. Thus with the choice $\eta = \frac{1}{2\alpha}$ all hypotheses of Theorem 5.2.2 are satisfied. $\square$

5.2.4 Instability

Quadratic Lyapunov function also can be used to develop instability criteria of various types. One example is the following result that, except for one value of t, does not involve a sign-definiteness assumption on $Q(t)$.

Theorem 5.2.4. *Suppose there exists an $n \times n$ matrix function $Q(t)$ that for all t is symmetric, continuously differentiable, and such that*

$$\|Q(t)\| \leq \rho, \qquad (5.33)$$

$$A^T(t)Q(t) + Q(t)A(t) + \dot{Q}(t) \leq -\nu I, \qquad (5.34)$$

where ρ and ν are finite positive constants. Also suppose there exists a t_a such that $Q(t_a)$ is not positive semidefinite. Then the linear state equation (5.19) is not uniformly stable.

Proof. Suppose $x(t)$ is the solution of Eq. (5.19) with $t_0 = t_a$ and $x_0 = x_a$ such that $x_a^T Q(t_a) x_a < 0$. Then, from Eq. (5.34),

$$x^T(t)Q(t)x(t) - x_0^T Q(t_0)x_0 = \int_{t_0}^{t} \frac{d}{d\sigma}[x^T(\sigma)Q(\sigma)x(\sigma)]d\sigma$$

$$\leq -\nu \int_{t_0}^{t} x^T(\sigma)x(\sigma)d\sigma$$

$$\leq 0, \quad t \geq t_0. \qquad \square$$

One consequence of this inequality, Eq. (5.33), and the choice of x_0 and t_0, is

$$-\rho\|x(t)\|^2 \leq x^T Q(t)x(t) \leq x_0^T Q(t_0)x_0 < 0, \quad t \geq t_0 \qquad (5.35)$$

and a further consequence is that

$$\nu \int_{t_0}^{t} x^T(\sigma)x(\sigma)d\sigma \leq x_0^T Q(t_0)x_0 - x^T(t)Q(t)x(t)$$

$$\leq |x^T(t)Q(t)x(t)| + |x_0^T Q(t_0)x_0| \qquad (5.36)$$

$$\leq 2|x^T(t)Q(t)x(t)|, \quad t \geq t_0.$$

Using Eqs. (5.33), (5.36) gives

$$\int_{t_0}^{t} x^T(\sigma)x(\sigma)d\sigma \leq \frac{2\rho}{\nu}\|x(t)\|^2, \quad t \geq t_0. \qquad (5.37)$$

The state equation can be shown to be not uniformly stable by proving that $x(t)$ is unbounded. This we do by a contradiction argument. Suppose that there exists a finite γ such that $\|x(t)\| \leq \gamma$, for all $t \geq t_0$. Then Eq. (5.37) gives

$$\int_{t_0}^{t} x^T(\sigma)x(\sigma)d\sigma \leq \frac{2\rho\gamma^2}{\nu}, \quad t \geq t_0$$

and the integrand, which is a continuously differentiable scalar function, must go to zero as $t \to 0$. Therefore $x(t)$ must also go to zero, and this implies that Eq. (5.35) is violated for sufficiently large t. The contradiction proves that $x(t)$ cannot be a bounded solution.

5.2.5 Time-invariant case

In the time-invariant case quadratic Lyapunov function with constant Q can be used to connect Theorem 5.2.2 with the familiar eigenvalue condition for exponential stability. If Q is symmetric and positive definite, then Eq. (5.26) is satisfied automatically. However, rather than specifying such a Q and checking to see if a positive ν exists such that Eq. (5.27) is satisfied, the approach can be reversed. Choose a positive definite matrix M, for example $M = \nu I$, where $\nu > 0$. If there exists a symmetric, positive-definite Q such that

$$QA + A^T Q = -M, \tag{5.38}$$

then all the hypotheses of Theorem 5.2.2 are satisfied. Therefore the associated linear state equation

$$\dot{x}(t) = Ax(t), \quad x(0) = x_0$$

is exponentially stable, all eigenvalues of A have negative real parts. Conversely the eigenvalues of A enter the existence question for solution of the Lyapunov equation (5.38).

Theorem 5.2.5. *Given an $n \times n$ matrix A, if M and Q are symmetric, positive-definite, $n \times n$ matrices satisfying Eq. (5.38), then all eigenvalues of A have negative real parts. Conversely if all eigenvalues of A have negative real parts, then for each symmetric $n \times n$ matrix M there exists a unique solution of Eq. (5.38) given by*

$$Q = \int_0^\infty e^{A^T t} M e^{At} dt. \tag{5.39}$$

Furthermore, if M is positive definite, then Q is positive definite.

Proof. For the converse, if all eigenvalues of A have negative real parts, it is obvious that the integral in Eq. (5.39) converges, so Q is well defined. To show that Q is a solution of Eq. (5.38), we calculate

$$\begin{aligned}
A^T Q + QA &= \int_0^\infty A^T e^{A^T t} M e^{At} dt + \int_0^\infty e^{A^T t} M e^{At} A \, dt \\
&= \int_0^\infty \frac{d}{dt} [e^{A^T t} M e^{At}] dt \\
&= e^{A^T t} M e^{At} \, |_0^\infty = -M.
\end{aligned}$$

To prove this solution is unique, suppose Q_a also is a solution. Then

$$(Q_a - Q)A + A^T (Q_a - Q) = 0. \tag{5.40}$$

But this implies

$$e^{A^T t}(Q_a - Q)Ae^{At} + e^{A^T t}A^T(Q_a - Q)e^{At} = 0, \quad t \geq 0$$

from which

$$\frac{d}{dt}[e^{A^T t}(Q_a - Q)e^{At}] = 0, \quad t \geq 0.$$

Integrating both sides from 0 to ∞ gives

$$0 = e^{A^T t}(Q_a - Q)e^{At} |_0^\infty = -(Q_a - Q).$$

That is, $Q_a = Q$. Now suppose that M is positive definite. Clearly Q is symmetric. To show it is positive definite simply note that for a zero $n \times 1$ vector x

$$x^T Q x = \int_0^\infty x^T e^{A^T t} M e^{At} x \, dt > 0, \tag{5.41}$$

since the integrand is a positive scalar function. (In detail, $e^{At} x \neq 0$ for $t \geq 0$, so positive definiteness of M shows that the integrand is positive for all $t \geq 0$.)   □

5.3 Input-output stability

In this section we address stability properties appropriate to the input-output behavior (zero-state response) of the linear state equation

$$\dot{x}(t) = A(t)x(t) + B(t)u(t),$$
$$y(t) = C(t)x(t). \tag{5.42}$$

That is, the initial state is set to zero, and attention is focused on the boundedness of the response to bounded inputs. There is no $D(t)u(t)$ term in Eq. (5.42) because a bounded $D(t)$ does not affect the treatment, while an unbounded $D(t)$ provides an unbounded response to an appropriate constant input. Of course the input-output behavior of Eq. (5.42) is specified by the impulse response

$$G(t, \sigma) = C(t)\Phi(t, \sigma)B(\sigma), \quad t \geq \sigma \tag{5.43}$$

and stability results are characterized in terms of boundedness properties of $\|G(t, \sigma)\|$. (Notice in particular that the weighting pattern is not employed.) For the time-invariant case, input-output stability also is characterized in terms of the transfer function of the linear state equation.

5.3.1 Uniform bounded-input bounded-output stability

Bounded-input, bounded-output stability is most simply discussed in terms of the largest value (over time) of the norm of the input signal, $\|u(t)\|$, in comparison to the largest value of the corresponding response norm $\|y(t)\|$. More precisely we use the standard notion of supremum. For example,

$$v = \sup_{t \geq t_0} \|u(t)\|$$

is defined as the smallest constant such that $u(t) \leq v$ for $t \geq t_0$. If no such bound exists, we write

$$\sup_{t \geq t_0} \|u(t)\| = \infty.$$

The basic notion is that the zero-state response should exhibit finite "gain" in terms of the input and output suprema.

Definition 5.3.1. The linear state equation (5.42) is called uniformly bounded-input, bounded-output stable if there exists a finite constant η such that for any t_0 and any input signal $u(t)$ the corresponding zero-state response satisfies

$$\sup_{t \geq t_0} \|y(t)\| \leq \eta \sup_{t \geq t_0} \|u(t)\|. \tag{5.44}$$

The adjective "uniform" does double duty in this definition. It emphasizes the fact that the same η works for all values of t_0, and that the same η works for all input signals. An equivalent definition based on the pointwise norms of $u(t)$ and $y(t)$ is explored.

Theorem 5.3.1. *The linear state equation (5.42) is uniformly bounded-input, bounded-output stable if and only if there exists a finite constant ρ such that for all t, τ with $t \geq \tau$,*

$$\int_\tau^t \|G(t, \sigma)\| d\sigma \leq \rho. \tag{5.45}$$

Proof. Assume first that such a ρ exists. Then for any t_0 and any input defined for $t \geq t_0$, the corresponding zero-state response of Eq. (5.42) satisfies

$$\|y(t)\| = \left\| \int_{t_0}^t C(t)\Phi(t, \sigma)B(\sigma)u(\sigma)d\sigma \right\|$$

$$\leq \int_{t_0}^t \|G(t, \sigma)\| \|u(\sigma)\| d\sigma, \quad t \geq t_0.$$

Replacing $\|u(\sigma)\|$ by its supremum over $\sigma \geq t_0$, and using Eq. (5.45),

$$\|y(t)\| \leq \int_{t_0}^t \|G(t, \sigma)\| d\sigma \sup_{t \geq t_0} \|u(t)\|$$

$$\leq \rho \sup_{t \geq t_0} \|u(t)\|, \quad t \geq t_0.$$

Therefore, taking the supremum of the left side over $t \geq t_0$, Eq. (5.44) holds with $\eta = \rho$, and the state equation is uniformly bounded-input, bounded-output stable. $\square$

Suppose now that Eq. (5.42) is uniformly bounded-input, bounded-output stable. Then there exists a constant η so that, in particular, the zero-state response for any t_0 and any input signal such that

$$\sup_{t \geq t_0} \|u(t)\| \leq 1$$

satisfies

$$\sup_{t \geq t_0} \|y(t)\| \leq \eta.$$

To set up a contradiction argument, suppose no finite ρ exists that satisfies Eq. (5.45). In other words for any given constant ρ there exist τ_ρ and $t_\rho > \tau_\rho$ such that

$$\int_{\tau_\rho}^{t_\rho} \|G(t_\rho, \sigma)\| d\sigma > \rho.$$

Taking $\rho = \eta$, that there exist τ_η, $t_\eta > \tau_\eta$, and indices i, j such that the i, j-entry of the impulse response satisfies

$$\int_{\tau_\eta}^{t_\eta} |G_{ij}(t_\eta, \sigma)| d\sigma > \eta. \tag{5.46}$$

With $t_0 = \tau_\eta$ consider the $m \times 1$ input signal $u(t)$ defined for $t \geq t_0$ as follows. Set $u(t) = 0$ for $t > t_\eta$, and for $t \in [t_0, t_\eta]$ set every component of $u(t)$ to zero except for the jth-component given by (the piecewise-continuous signal)

$$u_j(t) = \begin{cases} 1 & G_{ij}(t_\eta, t) > 0 \\ 0 & G_{ij}(t_\eta, t) = 0, \quad t \in [t_0, t_\eta] \\ -1 & G_{ij}(t_\eta, t) < 0. \end{cases}$$

This input signal satisfies $\|u(t)\| \leq 1$, for all $t \geq t_0$, but the ith-component of the corresponding zero-state response satisfies, by Eq. (5.46),

$$\begin{aligned} y_i(t_\eta) &= \int_{t_0}^{t_\eta} G_{ij}(t_\eta, \sigma) u_j(\sigma) d\sigma \\ &= \int_{t_0}^{t_\eta} |G_{ij}(t_\eta, \sigma)| d\sigma \\ &> \eta. \end{aligned}$$

Since $\|y(t_\eta)\| \geq \|y_i(t_\eta)\|$, a contradiction is obtained that completes the proof.

An alternate expression for the condition in Theorem 5.3.1 is that there exist a finite ρ such that for all t

$$\int_{-\infty}^{t} \|G(t, \sigma)\| d\sigma \leq \rho.$$

For a time-invariant linear state equation, $G(t, \sigma) = G(t, -\sigma)$, and the impulse response customarily is written as $G(t) = Ce^{At}B$, $t \geq 0$. Then a change of integration variable shows that a necessary and sufficient condition for uniform bounded-input, bounded-output stability for a time-invariant state equation is finiteness of the integral

$$\int_0^\infty \|G(t)\| dt. \tag{5.47}$$

5.3.2   Relation to uniform exponential stability

We now turn to establishing connections between uniform bounded-input, bounded-output stability and the property of uniform exponential stability of the zero-input response.

Lemma 5.3.1. *Suppose the linear state equation (5.42) is uniformly exponentially stable, and there exist finite constant β and μ such that for all t*

$$\|B(t)\| \leq \beta, \quad \|C(t)\| \leq \mu. \tag{5.48}$$

Then the state equation also is uniformly bounded-input, bounded-output stable.

Proof. Using the transition matrix bound implied by uniform exponential stability,

$$\int_\tau^t \|G(t,\sigma)\|d\sigma \leq \int_\tau^t \|C(t)\|\|\Phi(t,\sigma)\|\|B(\sigma)\|d\sigma$$
$$\leq \mu\beta \int_\tau^t \gamma e^{-\lambda(t-\sigma)}d\sigma$$
$$\leq \frac{\mu\beta\gamma}{\lambda}$$

for all t, τ with $t \geq \tau$. Therefore the state equation is uniformly bounded-input, bounded-output stable by Theorem 5.3.1. $\qquad\square$

That coefficient bounds as in Eq. (5.48) are needed to obtain the implication in Lemma 5.3.1 should be clear. However, the simple proof might suggest that uniform exponential stability is a needlessly strong condition for uniform bounded-input, bounded-output stability.

In developing implication of uniform bounded-input, bounded-output stability for uniform exponential stability, we need to strengthen the usual controllability and observability properties. Specifically it will be assumed that these properties are uniform in time in special way. For simplicity, admittedly a commodity in short supply for the next few pages, the development is subdivided into two parts. First we deal with linear state equations where the output is precisely the state vector ($C(t)$ is the $n \times n$ identity). In this instance the natural terminology is uniform bounded-input, bounded-state stability.

Theorem 5.3.2. *Suppose for the linear state equation*

$$\dot{x}(t) = A(t)x(t) + B(t)u(t),$$
$$y(t) = x(t),$$

there exist finite positive constants $\alpha, \beta, \varepsilon,$ and δ such that for all t

$$\|A(t)\| \leq \alpha, \quad \|B(t)\| \leq \beta, \quad \varepsilon I \leq W(t-\delta,t). \tag{5.49}$$

Then the state equation is uniformly bounded-input, bounded-state stable if and only if it is uniformly exponentially stable.

Proof. One direction of proof is supplied by Lemma 5.3.1, so assume the linear state equation (5.42) is uniformly bounded-input, bounded-state stable. Applying Theorem 5.3.1 with $C(t) = I$, there exists a finite constant ρ such that

$$\int_\tau^t \|\Phi(t,\sigma)B(\sigma)\|d\sigma \leq \rho \tag{5.50}$$

for all t, τ such that $t \geq \tau$. We next show that this implies existence of a finite constant ψ such that

$$\int_\tau^t \|\Phi(t,\sigma)\| d\sigma \leq \psi$$

for all t, τ such that $t \geq \tau$, and thus conclude uniform exponential stability.

First, since $A(t)$ is bounded, corresponding to the constant δ in Eq. (5.49) there exist a finite constant κ such that

$$\|\Phi(t,\sigma)\| \leq \kappa, \quad |t - \sigma| \leq \delta. \tag{5.51}$$

Second, the lower bound on the controllability Gramian in Eq. (5.49) gives

$$W^{-1}(t - \delta, t) \leq \frac{1}{\varepsilon} I$$

for all t, and therefore

$$\|W^{-1}(t - \delta, t)\| \leq \frac{1}{\varepsilon}$$

for all t. In particular these bounds show that

$$\|B^T(\gamma)\Phi^T(\sigma - \delta, \gamma)W^{-1}(\sigma - \delta, \sigma)\| \leq \|B^T(\gamma)\|\|\Phi^T(\sigma-\delta, \gamma)\|\|W^{-1}(\sigma-\delta,\sigma)\|$$
$$\leq \frac{\beta\kappa}{\varepsilon}$$

for all σ, γ satisfying $|\sigma - \delta - \gamma| \leq \delta$. Therefore writing

$$\Phi(t,\sigma - \delta) = \Phi(t, \sigma - \delta)W(\sigma - \delta, \sigma)W^{-1}(\sigma - \delta, \sigma)$$
$$= \int_{\sigma-\delta}^t \Phi(t,\gamma)B(\gamma)B^T(\gamma)\Phi^T(\sigma - \delta, \gamma)W^{-1}(\sigma - \delta, \sigma)d\gamma$$

we obtain, since $\sigma - \delta \leq \gamma \leq \sigma$ implies $|\sigma - \delta - \gamma| \leq \delta$,

$$\|\Phi(t,\sigma - \delta)\| \leq \frac{\beta\kappa}{\varepsilon}\int_{\sigma-\delta}^\sigma \|\Phi(t,\gamma)B(\gamma)\|d\gamma.$$

Then

$$\int_\tau^t \|\Phi(t,\sigma-\delta)\| d(\sigma-\delta) \leq \frac{\beta\kappa}{\varepsilon}\int_\tau^t \left[\int_{\sigma-\delta}^\sigma \|\Phi(t,\gamma)B(\gamma)\|d\gamma\right]d(\sigma-\delta). \tag{5.52}$$

The proof can be completed by showing that the right side of Eq. (5.45) is bounded for all t, τ such that $t \geq \tau$. $\qquad\square$

In the inside integral on the right side of Eq. (5.45), change the integration variable from γ to $\xi = \gamma - \sigma + \delta$, and then interchange the order of integration to write the right side of Eq. (5.45) as

$$\frac{\beta\kappa}{\varepsilon} \int_0^\delta \left[\int_\tau^t \|\Phi(t, \xi + \sigma - \delta)B(\xi + \sigma - \delta)\| d(\sigma - \delta) \right] d\xi.$$

In the inside integral in this expression, change the integration variable from $\sigma - \delta$ to $\zeta = \xi + \sigma - \delta$ to obtain

$$\frac{\beta\kappa}{\varepsilon} \int_0^\delta \left[\int_{\tau+\xi}^{t+\xi} \|\Phi(t, \zeta)B(\zeta)\| d\zeta \right] d\xi. \tag{5.53}$$

Since $0 \le \xi \le \delta$ we can use Eqs. (5.50), (5.51) with the composition property to bound the inside integral in Eq. (5.53)) as

$$\int_{\tau+\xi}^{t+\xi} \|\Phi(t, \zeta)B(\zeta)\| d\zeta \le \|\Phi(t, t+\xi)\| \int_{\tau+\xi}^{t+\xi} \|\Phi(t+\xi, \zeta)B(\zeta)\| d\zeta$$

$$\le \kappa\rho.$$

Therefore Eq. (5.52) becomes

$$\int_\tau^t \|\Phi(t, \sigma - \delta)\| d(\sigma - \delta) \le \frac{\beta\kappa}{\varepsilon} \int_0^\delta \kappa\rho \, d\xi$$

$$\le \frac{\beta\kappa^2\rho\delta}{\varepsilon}.$$

This holds for all t, τ such that $t \ge \tau$, so uniform exponential stability of linear state equation with $C(t) = I$.

To address the general case, where $C(t)$ is not an identity matrix, recall that the observability Gramian for the state equation (5.42) is defined by

$$M(t_0, t_f) = \int_{t_0}^{t_f} \Phi^T(t, t_0)C^T(t)C(t)\Phi(t, t_0)dt. \tag{5.54}$$

Theorem 5.3.3. *Suppose that for the linear state equation (5.42) there exist finite positive constants $\alpha, \beta, \mu, \varepsilon_1, \delta_1, \varepsilon_2,$ and δ_2 such that*

$$\|A(t)\| \le \alpha, \quad \|B(t)\| \le \beta, \quad \|C(t)\| \le \mu,$$

$$\varepsilon_1 I \le W(t - \delta_1, t), \quad \varepsilon_2 I \le M(t, t + \delta_2) \tag{5.55}$$

for all t. Then the state equation is uniformly bounded-input, bounded-output stable if and only if it is uniformly exponentially stable.

Proof. Again uniform exponential stability implies uniform bounded-input, bounded-output stability by Lemma 5.3.1. So suppose that Eq. (5.42) is uniformly bounded-input, bounded-output stable, and η is such that the zero-state response satisfies

$$\sup_{t \geq t_0} \|y(t)\| \leq \eta \sup_{t \geq t_0} \|u(t)\| \tag{5.56}$$

for all inputs $u(t)$. We will show that the associated state equation with $C(t) = I$, namely,

$$\begin{aligned} \dot{x}(t) &= A(t)x(t) + B(t)u(t) \\ y_a(t) &= x(t) \end{aligned} \tag{5.57}$$

also is uniformly bounded-input, bounded-state stable. To set up a contradiction argument, assume the negation. Then for the positive constant $\sqrt{\frac{\eta^2 \delta_2}{\varepsilon_2}}$ there exists a t_0, $t_a > t_0$, and bounded input signal $u_b(t)$ such that

$$\|y_a(t_a)\| = \|x(t_a)\| > \sqrt{\frac{\eta^2 \delta_2}{\varepsilon_2}} \sup_{t \geq t_0} \|u_b(t)\|. \tag{5.58}$$

Furthermore we can assume that $u_b(t)$ satisfies $u_b(t) = 0$ for $t > t_a$. Applying this input to Eq. (5.42), keeping the same initial time t_0, the zero-state response satisfies

$$\begin{aligned} \delta_2 \sup_{t_a \leq t \leq t_a + \delta_2} \|y(t)\|^2 &\geq \int_{t_a}^{t_a + \delta_2} \|y(t)\|^2 dt \\ &= \int_{t_a}^{t_a + \delta_2} x^T(t_a)\Phi^T(t, t_a)C^T(t)C(t)\Phi(t, t_a)x(t_a)dt \\ &= x^T(t_a)M(t_a, t_a + \delta_2)x(t_a). \end{aligned}$$

Invoking the hypothesis on the observability Gramian, and then Eq. (5.58),

$$\begin{aligned} \delta_2 \sup_{t_a \leq t \leq t_a + \delta_a} \|y(t)\|^2 &\geq \varepsilon_2 \|x(t_a)\|^2 \\ &> \eta^2 \delta_2 (\sup_{t \geq t_0} \|u_b(t)\|)^2. \end{aligned}$$

Using elementary properties of the supremum, including

$$(\sup_{t_a \leq t \leq t_a + \delta_2} \|y(t)\|)^2 = \sup_{t_a \leq t \leq t_a + \delta_2} \|y(t)\|^2$$

yields

$$\sup_{t \geq t_0} \|y(t)\| > \eta \sup_{t \geq t_0} \|u_b(t)\|. \tag{5.59}$$

$\square$

Thus we have shown that the bounded input $u_b(t)$ is such that bound (5.56) for uniform bounded-input, bounded-output stability of Eq. (5.42) is violated. This contradiction implies Eq. (5.57) is uniformly bounded-input, bounded-state stable. Then by Theorem 5.3.2 the state equation (5.57) is uniformly exponentially stable and, hence, Eq. (5.42) also is uniformly exponentially stable.

5.3.3 Time-invariant case

Complicated and seemingly contrived manipulations in the proofs of Theorems 5.3.2 and 5.3.3 motivate separate consideration of the time-invariant case. In the time-invariant setting, simpler characterizations of stability properties, and of controllability and observability, yield more straightforward proofs. For the linear state equation

$$\dot{x}(t) = Ax(t) + Bu(t),$$
$$y(t) = Cx(t) \tag{5.60}$$

the main task in proving an analog of Theorem 5.3.3 is to show that controllability, observability, and finiteness of

$$\int_0^\infty \|Ce^{At}B\|dt \tag{5.61}$$

imply finiteness of

$$\int_0^\infty \|e^{At}\|dt.$$

Theorem 5.3.4. *Suppose the time-invariant linear state equation (5.60) is controllable and observable. Then the state equation is uniformly bounded-input, bounded-output stable if and only if it is exponentially stable.*

Proof. Clearly exponential stability implies uniform bounded-input, bounded-output stability since

$$\int_0^\infty \|Ce^{At}B\|dt \le \|C\|\|B\| \int_0^\infty \|e^{At}\|dt.$$

Conversely, suppose Eq. (5.43) is uniformly bounded-input, bounded-output stable. Then Eq. (5.61) is finite, and this implies

$$\lim_{t\to\infty} Ce^{At}B = 0. \tag{5.62}$$

Using a representation for the matrix exponential, we can write the impulse response in the form

$$Ce^{At}B = \sum_{k=1}^l \sum_{j=1}^{\sigma_k} G_{kj} \frac{t^{j-1}}{(j-1)!} e^{\lambda_k t}, \tag{5.63}$$

where $\lambda_1, \ldots, \lambda_l$ are the distinct eigenvalues of A, and the G_{kj} are $p \times m$ constant matrices. Then

$$\frac{d}{dt} Ce^{At}B = \sum_{k=1}^l \left[G_{k1}\lambda_k + \sum_{j=2}^{\sigma_k} G_{kj} \left(\frac{\lambda_k t^{j-1}}{(j-1)!} + \frac{t^{j-2}}{(j-2)!} \right) \right] e^{\lambda_k t}.$$

If we suppose that this function does not go to zero, then from a comparison with Eq. (5.63) we arrive at a contradiction with Eq. (5.62). Therefore

$$\lim_{t \to \infty} \left(\frac{d}{dt} C e^{At} B \right) = 0.$$

That is,

$$\lim_{t \to \infty} C A e^{At} B = \lim_{t \to \infty} C e^{At} A B = 0.$$

This reasoning can be repeated to show that any time derivative of the impulse response goes to zero as $t \to \infty$. Explicitly,

$$\lim_{t \to \infty} C A^i e^{At} A^j B = 0; \quad i, j = 0, 1, \dots.$$

These data implies

$$\lim_{t \to \infty} \begin{bmatrix} C \\ CA \\ \vdots \\ CA^{n-1} \end{bmatrix} e^{At} [B \; AB \; \dots \; A^{n-1}B] = 0. \tag{5.64}$$

Using the controllability and observability hypotheses, select n linearly independent columns of the controllability matrix to form an invertible matrix W_a, and n linearly independent rows of the observability matrix to form an invertible M_a. Then, from Eq. (5.64),

$$\lim_{t \to \infty} M_a e^{At} W_a = 0.$$

Therefore

$$\lim_{t \to \infty} e^{At} = 0.$$

For some purposes it is useful to express the condition for uniform bounded-input, bounded-output stability of Eq. (5.60) in terms of the transfer function $G(s) = C(sI - A)^{-1}B$. We use the familiar terminology that a pole of $G(s)$ is a (complex, in general) value of s, say s_0, such that for some i, j, $|G_{ij}(s_0)| = \infty$.

If each entry of $G(s)$ has negative-real-part poles, then a partial-fraction-expansion computation shows that each entry of $G(t)$ has a "sum of t-multiplied exponentials" form, with negative-real-part exponents. Therefore

$$\int_0^{\infty} \|G(t)\| dt \tag{5.65}$$

is finite, and any realization of $G(s)$ is uniformly bounded-input, bounded-output stable. On the other hand if Eq. (5.65) is finite, then the exponential terms in any entry of $G(t)$ must have negative real parts. (Write a general entry in terms of distinct exponentials, and use a contradiction argument.) But then every entry of $G(s)$ has negative-real-part poles.

Supplying this reasoning with a little more specificity proves a standard result.   □

Theorem 5.3.5. *The time-invariant linear state equation (5.60) is uniformly bounded-input, bounded-output stable if and only if all poles of the transfer function $G(s) = C(sI - A)^{-1}B$ have negative real parts.*

For the time-invariant linear state equation (5.60), the relation between input-output stability and internal stability depends on whether all distinct eigenvalues of A appear as poles of $G(s) = C(sI - A)^{-1}B$. Controllability and observability guarantee that this is the case. (Unfortunately, eigenvalues of A sometimes are called "poles of A," a loose terminology that at best obscures delicate distinctions.)

Fundamental approaches to control system analysis

6

The main purpose of this introductory chapter is to familiarize the reader with such basic theories on control system structure and behavior as the polynomial matrix description (PMD) theory, behavioral theory, and chain-scattering representation (CSR) approaches. This chapter also serves to prepare for the representation of our results in the succeeding chapters.

6.1 PMD theory of linear multivariable control systems

The main aim of this section is to briefly introduce the background and preliminary results in PMD theory, which are needed in the sequel of this book. Regarding the related issue of determination of finite and infinite frequency structure of a rational matrix and the issue of the resolvent decomposition and solution of regular PMD, this book will present its contributions in Chapters 7–10. The main references to the following introduction are Rosenbrock [5], Vardulakis [13], and Kailath [32].

The initial aim [5] of PMD theory was to describe various time domain results of Kalman's state space theory into a powerful algebraic language by using the existing results in matrix theory. This led to a better understanding of the mathematical structure of linear multivariable systems by generalizing the classical single-input single-output transfer function approaches to multivariable case. It finally resulted in various synthesis techniques for multivariable feedback systems. So far PMD theory has been established as a very successful and still developing area in the linear multivariable control system theory.

Throughout this book, $\Re[s]^{p \times q}$ denotes the set of $p \times q$ polynomial matrices, while $\Re(s)^{p \times q}$ denotes the set of $p \times q$ rational matrices. Regular (PMDs) or *linear nonhomogeneous matrix differential equations* (LNHMDEs) are described by

$$\Sigma : \begin{cases} A(\rho)\beta(t) & = B(\rho)u(t), \\ y(t) & = C(\rho)\beta(t) + D(\rho)u(t), \quad t \geq 0, \end{cases} \tag{6.1}$$

where $\rho := d/dt$ is the differential operator, $A(\rho) \in \Re[\rho]^{r \times r}$, $\mathrm{rank}_{\Re[\rho]}A(\rho) = r$, $B(\rho) \in \Re[\rho]^{r \times m}$, $C(\rho) \in \Re[\rho]^{p \times r}$, $D(\rho) \in \Re[\rho]^{p \times m}$, $\beta(t) : [0, +\infty) \to \Re^r$ is the *pseudo-state* of the PMDs, $u(t) : [0, +\infty) \to \Re^m$ is a p times piecewise continuously differentiable function called the *input*, $y(t) : [0, +\infty) \to \Re^p$ is the output vector of the PMD.

A Generalized Framework of Linear Multivariable Control. http://dx.doi.org/10.1016/B978-0-08-101946-7.00006-8

Taking the Laplace transform of Eq. (6.1) and assuming "zero initial conditions," i.e., that

$$u(0)^{(i)} = 0, \quad \beta(0)^{(i)} = 0, \quad y(0)^{(i)} = 0,$$

Eq. (6.1) can be written as

$$\begin{cases} A(s)\hat{\beta}(s) & = B(s)\hat{u}(s), \\ \hat{y}(s) & = C(s)\hat{\beta}(s) + D(s)\hat{u}(s), \end{cases} \tag{6.2}$$

where $\hat{\beta}(s) := \mathcal{L}\{\beta(t)\}$, $\hat{u}(s) := \mathcal{L}\{u(t)\}$, $\hat{y}(s) := \mathcal{L}\{y(t)\}$ are Laplace transforms of $\beta(t)$, $u(t)$, and $y(t)$, respectively. From Eq. (6.2) we have

$$\hat{y}(s) = [C(s)A(s)^{-1}B(s) + D(s)]\hat{u}(s). \tag{6.3}$$

Definition 6.1.1. *The rational matrix*

$$H(s) := C(s)A(s)^{-1}B(s) + D(s) \in \Re(s)^{p \times m}$$

is called the transfer function matrix *of the system* $\sum$.

A polynomial matrix $T(s) \in \Re[s]^{p \times p}$ is called $\Re[s]$-*unimodular* or simply *unimodular* if there exists a matrix $\hat{T}(s) \in \Re[s]^{p \times p}$ such that $T(s)\hat{T}(s) = I_p$, where I_p denotes the $p \times p$ identity matrix, equivalently if $\det T(s) = c \in \Re, c \neq 0$.

Definition 6.1.2. *The* degree *of a polynomial matrix* $T(s) \in \Re[s]^{p \times m}$, *denoted by* deg $T(s)$, *is defined as the maximum degree among the degrees of all its maximum order (nonzero) minors.*

Corollary 6.1.1. *If $T(s)$ is a nonsingular square polynomial matrix, then*

$$\deg T(s) = \deg(\det(T(s))),$$

where $\det(\cdot)$ denotes the determinant of the indicated matrix.

Corollary 6.1.2. *$T(s) \in \Re[s]^{p \times p}$ is unimodular if and only if* deg $T(s) = 0$.

Two rational matrices $T_1(s)$, $T_2(s) \in \Re[s]^{p \times m}$ are called equivalent if there exist unimodular matrices $T_L(s) \in \Re[s]^{p \times p}$, $T_R(s) \in \Re[s]^{m \times m}$ such that

$$T_L(s)T_1(s)T_R(s) = T_2(s).$$

Any rational matrix is equivalent to its *Smith-McMillan form*, which is a canonical form of a matrix.

Theorem 6.1.1 ([5] Smith-McMillan form of a rational matrix in C). Let $T(s) \in \Re[s]^{p \times m}$ with $\text{rank}_{\Re[s]}T(s) = r$, $r \leq \min\{p, m\}$. Then $T(s)$ is equivalent to a diagonal matrix $S^C_{T(s)}$ having the form

$$S^C_{T(s)} := \text{block diag} \left[\frac{\varepsilon_1(s)}{\psi_1(s)}, \frac{\varepsilon_2(s)}{\psi_2(s)}, \ldots, \frac{\varepsilon_r(s)}{\psi_r(s)}, 0_{p-r,m-r} \right], \tag{6.4}$$

where $\varepsilon_i(s)$, $\Psi_i(s) \in \Re[s]$ are monic and coprime such that $\varepsilon_i(s)$ divides $\varepsilon_{i+1}(s)$, and $\Psi_{i+1}(s)$ divides $\Psi_i(s), i = 1, 2, \ldots, r - 1$.

If $T(s) \in \Re[s]^{p \times m}$ then $\Psi_i(s) = 1, \forall i \in \mathbf{r}$, that is, $S^C_{T(s)}$ is also a polynomial matrix and it is called the *Smith form* of $T(s)$. Otherwise, i.e., if $T(s)$ is nonpolynomial, for some i and j, $\Psi_i(s)$ are nonconstant, that is $S^C_{T(s)}$ is also nonpolynomial and it is called *McMillan form* of $T(s)$.

The *zeroes* of $T(s)$ are defined as the zeros of the polynomials $\varepsilon_i(s), i \in \mathbf{r}$. The *poles* of $T(s)$ are defined as the zeros of the polynomials $\Psi_i(s), i \in \mathbf{r}$.

Vardulakis et al. [87] introduced the concept of the Smith-McMillan form at infinity of a rational matrix. The main definitions are briefly presented here.

Definition 6.1.3. $T(s) \in \Re[s]^{p \times m}$ *will be called* proper *if* $\lim_{s \to \infty} T(s)$ *exists. If the limit is zero then* $T(s)$ *will be called* strictly proper, *while if this limit is nonzero* $T(s)$ *will be called* exactly proper.

Let $\Re_{pr}(s)$ denote the ring of proper rational functions.

Definition 6.1.4. The $m \times m$ rational matrix $W(s) \in \Re_{pr}^{p \times m}(s)$ is said to be *biproper* if and only if

1. $\lim_{s \to \infty} W(s) = W_\infty \in \Re^{m \times m}$ exists, and
2. $\det W_\infty \neq 0$.

The p × m rational matrices $T_1(s)$ and $T_2(s)$ are said to be *equivalent at infinity* if there exist biproper matrices $W(s) \in \Re_{pr}^{p \times p}(s), V(s) \in \Re_{pr}^{m \times m}(s)$ such that

$$W(s)T_1(s)V(s) = T_2(s).$$

Since $W(s)$ and $V(s)$ are biproper, it can be seen from Definition 6.1.4 that $W(s)$ and $V(s)$ possess neither poles nor zeros at infinity. It therefore follows intuitively from this that $T_1(s)$ and $T_2(s)$ have an identical pole-zero structure at infinity. A canonical form for a rational matrix under the equivalence at infinity is its Smith-McMillan form at infinity, $S^\infty_{T(s)}(s)$.

Theorem 6.1.2 (Smith-McMillan form of a rational matrix at $s = \infty$). Let $T(s) \in \Re^{p \times m}(s)$ with $\text{rank}_{\Re[s]} T(s) = r$. Then $T(s)$ is equivalent at $s = \infty$ to a diagonal matrix having the form

$$S^\infty_{T(s)}(s) = \text{block diag}[s^{q_1}, s^{q_2}, \ldots, s^{q_k}, 1/s^{\hat{q}_{k+1}}, \ldots, 1/s^{\hat{q}_r}, 0_{p-r,m-r}], \tag{6.5}$$

where $1 \leq k \leq r$ and

$$q_1 \geq q_2 \geq \cdots \geq q_k \geq 0$$

$$\hat{q}_r \geq \hat{q}_{r-1} \geq \cdots \geq \hat{q}_{k+1} \geq 0$$

are respectively the orders of its *poles* and *zeros* at $s = \infty$.

A reduction approach for computing Smith-McMillan form at ∞ of a rational matrix is suggested in [87].

Any rational function can be represented as a ratio of coprime polynomials, this can be generalized to the matrix case.

Definition 6.1.5. Let $T(s) \in \Re(s)^{p \times m}$ with $\text{rank}_{\Re(s)} T(s) = r, 1 \leq r \leq \min\{p, m\}$. Then there exist nonunique pairs: $A_1(s) \in \Re(s)^{p \times p}, B_1(s) \in \Re(s)^{p \times m}$ and left coprime, $B_2(s) \in \Re[s]^{p \times m}, A_2(s) \in \Re[s]^{m \times m}$ and right coprime such that

$$T(s) = A_1(s)^{-1} B_1(s) = B_2(s) A_2(s)^{-1}. \tag{6.6}$$

A representation of a rational matrix $T(s)$ given by Eq. (6.6) is called a left (right) coprime *polynomial matrix fraction description* (MFD) of $T(s)$.

Let $A(s) \in R[s]^{r \times r}$, $\text{rank}_{\Re(s)} A(s) = r$. Then by Theorem 6.1.2 $A(s)$ is equivalent at $s = \infty$ to its Smith-McMillan form $S_{A(s)}^{\infty}(s)$ having the form

$$S_{A(s)}^{\infty}(s) = \text{diag}[s^{q_1}, s^{q_2}, \ldots, s^{q_k}, 1/s^{\hat{q}_{k+1}}, \ldots, 1/s^{\hat{q}_r}], \tag{6.7}$$

where $1 \leq k \leq r$ and

$$q_1 \geq q_2 \geq \cdots \geq q_k \geq 0,$$

$$\hat{q}_r \geq \hat{q}_{r-1} \geq \cdots \geq \hat{q}_{k+1} \geq 0,$$

If $A(s)$ has at least one zero at $s = \infty$, let

$$A(s)^{-1} = H_k s^k + H_{k-1} s^{k-1} + \cdots + H_1 s + H_0 + H_{-1} s^{-1} + H_{-2} s^{-2} + \cdots \tag{6.8}$$

be the Laurent expansion of $A(s)^{-1}$ at $s = \infty$, where $k > 0$, $H_k \neq 0$. Then one has

$$k = \hat{q}_r. \tag{6.9}$$

Let

$$A(s) = A_0 + A_1 s + \cdots + A_{q_1} s^{q_1} \in R^{r \times r}[s]. \tag{6.10}$$

$n := \deg \det(A(s))$ and $\lambda_0 \in C$ such that $\det(A(\lambda_0)) = 0$. The sequence of r-dimensional vectors $x_0, x_1, \ldots, x_m (x_0 \neq 0)$ for which the following equalities hold

$$\sum_{i=0}^{q} \frac{1}{i!} A^i(\lambda_0) x_{q-i} = 0, \quad q = 0, 1, 2, \ldots, m$$

is called *a Jordan Chain* of length $(m+1)$ for $A(s)$ corresponding to λ_0. Now let

$$S_{A(s)}^{C}(s) = \text{diag}[1, 1, \ldots, 1, f_k(s), f_{k+1}(s), \ldots, f_r(s)]$$

$1 \leq k \leq r$ be the Smith Form of $A(s)$.

$$f_j(s)/f_{j+1}(s), \quad j = k, k+1, \ldots, r-1.$$

Let

$$f_j(s) = \prod_{i=1}^{l} (s - \lambda_i)^{\sigma_{ij}}, \quad j = k, k+1, \ldots, r$$

be the decomposition of the invariant polynomials $f_j(s)$ into irreducible elementary divisors over C, i.e., assure that

$$\det(A(s)) = 0$$

has l distinct zeroes

$$\lambda_1, \lambda_2, \ldots, \lambda_l \in R,$$

where

$$0 \leq \sigma_{ik} \leq \sigma_{i,k+1} \leq \cdots \leq \sigma_{ir}, i \in \mathbf{I}$$

are the partial multiplicities of the eigenvalues λ_i. Let

$$x_{j0}^i, x_{j1}^i, \ldots, x_{j,\sigma_{ij}-1}^i \in R^r, (x_{j0}^j \neq 0),$$

$$i \in \mathbf{I}, \quad j = k, k+1, \ldots, r$$

be the Jordan Chain of lengths σ_{ij} corresponding to the eigenvalue λ_i of $A(s)$ and consider the matrices

$$C_i = [x_{k,0}^i, \ldots, x_{k,\sigma_{i,k}-1}^i] \cdots [x_{r,0}^i, \ldots, x_{r,\sigma_{i,r}-1}^i] \in R^{r \times m_i},$$

where $m_i = \sum_{j=k}^r \sigma_{ij}, i \in \mathbf{I}$ and

$$J_i = \text{block diag}[J_{ik}, J_{i,k+1}, \ldots, J_{ir}] \in R^{m_i \times m_i}, i \in \mathbf{I},$$

where

$$J_{ij} = \begin{bmatrix} \lambda_i & 1 & 0 & \cdots & 0 \\ 0 & \lambda_i & 1 & \cdots & 0 \\ \cdot & \cdot & \cdot & \cdots & \cdot \\ 0 & 0 & 0 & \cdots & 1 \\ 0 & 0 & 0 & \cdots & \lambda_i \end{bmatrix} \in R^{\sigma_{ij} \times \sigma_{ij}}$$

$$i \in \mathbf{I}, \quad j = k, k+1, \ldots, r.$$

Definition 6.1.6. The matrix pair (C, J) with

$$C = [C_1, C_2, \ldots, C_l] \in R^{r \times n}$$

$$J = \text{block diag}[J_1, J_2, \ldots, J_l] \in R^{n \times n}$$

where $n := m_1 + m_2 + \cdots + m_l = \deg \det(A(s))$ is called the *finite Jordan pair* of $A(s)$.

Proposition 6.1.1 ([17]). *The matrices C, J satisfy*

$$A_{q_1} C J^{q_1} + A_{q_1-1} C J^{q_1-1} + \cdots + A_0 C = 0$$

$$Q_n = \begin{bmatrix} C \\ CJ \\ \cdot \\ \cdot \\ \cdot \\ CJ^{n-1} \end{bmatrix} \in R^{m \times n}, \quad \text{rank } Q_n = n.$$

Definition 6.1.7. Let $T(s) \in R_{pr}(s)^{p \times m}$, then a quadruple (A, B, C, E) such that

$$T(s) = C(sI_n - A)^{-1}B + E$$

is called a realization of the proper rational matrix $T(s)$.

Definition 6.1.8. A realization (A, B, C, E) of a $T(s) \in R_{pr}(s)^{p \times m}$ is called a minimal realization of $T(s)$ if

$$n = \delta_M(T(s)),$$

where $\delta_M(T(s))$ is the *McMillan* degree of $T(s)$.

Definition 6.1.9. Let $T(s) \in R[s]^{p \times m}$. Then a triple of matrices $C_\infty \in R^{p \times \mu}, A_\infty \in R^{\mu \times \mu}, B_\infty \in R^{\mu \times m}$ such that

$$T(s) = C_\infty(I_\mu - sA_\infty)^{-1}B_\infty \tag{6.11}$$

is called a realization of the polynomial matrix $T(s)$.

A realization of $T(s) \in R[s]^{p \times m}$ can always be obtained from a realization of the strictly proper rational matrix $\tilde{T}(w) := (1/w)T(1/w) \in R_{pr}^{p \times m}(w)$. Because if $C_\infty \in R^{p \times \mu}, A_\infty \in R^{\mu \times \mu}, B_\infty \in R^{\mu \times m}$, is a realization of $\tilde{T}(w)$, i.e., if

$$\frac{1}{w}T\left(\frac{1}{w}\right) = C_\infty(wI_\mu - A_\infty)^{-1}B_\infty, \tag{6.12}$$

then Eq. (6.12) by the substitution $\frac{1}{w} = s$ gives Eq. (6.11).

Proposition 6.1.2 (Vardulakis [13]). *Let*

$$T(s) = T_0 + T_1 s + \cdots + T_{q_1} s^{q_1} \in R^{p \times m}[s], \quad \text{rank}_{\Re(s)} T(s) = r,$$

its Smith-McMillan form at infinity is

$$S_{T(s)}^\infty(s) = block\ diag[s^{q_1}, s^{q_2}, \ldots, s^{q_k}, 1/s^{\hat{q}_{k+1}}, \ldots, 1/s^{\hat{q}_r}, 0_{p-r,m-r}]$$

are respectively the orders of its poles and zeros at $s = \infty$. Then

1. The McMillan degree of $\tilde{T}(w) := (1/w)T(1/w) \in R_{pr}^{p \times m}(w)$ is given by

$$\mu = \delta_M\left[\frac{1}{w}T\left(\frac{1}{w}\right)\right] = \sum_{i=1}^{k}(q_i + 1) = q_1 + q_2 + \cdots + q_k + k.$$

2. If $(C_\infty, J_\infty, B_\infty)$ is a minimal realization of $T(s)$ with $J_\infty \in R^{\mu \times \mu}$ in Jordan form, then

$$J_\infty = block\ diag[J_{\infty 1}, J_{\infty 2}, \ldots, J_{\infty k}] \in R^{\mu \times \mu},$$

where

$$J_{\infty i} = \begin{bmatrix} 0 & 1 & 0 & \cdots & 0 \\ 0 & 0 & 1 & \cdots & 0 \\ \cdot & \cdot & \cdot & \cdots & \cdot \\ 0 & 0 & 0 & \cdots & 1 \\ 0 & 0 & 0 & \cdots & 0 \end{bmatrix} \in R^{(q_i+1) \times (q_i+1)}, \quad i = 1, 2, \ldots, k.$$

Crucial to the issue of the solution of regular PMDs, which we will discuss later, is the concept of an *infinite Jordan pair* of a regular polynomial matrix, i.e., a Jordan pair C_∞, J_∞ which corresponds to its zeros at $s = \infty$.

Definition 6.1.10 ([17]). Let $A(s) = A_0 + A_1 s + \cdots + A_{q_1} s^{q_1} \in \Re[s]^{r \times r}$, $\text{rank}_{\Re[s]} A(s) = r$, $q_1 \geq 1$. Then a pair

$$C_\infty \in R^{r \times v}, \quad J_\infty = block\ diag\left[J_{\infty 1}, J_{\infty 2}, \ldots, J_{\infty \zeta}\right] \in R^{v \times v}$$

$$J_{\infty i} = \begin{bmatrix} 0 & 1 & 0 & \cdots & 0 \\ 0 & 0 & 1 & \cdots & 0 \\ . & . & . & \cdots & . \\ 0 & 0 & 0 & \cdots & 1 \\ 0 & 0 & 0 & \cdots & 0 \end{bmatrix} \in R^{v_i \times v_i}, \quad i = 1, 2, \ldots, \zeta$$

$v := \sum_{i=1}^{\zeta} v_i, v_i, \zeta \in \mathbb{Z}^+$is called an *infinite Jordan pair* of $A(s)$ if it is a (finite) Jordan pair of the "dual" polynomial matrix

$$\tilde{A}(w) := w^{q_1} A \left(\frac{1}{w} \right) = A_0 w^{q_1} + A_1 w^{q_1-1} + \cdots + A_{q_1} \in \Re[w]^{r \times r}$$

corresponding to its zero at $w = 0$, or equivalently if and only if (by Proposition 6.1.1)

$$A_0 C_\infty J_\infty^{q_1} + A_1 C_\infty J_\infty^{q_1-1} + \cdots + A_{q_1} C_\infty = 0,$$

and

$$\text{rank } Q_v^\infty = \text{rank} \begin{bmatrix} C_\infty \\ C_\infty J_\infty \\ \vdots \\ C_\infty J_\infty^{v-1} \end{bmatrix} = v.$$

Regarding the resolvent decomposition of a regular polynomial matrix which is closely related to the solution of regular PMD, Vardulakis [13] gave the following result.

Theorem 6.1.3. *Let $A(s) = A_0 + A_1 s + \cdots + A_{q_1} s^{q_1} \in \Re[s]^{r \times r}$, $rank_{\Re[s]} A(s) = r$, $q_1 \geq 1$ with Smith-McMillan form at $s = \infty S_{A(s)}^\infty$ given by Eq. (6.7). Write*

$$A^{-1}(s) = H_{pol}(s) + H_{sp}(s),$$

where $H_{pol}(s) \in \Re[s]^{r \times r}$ and $H_{sp}(s) \in \Re_{pr}^{r \times r}(s)$ is strictly proper.

Let $n = \deg(\det(A(s))) = \delta_M(H_{sp}(s))$, $v = \sum_{i=k+1}^r (\hat{q}_i + 1)$. Let $C \in \Re^{r \times n}, J \in \Re^{n \times n}, B \in \Re^{n \times r}$ be a minimal realization of $H_{sp}(s)$ and $C_\infty \in \Re^{r \times v}, J_\infty \in \Re^{v \times v}, B_\infty \in \Re^{v \times r}$ be a minimal realization of $H_{pol}(s)$. Then C, J is a finite Jordan pair of $A(s)$ and C_∞, J_∞ is an infinite Jordan pair of $A(s)$. Furthermore, $A(s)^{-1}$ can be written as

$$A(s)^{-1} = [C, C_\infty] \left[\begin{array}{c|c} sI_n - J & O_{n,v} \\ \hline O_{v,n} & I_v - sJ_\infty \end{array} \right]^{-1} \begin{bmatrix} B \\ B_\infty \end{bmatrix}. \tag{6.13}$$

6.2 Behavioral approach in systems theory

The purpose of this section is to briefly introduce behavioral theory. Based on these preliminary results we will present a new approach *realization of behavior* in Chapter 12. The main references to the following introduction are [26, 27, 88, 89].

Both the transfer function and the state space approaches view a system as a signal processor that accepts inputs and transfers them into outputs. In the transfer function approach, this processor is described through the way in which exponential inputs are transformed into exponential outputs. In the state space approach, this processor involves the state as an intermediate variable, but the ultimate aim remains to describe how inputs lead to outputs. This input-output point of view has played an important role in system control theory. However, the starting point of behavioral theory is fundamentally different. As claimed in [27, 28], such a starting point is more suited to modeling and more suitable for actual applications in certain circumstances.

In the behavioral approach a mathematical model is viewed as a subset of a universum of possibilities. Before one accepts a mathematical model as a description of reality, all outcomes in the universum are in principle possible. After one accepts the mathematical model as a convenient description, one can declare that only outcomes in a certain subset are possible. This subset is called the *behavior* of the mathematical model. Starting from this perspective, one arrives at the notion of a dynamical system as simply a subset of time-trajectories, as a family of time signals taking on values in a suitable signal space.

It is in terms of the time trajectories of a specific system that all the concepts in behavioral theory are put forward. Linear time-invariant differential systems have such a nice structure that they fall into the scope of behavioral approach immediately. When one has a set of variables that can be described by such a system, then there is a transparent way of describing how trajectories in the behavior are generated. Some of the variables, it turns out, are free, i.e., unconstrained. They can thus be viewed as unexplained by the model and imposed on the system by the environment. These variables are called *inputs*. Once these variables are determined, the remaining variables called *outputs* are not yet completely specified, because of the possible trajectories, which are dependent on the past history of the system. This means that the outputs are still dependent on the initial conditions of the system. To formulate this relationship between the outputs, the inputs and the initial conditions of the system, one thus has to use the concept of *state*.

When one models an *interconnected* physical system, then unavoidably auxiliary variables, in addition to the variables modeled, will appear in the model. In order to distinguish them from the *manifest variables*, which are the variables whose behavior the model aims to describe, these auxiliary variables are called *latent* variables. The interaction between manifest and latent variables is one of the themes in this book. In this book a new approach, *realization of behavior*, will be presented to expose this interaction. In [27] it was shown how to eliminate latent variables and how to introduce state variables. Thus a system of linear differential equations containing latent variables can be transformed in an equivalent system in which these latent variables have been eliminated.

The basic idea in our approach of realization of behavior is, however, to find an ARMA representation for a given frequency behavior description such that the known frequency behavior is completely recovered to the corresponding dynamical behavior. From this point of view, realization of behavior is seen to be a converse procedure to the above latent variable eliminating process. Such a realization approach is believed to be highly significant in modeling dynamical system in some real cases where the system behavior is conveniently described in the frequency domain. Since no numerical computation is required of the procedure, the realization of behavior is believed to be particularly suitable for situations in which the coefficients are symbolic rather than numerical.

Definition 6.2.1 ([89]). Let $\mathbb{U}$ be a universum, $\mathbb{E}$ a set, and $f_1, f_2 : \mathbb{U} \to \mathbb{E}$. The mathematical model $(\mathbb{U}, \mathfrak{B})$ with $\mathfrak{B} = \{u \in \mathbb{U} | f_1(u) = f_2(u)\}$ is said to be described by *behavioral equations* and is denoted by $(\mathbb{U}, \mathbb{E}, f_1, f_2)$. The set $\mathbb{E}$ is called the *equating space*. We also call $(\mathbb{U}, \mathbb{E}, f_1, f_2)$ a *behavioral equation representation* of $(\mathbb{U}, \mathfrak{B})$.

Latent variables appear frequently in system modeling practice, for which examples abound and are provided in [27, 89]. The need to use latent variables is recognized from the situations where for mathematical reasons they are unavoidably involved in expressing the basic laws in the modeling process. For example, *state variables* are needed in system theory in order to express the memory of a dynamical system, internal voltages and currents are needed in electrical circuits in order to express the external port behavior, momentum is needed in Hamiltonian mechanics in order to describe the evolution of the position, prices are needed in economics in order to explain the production and exchange of economic goods, etc.

Definition 6.2.2 ([89]). A *mathematical model with latent variables* is defined as a triple $(\mathbb{U}, \mathbb{U}_l, \mathfrak{B}_f)$ with $\mathbb{U}$ the *universum* of manifest variables, $\mathbb{U}_l$ the universum of *latent variables*, and $\mathfrak{B}_f \subseteq \mathbb{U} \times \mathbb{U}_l$ the *full behavior*. It defines the *manifest mathematical* model $(\mathbb{U}, \mathfrak{B})$ with $\mathfrak{B} := \{u \in \mathbb{U} | \exists l \in \mathbb{U}_l \text{ such that } (u, l) \in \mathfrak{B}_f\}$; $\mathfrak{B}$ is called the *manifest behavior* (or the *external behavior*) or simply the *behavior*. We call $(\mathbb{U}, \mathbb{U}_l, \mathfrak{B}_f)$ a latent *variable representation* of $(\mathbb{U}, \mathfrak{B})$.

Now by applying the above ideas the following basic description about *dynamical system* can be set up in a language of *behavioral theory*.

Definition 6.2.3 ([27]). A *dynamical system* Σ is defined as a triple

$$\Sigma = (\mathbb{T}, \mathbb{W}, \mathfrak{B}),$$

with $\mathbb{T}$ a subset of $\mathbb{R}$, called the *time axis*, $\mathbb{W}$ a set called the *signal space*, and $\mathfrak{B}$ is a subset of $\mathbb{W}^T$ called the *behavior*.

Now consider the following class of dynamical systems with latent variables

$$R\left(\frac{d}{dt}\right) w = M\left(\frac{d}{dt}\right) l, \qquad (6.14)$$

where $w : \mathbb{R} \to \mathbb{R}^q$ is the trajectory of the manifest variables, whereas $l : \mathbb{R} \to \mathbb{R}^d$ is the trajectory of the latent variables. The equating space is $\mathbb{R}^g$, and the behavioral equations are parameterized by the two polynomial matrices $R(\xi) \in \mathbb{R}^{g \times q}[\xi]$ and $M(\xi) \in \mathbb{R}^{g \times d}[\xi]$.

The question of *elimination of latent variables* is, what sort of behavioral equation does Eq. (6.14) imply about the manifest variable w alone? In particular, we wonder whether the relations imposed on the manifest variable w by the full behavioral equations (6.14) can themselves be written into the form of a system of differential equations. In other words, this question is to ask whether or not the set

$$\mathfrak{B} = \left\{ w \in \mathfrak{L}_1^{loc}(\mathbb{R}, \mathbb{R}^q) | \exists l \in \mathfrak{L}_1^{loc}(\mathbb{R}, \mathbb{R}^d) \text{ satisfies } R\left(\frac{d}{dt}\right) w = M\left(\frac{d}{dt}\right) l \text{ weakly} \right\}$$

(6.15)

can be written as the (weak) solution set of a system of linear differential equations. The above question is very important in situations where one has to introduce more variables in order to obtain a model of the relation between certain variables in a system of some complexity, after one proposes this model those auxiliary variables in which one is not interested may be eliminated by manipulating the equations (model).

An appealing and insightful answer to the above question is the following *latent variable elimination procedure* [27].

Theorem 6.2.1. *Denote the full behavior of Eq. (6.14) by*

$$\mathfrak{B}_f = \left\{ (w, l) \in \mathfrak{L}_1^{loc}(\mathbb{R}, \mathbb{R}^q \times \mathbb{R}^d)) | R\left(\frac{d}{dt}\right) w = M\left(\frac{d}{dt}\right) l \text{ weakly} \right\}.$$

(6.16)

Let the unimodular matrix $U(\xi) \in \mathbb{R}^{g \times g}[\xi]$ be such that

$$U(\xi)M(\xi) = \left[\begin{array}{c} 0 \\ M''(\xi) \end{array} \right], \quad U(\xi)R(\xi) = \left[\begin{array}{c} R'(\xi) \\ R''(\xi) \end{array} \right],$$

(6.17)

with $M''(\xi)$ of full row rank. Then the C^∞ part of the manifest behavior $\mathfrak{B}$, denoted by $\mathfrak{B} \cap C^\infty(\mathbb{R}, \mathbb{R}^q)$ with $\mathfrak{B}$ given by Eq. (6.15), consists of the C^∞ solutions of

$$R'\left(\frac{d}{dt}\right) w = 0.$$

(6.18)

6.3 Chain-scattering representations

The main aim of this section is to briefly introduce some background and preliminary results on the CSR, which have been widely used in circuit theory, signal processing, and $\mathcal{H}_\infty$ control theory. In Chapters 11 and 12 we will present some generalizations to these results. The main references to the following introduction are [21, 22, 90].

There are two basic interconnections widely used in circuit theory and signal processing. One is the series-parallel interconnection which is shown in Fig. 6.2, the other one is the cascade connection which is shown in Fig. 6.2.

Consider two well-defined n-ports in Fig. 6.1, both having the same numbers of series ports on the one hand, and of shunt ports on the other, and designate by H_1 and H_2 their hybrid matrices. It has been proved [21] that *if the shunt ports are paralleled (port by port) and if the series ports are connected in series, the resulting n-port has the hybrid matrix*

$$H = H_1 + H_2.$$

(6.19)

Generally one comes to the conclusion that *impedance matrices* (when they exist) *add up for series connections at all ports, and admittance matrices add up for parallel connections.*

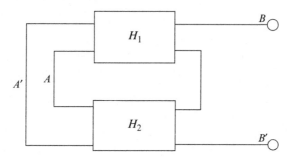

Fig. 6.1 Parallel connections.

Fig. 6.2 Cascade interconnection.

In addition to the above series-parallel interconnection, it is often convenient to consider a *cascade* (or chain) *connection* of two subnetworks which is shown in Fig. 6.2, where the output ports of the first subnetwork are identical to the input ports of the second subnetwork. Let x_a, x_b, and x_c be the electrical variables at the input ports of the first subnetwork, at the interconnected ports, and at the output ports of the second subnetwork, respectively. If the equations of the first subnetwork can be written into

$$x_a = K_1 x_b, \tag{6.20}$$

and the equations of the second subnetwork into a similar form

$$x_b = K_2 x_c, \tag{6.21}$$

the elimination of internal variables in the interconnected subnetwork is immediate, and the final equations are obtained as

$$x_a = K_1 K_2 x_c. \tag{6.22}$$

The above equations thus describe the input-output relationship in the cascade connection.

If one writes Eq. (6.20) explicitly into

$$\begin{bmatrix} v_a \\ i_a \end{bmatrix} = \begin{bmatrix} A & B \\ C & D \end{bmatrix} \begin{bmatrix} v_b \\ -i_b \end{bmatrix} := K \begin{bmatrix} v_b \\ -i_b \end{bmatrix} \tag{6.23}$$

the output vector x_b (with a sign change in i_b) of Eq. (6.23) is the input vector to a second subnetwork in cascade with the first. The matrix K appearing in Eq. (6.23) is naturally called the *chain matrix* of the $2n$-port, and one has the following theorem.

Theorem 6.3.1 ([21]). *The chain matrix of a cascade connection of 2n-port is the product of the individual chain matrices in the order of connection.*

The above *cascade structure* is the most salient characteristic feature of the chain matrix. In terms of control, the chain matrix represents the feedback simply as a multiplication of a matrix. This property makes the analysis of closed-loop systems very simple and makes the role of factorization clearer. Based on this, Kimura [22] brought forward the use of chain-scattering matrix in control system design. Another remarkable property of the CSR is the symmetry (duality) between the CSR and its inverse. This property is also regarded to be quite relevant to control system design.

A serious disadvantage of the CSR is, however, that it only exists for the special plants that satisfy certain full rank conditions. In order to obtain the CSR of the general plants that do not satisfy the full rank conditions, one is forced to augment the plants. Such augmentation is, however, irrelevant to and has no implication in control system design.

This book will present a generalization of the CSR to the case of general plants. Through the notion of input-output consistency, the conditions under which the generalized CSR and the dual generalized CSR exist will be proposed. The generalized chain-scattering matrices will be formulated into a general parameterized form by using the generalized inverse of matrices. The algebraic system properties such as the cascade structure and the symmetry (duality) property of this approach will be exploited completely.

Consider a plant P (Fig. 6.3) with two kinds of inputs (w, u) and two kinds of outputs (z, y) represented by

$$\begin{bmatrix} z \\ y \end{bmatrix} = P \begin{bmatrix} w \\ u \end{bmatrix} = \begin{bmatrix} P_{11} & P_{12} \\ P_{21} & P_{22} \end{bmatrix} \begin{bmatrix} w \\ u \end{bmatrix}, \tag{6.24}$$

where P_{ij} $(i,j = 1,2)$ are all rational matrices with dimensions $m_i \times k_j$ $(i,j = 1,2)$.

If P_{21} is invertible, then one has the CSR of P as

$$\begin{bmatrix} z \\ w \end{bmatrix} = CHAIN(P) \begin{bmatrix} u \\ y \end{bmatrix}, \tag{6.25}$$

where

$$CHAIN(P) = \begin{bmatrix} P_{12} - P_{11}P_{21}^{-1}P_{22} & P_{11}P_{21}^{-1} \\ -P_{21}^{-1}P_{22} & P_{21}^{-1} \end{bmatrix}. \tag{6.26}$$

If P represents a usual input-output relation of a system, CHAIN(P) represents the characteristic of power ports which in turn reflects physical structure of the plant. The

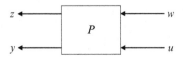

Fig. 6.3 Input-output representation.

Fig. 6.4 Chain-scattering representation.

CSR describes the plant as a wave scatterer between (u, z)-wave and the (w, y)-wave that travel oppositely to each other (Fig. 6.4).

The main reason of using the CSR lies in its ability to represent the feedback connection as a cascade one. The cascade connection of two CSRs, G_1 and G_2, is actually a feedback connection because the loops across the two systems, G_1 and G_2, exists. The resulting CSR is just $G_1 G_2$ of the two representation. This property thus greatly simplifies the analysis and synthesis of feedback connection. This can be seen by eliminating the intermediate variables (z_1, w_1) from the following relations:

$$\begin{bmatrix} z \\ w \end{bmatrix} = G_1 \begin{bmatrix} z_1 \\ w_1 \end{bmatrix}, \quad \begin{bmatrix} z_1 \\ w_1 \end{bmatrix} = G_2 \begin{bmatrix} z_2 \\ w_2 \end{bmatrix}. \tag{6.27}$$

If $G_i = CHAIN(P_i)$, $i = 1, 2$, the cascade connection represents the feedback connection represented in Fig. 6.5. This connection is also termed a *star product* in Redheffer [91]. The use of CSR simply represents this connection by the product of the two individual representations.

Another interesting property of CSR is that its inverse (if exists) is dually represented as

$$H = (CHAIN(P))^{-1} = \begin{bmatrix} P_{12}^{-1} & -P_{12}^{-1} P_{11} \\ P_{22} P_{12}^{-1} & P_{21} - P_{22} P_{12}^{-1} P_{11} \end{bmatrix}. \tag{6.28}$$

The representation (6.28) exists if P_{12} is invertible. It is called the *dual CSR* of P which is denoted by

$$DCHAIN(P) = \begin{bmatrix} P_{12}^{-1} & -P_{12}^{-1} P_{11} \\ P_{22} P_{12}^{-1} & P_{21} - P_{22} P_{12}^{-1} P_{11} \end{bmatrix}. \tag{6.29}$$

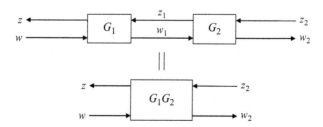

Fig. 6.5 Cascade property of chain-scattering representation.

The duality between $CHAIN(P)$ and $DCHAIN(P)$ is represented in the following identity

$$\begin{bmatrix} 0 & I \\ -I & 0 \end{bmatrix} CHAIN(P^T) \begin{bmatrix} 0 & -I \\ I & 0 \end{bmatrix} = [DCHAIN(P^T)]^T. \tag{6.30}$$

Now, we look at the realizations of the CSR and the dual CSRs. Let

$$P = \left[\begin{array}{c|cc} A & B_1 & B_2 \\ \hline C_1 & D_{11} & D_{12} \\ C_2 & D_{21} & D_{22} \end{array} \right] \tag{6.31}$$

be a state-space realization of the plant P. In order to obtain the realization of the CSR and the dual CSRs, one may write out the following state equation of the plant P explicitly

$$x = Ax + B_1 w + B_2 u \tag{6.32}$$
$$z = C_1 x + D_{11} w + D_{12} u \tag{6.33}$$
$$y = C_2 x + D_{21} w + D_{22} u. \tag{6.34}$$

In order that state space representations of $CHAIN(P)$ and $DCHAIN(P)$ exist, one must assume that D_{21}^{-1} and D_{12}^{-1} exist. In that case, one can solve Eq. (6.34) for w yielding

$$w = D_{21}^{-1}(-C_2 x - D_{22} u + y). \tag{6.35}$$

By substituting this relation into Eqs. (6.32), (6.33), one has a realization of $CHAIN(P)$ as

$$CHAIN(P) = \left[\begin{array}{c|cc} A - B_1 D_{21}^{-1} C_2 & B_2 - B_1 D_{21}^{-1} D_{22} & B_1 D_{21}^{-1} \\ \hline C_1 - D_{11} D_{21}^{-1} C_2 & D_{12} - D_{11} D_{21}^{-1} D_{22} & D_{11} D_{21}^{-1} \\ -D_{21}^{-1} C_2 & -D_{21}^{-1} D_{22} & D_{21}^{-1} \end{array} \right]. \tag{6.36}$$

Similarly, $DCHAIN(P)$ is given by

$$DCHAIN(P) = \left[\begin{array}{c|cc} A - B_2 D_{12}^{-1} C_1 & B_2 D_{12}^{-1} & B_1 - B_2 D_{12}^{-1} D_{11} \\ \hline -D_{12}^{-1} C_1 & D_{12}^{-1} & -D_{12}^{-1} D_{11} \\ C_2 - D_{22} D_{12}^{-1} C_1 & D_{22} D_{12}^{-1} & D_{21} - D_{22} D_{12}^{-1} D_{11} \end{array} \right]. \tag{6.37}$$

6.4 Conclusions

This chapter has briefly introduced certain fundamental approaches to control system analysis, specifically the PMD theory, the behavioral approach and the CSR. Based on these, this book will present its main contributions in the following chapters. In the PMD theory, our interests will be focused on certain fundamental issues as determination of the finite and infinite frequency structure of any rational matrix, analysis of the resolvent decomposition of a regular polynomial matrix and formulations of the solution of a regular PMD.

Concerning the chain-scattering representation, this book will provide a generalization to the known approach. Such a generalized CSR is believed to be useful in circuit theory and $\mathcal{H}_\infty$ control theory. Related to the behavior theory, this book will present a new notion: realization of behavior. Realization of behavior is seen to be a converse procedure to the latent variable elimination theorem [27]. Finally, two key notions in control system analysis, well-posedness and internal stability, will be discussed.

Determination of finite and infinite frequency structure of a rational matrix

7

7.1 Introduction

The finite and infinite frequency structures of a rational matrix are of great importance, particularly in the area of systems analysis and design. The classical methods of determining such structure are not stable in numerical computation, because the methods are generally based on unimodular matrix transformations, which result in an extraordinarily large number of polynomial manipulations. Van Dooren et al. [1] presented a method for determining the Smith-McMillan form of a rational matrix from its Laurent expansion about a particular finite point. Subsequently, Verghese and Kailath [2], and Pugh et al. [3] extended this theory to produce a method of determining the infinite pole and zero structures of a rational matrix from its Laurent expansion about the point at infinity. A detailed reduction approach that can be used in computing the Smith-McMillan form at ∞ of a rational matrix is given by Vardulakis et al. [87]. For matrix pencils, a Toeplitz-based theory of computing Kronecker invariants with geometric interpretations has been developed in [92–95].

In this chapter a novel method is developed that determines the finite and infinite frequency structure of any rational matrix. For a polynomial matrix, a natural relationship between the rank information of the Toeplitz matrices and the multiplicities of the corresponding *irreducible elementary divisors* (IREDs) in its Smith form is established. These special Toeplitz matrices correspond to the derivatives at some finite zero of that matrix. This relationship proves to be fundamental and efficient in the study of the finite frequency structure of a polynomial matrix. For a polynomial matrix, this technique can be employed to find its infinite frequency structure by examining the finite frequency structure of the *dual* of what will be termed its *companion polynomial matrix*. It can then be extended to find the finite frequency structure of a rational matrix via *polynomial matrix fraction description* (MFD). The proposed method is neat and numerically advantageous when compared with the cumbersome and unstable classical procedures based on elementary transformations with unimodular matrices, though this method has the same numerical problems with those of computing the Jordan form.

It should also be noted that the usual methods for computing zeros introduce errors and how these errors affect the numerical computations is not easy to assess at this stage. Compared to the methods of Van Dooren et al. [1], Verghese and Kailath [2], and Pugh et al. [3], the present approach, which is based on analyzing the *nullity* of the Toeplitz matrices formed from the derivatives of the polynomial matrix rather than the Laurent expansion, will be more straightforward and much simpler. For it is easier and

A Generalized Framework of Linear Multivariable Control. http://dx.doi.org/10.1016/B978-0-08-101946-7.00007-X

more direct to obtain the derivatives of a polynomial matrix than to obtain its Laurent expansion. These relevant Toeplitz matrices are based on the information given by the *Smith zeros* [4, 5]) of the system. These are relatively easy to compute due to the fact that several numerical algorithms [6] have been proposed to find the location of these Smith zeros. Moreover, the proposed procedure will terminate after a minimal number of steps as soon as the set stop criterion is satisfied, which thus represents another numerical advantage over Van Dooren et al. [1], Verghese and Kailath [2], and Pugh et al. [3].

7.2 The Toeplitz rank information

In this section, we are to develop a fundamental algorithm that uses only the rank information of the Toeplitz matrices to propose the information of the IREDs in the Smith form of a polynomial matrix. Such Toeplitz matrices are formed from the derivatives of the polynomial matrix at its Smith zeros [4, 5]. Since several numerical algorithms have been proposed [6] for the computation of the zeros of a system, the Toeplitz matrices that are needed in this algorithm are relatively easy to obtain. Related properties of Kernels of the Toeplitz pencils can be found in the works of Gohberg et al. [17].

Let $A(s) \in \Re[s]^{p \times m}$, $\text{rank}_{\Re[s]}A(s) = r$, $r \leq \min\{p, m\}$. Then there exist unimodular matrices $U_L(s) \in \Re[s]^{p \times p}$, $U_R(s) \in \Re[s]^{m \times m}$ such that

$$U_L(s)A(s)U_R(s) = S_{A(s)}^C, \tag{7.1}$$

where

$$S_{A(s)}^C := \begin{bmatrix} 1 & & & & & \\ & \ddots & & & & \\ & & 1 & & & 0_{r,m-r} \\ & & & f_k(s) & & \\ & & & & \ddots & \\ & & & & & f_r(s) & \\ \hline & & 0_{p-r,r} & & & 0_{p-r,m-r} \end{bmatrix}$$

is the Smith form of $A(s)$. $f_i(s) \in \Re[s]$ are the invariant polynomials of $A(s)$ and $f_j(s)/f_{j+1}(s), j = k, k+1, \ldots, r-1$. Assume that A(s) has l distinct eigenvalues (finite zeros) $\lambda_1, \lambda_2, \ldots, \lambda_l$, where for simplicity of notation we assume that $\lambda_i \in \Re$, $i \in \mathbf{I} := \{1, 2, \ldots, l\}$, each invariant polynomial $f_k(s), \ldots, f_r(s)$ can be decomposed into IREDs over $\Re$, i.e., let

$$f_k(s) = (s - \lambda_1)^{p_{11}} (s - \lambda_2)^{p_{21}} \cdots (s - \lambda_l)^{p_{l1}}$$
$$f_{k+1}(s) = (s - \lambda_1)^{p_{12}} (s - \lambda_2)^{p_{22}} \cdots (s - \lambda_l)^{p_{l2}}$$
$$\vdots$$
$$f_r(s) = (s - \lambda_1)^{p_{1k(1)}} (s - \lambda_2)^{p_{2k(2)}} \cdots (s - \lambda_l)^{p_{lk(l)}}, \tag{7.2}$$

where

$$0 < p_{i1} \leq p_{i2} \leq \cdots p_{ik(i)}, \quad i = 1, 2, \ldots, l$$

are the partial multiplicities of the eigenvalue $\lambda_i . \lambda_i \in \mathbf{I}$. Eq. (7.2) can be written as

$$f_k(s) = (s - \lambda_i)^{p_{i1}} \hat{f}_k(s)$$
$$f_{k+1}(s) = (s - \lambda_i)^{p_{i2}} \hat{f}_{k+1}(s)$$
$$\vdots$$
$$f_r(s) = (s - \lambda_i)^{p_{ik(i)}} \hat{f}_r(s), \tag{7.3}$$

where

$$\hat{f}_j(\lambda_i) \neq 0, \quad j = k, k+1, \ldots, r.$$

We denote

$$\tau_i := \sum_{j=1}^{k(i)} p_{ij}, \quad n := \sum_{i=1}^{l} \tau_i \tag{7.4}$$

τ_i is the *local degree* of $A(s)$ with respect to the finite zero λ_i, n is the degree of $A(s)$. It can easily be seen that the Smith form of a polynomial matrix is completely determined by the scalars

$$
\begin{array}{cccccc}
\lambda_1, & \tau_1, & p_{11}, & p_{12}, & \cdots, & p_{1k(1)}; \\
\lambda_2, & \tau_2, & p_{21}, & p_{22}, & \cdots, & p_{2k(2)}; \\
& \vdots & & & & \\
\lambda_l, & \tau_l, & p_{l1}, & p_{l2}, & \cdots, & p_{lk(l)};
\end{array}
\tag{7.5}
$$

which are sometimes called the *Segre characteristics* of $A(s)$, the properties of which are studied in detail in [92, 93]. The information in Eq. (7.5) is somewhat redundant since the constants always satisfy Eq. (7.4).

It should be noted that $k(i)$ denotes the total number of the IREDs that appear in the Smith form of $A(s)$ in the form of $(s - \lambda_i)^m$ with $m \geq 1$. A basic result concerning this is as follows, which can be deducted from the known properties [17] of geometric multiplicity of the eigenvalue.

Theorem 7.2.1. *Let $A(s) \in \Re[s]^{p \times m}$, $\text{rank}_{\Re[s]} A(s) = r$. Then for any finite zero λ_i of $A(s)$,*

$$k(i) = \text{nullity of } A(\lambda_i) = r - \text{rank } A(\lambda_i). \tag{7.6}$$

The following analysis, which results in a related Toeplitz matrix, follows from Gohberg et al. [17]. Let $u_j(s) \in \Re[s]^{m \times 1}$ $(j = 1, 2, \ldots, m)$, $v_j(s) \in \Re[s]^{p \times 1}$ $(j = 1, 2, \ldots, p)$ be the columns of $U_R(s)$ and $U_L(s)^{-1}$, then from Eqs. (7.1), (7.3), we have

$$A(s)u_k(s) = v_k(s)(s - \lambda_i)^{p_{i1}}\hat{f}_k(s)$$
$$A(s)u_{k+1}(s) = v_{k+1}(s)(s - \lambda_i)^{p_{i2}}\hat{f}_{k+1}(s)$$
$$\vdots$$
$$A(s)u_r(s) = v_r(s)(s - \lambda_i)^{p_{ik(i)}}\hat{f}_r(s), \tag{7.7}$$

which for $s = \lambda_i$ gives

$$A(\lambda_i)u_j(\lambda_i) = 0, \quad j = k, k+1, \ldots, r, \ i \in \mathbf{I}. \tag{7.8}$$

Taking the first derivative of Eq. (7.7) with respect to s we have

$$A(s)^{(1)}u_k(s) + A(s)u_k^{(1)}(s) = v_k^{(1)}(s - \lambda_i)^{p_{i1}}\hat{f}_k(s) + v_k(s)p_{i1}(s - \lambda_i)^{p_{i1}-1}\hat{f}_k(s)$$
$$+ v_k(s)(s - \lambda_i)^{p_{i1}}\hat{f}_k(s)^{(1)}$$
$$A(s)^{(1)}u_{k+1}(s) + A(s)u_{k+1}^{(1)}(s) = v_{k+1}^{(1)}(s - \lambda_i)^{p_{i2}}\hat{f}_{k+1}(s) + v_{k+1}(s)p_{i2}(s - \lambda_i)^{p_{i2}-1}\hat{f}_{k+1}(s)$$
$$+ v_{k+1}(s)(s - \lambda_i)^{p_{i2}}\hat{f}_{k+1}(s)^{(1)}$$
$$\vdots$$
$$A(s)^{(1)}u_r(s) + A(s)u_r^{(1)}(s) = v_r^{(1)}(s - \lambda_i)^{p_{ik(i)}}\hat{f}_r(s) + v_r(s)p_{ik(i)}(s - \lambda_i)^{p_{ik(i)}-1}\hat{f}_r(s)$$
$$+ v_r(s)(s - \lambda_i)^{p_{ik(i)}}\hat{f}_r(s)^{(1)} \tag{7.9}$$

which for $s = \lambda_i$ and $p_{ij} > 1$ gives

$$A(\lambda_i)^{(1)}u_j(\lambda_i) + A(\lambda_i)u_j^{(1)}(\lambda_i) = 0, \quad j = k, \ldots, r; \ i \in \mathbf{I}. \tag{7.10}$$

In a similar manner and taking the derivative of Eq. (7.9), i.e., the second derivative of Eq. (7.7), and evaluating at $s = \lambda_i$ we obtain for $j = k, k+1, \ldots, r; \ i \in \mathbf{I}$

$$\frac{1}{2!}A(\lambda_i)^{(2)}u_j(\lambda_i) + A(\lambda_i)^{(1)}u_j^{(1)}(\lambda_i) + A(\lambda_i)\frac{1}{2!}u_j^{(2)}(\lambda_i) = 0. \tag{7.11}$$

Generally, for any IRED $(s - \lambda_i)^{p_{ij}}(i \in \mathbf{I}, \ j = 1, 2, \ldots, k(i))$ in $S_{A(s)}^C$, we have

$$\begin{bmatrix} A(\lambda_i) & & & \\ A(\lambda_i)^{(1)} & A(\lambda_i) & & \\ \vdots & \vdots & \ddots & \\ \frac{1}{(p_{ij}-1)!}A(\lambda_i)^{(p_{ij}-1)} & \frac{1}{(p_{ij}-2)!}A(\lambda_i)^{(p_{ij}-2)} & \cdots & A(\lambda_i) \end{bmatrix} \begin{bmatrix} u_j(\lambda_i) \\ \frac{1}{1!}u_j^{(1)}(\lambda_i) \\ \vdots \\ \frac{1}{(p_{ij}-1)!}u_j^{(p_{ij}-1)}(\lambda_i) \end{bmatrix}$$
$$= 0, \quad j = k, k+1, \ldots, r. \tag{7.12}$$

For simplicity, we denote the Toeplitz matrix as follows:

$$
T_\mu(\lambda_i) := \begin{bmatrix} A(\lambda_i) & & & \\ A(\lambda_i)^{(1)} & A(\lambda_i) & & \\ \vdots & \vdots & \ddots & \\ \frac{1}{\mu!}A(\lambda_i)^{(\mu)} & \frac{1}{(\mu-1)!}A(\lambda_i)^{(\mu-1)} & \cdots & A(\lambda_i) \end{bmatrix} \quad \mu = 1,2,\dots . \quad (7.13)
$$

Proposition 7.2.1. *For any IRED* $(s - \lambda_i)^{p_{ij}}(i \in I, \quad j = 1,2,\dots,k(i))$ *in* $S^C_{A(s)}$,

$$
Ker\, T_{p_{ij}-1}(\lambda_i) = Span \left\{ \begin{bmatrix} u_j(\lambda_i) \\ u_j^{(1)}(\lambda_i) \\ \vdots \\ u_j^{(p_{ij}-1)}(\lambda_i) \end{bmatrix} ; j = k,k+1,\dots,r \right\}. \quad (7.14)
$$

Proof. This follows directly by examining Eq. (7.12). From Vardulakis [13], the sequence

$$
u_j(\lambda_i), u_j^{(1)}(\lambda_i), \dots, u_j^{(p_{ij}-1)}(\lambda_i); \quad j = k,k+1,\dots,r
$$

forms a Jordan chain thus is linearly independent. Subsequently

$$
\begin{bmatrix} u_j(\lambda_i) \\ u_j^{(1)}(\lambda_i) \\ \vdots \\ u_j^{(p_{ij}-1)}(\lambda_i) \end{bmatrix} ; \quad j = k,k+1,\dots,r,
$$

are linearly independent. Considering in Eq. (7.12) it is seen that all of them are in the Kernel space of $T_{p_{ij}-1}(\lambda_i)$. The above two facts lead to the required assertion. □

Theorem 7.2.2. *Let* $A(s) \in R[s]^{p \times m}$, $rank_{\Re[s]}A(s) = r$. *Let* $n_\mu(i)$ *denote the nullity of* $T_\mu(\lambda_i)$, *and* $k_{\mu+1}(i)$ *denote the number of the IREDs in* $S^C_{A(s)}$ *of the form of* $(s - \lambda_i)^m$ *with* $m \geq \mu + 1$, *then*

$$
\begin{aligned} k_{\mu+1}(i) &= n_\mu(i) - n_{\mu-1}(i) \\ &= r - rank\, T_\mu(\lambda_i) + rank\, T_{\mu-1}(\lambda_i); \quad i \in I, \; \mu = 1,2,\dots . \end{aligned} \quad (7.15)
$$

Proof. Let $p_h(i)$ be the number of the IREDs in the form of $(s - \lambda_i)^h$, as above $k(i)$ denote the total number of the IREDs. It follows that the number of $(s - \lambda_i)^m$ with $m \geq h$ is

$$
k_h(i) = k(i) - (p_1(i) + p_2(i) + \cdots + p_{h-1}(i)).
$$

By Proposition 7.2.1 and Theorem 7.2.1, we know

$$
2k(i) - p_1(i) = n_1(i) = 2r - rank\, T_1(\lambda_i),
$$

so

$$
\begin{aligned}
k(i) - p_1(i) &= 2r - \text{rank } T_1(\lambda_i) - k(i) \\
&= (2r - \text{rank } T_1(\lambda_i)) - (r - \text{rank } A(\lambda_i)) \\
&= n_1(i) - n_0(i) \\
&= r - \text{rank } T_1(\lambda_i) + \text{rank } T_0(\lambda_i).
\end{aligned}
$$

In general, we find that

$$
\begin{aligned}
hk(i) - (h-1)p_1(i) - (h-2)p_2(i) - \cdots - p_{h-1}(i) &= n_{h-1}(i) \\
&= hr - \text{rank } T_{h-1}(i),
\end{aligned}
$$

and

$$
\begin{aligned}
ccl(h+1)k(i) - hp_1(i) - (h-1)p_2(i) - \cdots - 2p_{h-1}(i) - p_h(i) &= n_h(i) \\
&= (h+1)r - \text{rank } T_h(i),
\end{aligned}
$$

so

$$
\begin{aligned}
k(i) - (p_1(i) + p_2(i) + \cdots + p_{h-1}(i) + p_h(i)) &= k_{h+1}(i) \\
&= n_h(i) - n_{h-1}(i) \\
&= r - \text{rank } T_h(i) + \text{rank } T_{h-1}(i),
\end{aligned}
$$

which is the desired result.   □

The above theorem has in fact provided an extension to the related findings of Karcanias and Kalogeropoulos [94] from the matrix pencils to a polynomial matrix. Thus a direct relationship between the rank information of the Toeplitz matrices $T_\mu(\lambda_i)$ ($\mu = 0, 1, \ldots; i \in \mathbf{I}$) and the multiplicities of the IREDs in $S_{A(s)}^C$ has been established, which is fundamental in our approach to determine the finite and infinite frequency structure of a rational matrix.

Example 7.2.1. Let

$$
A(s) = \begin{bmatrix} s+1 & s+1 \\ 0 & (s+2)^2 \end{bmatrix},
$$

from $\det A(s) = (s+1)(s+2)^2 = 0$, we know that $A(s)$ has two finite zeros $\lambda_1 = -1, \lambda_2 = -2$. $\mathbf{I} = \{1, 2\}$, $r = 2$.

$$
A(-1) = \begin{bmatrix} 0 & 0 \\ 0 & 1 \end{bmatrix}, \quad A^{(1)}(-1) = \begin{bmatrix} 1 & 1 \\ 0 & 2 \end{bmatrix};
$$

$$
A(-2) = \begin{bmatrix} -1 & -1 \\ 0 & 0 \end{bmatrix}, \quad A^{(1)}(-2) = \begin{bmatrix} 1 & 1 \\ 0 & 0 \end{bmatrix}, \quad \tfrac{1}{2!}A^{(2)}(-2) = \begin{bmatrix} 0 & 0 \\ 0 & 1 \end{bmatrix}.
$$

So the Toeplitz matrices and the rank information are

$$T_0(-1) = A(-1), \quad \text{rank } T_0(-1) = 1, \quad n_0(1) = 2 - 1 = 1,$$

$$T_1(-1) = \begin{bmatrix} A(-1) \\ A^{(1)}(-1) & A(-1) \end{bmatrix}, \quad \text{rank } T_1(-1) = 3,$$

$$n_1(1) = 2 \times 2 - 3 = 1;$$

$$T_0(-2) = A(-2), \quad \text{rank } T_0(-2) = 1, \quad n_0(2) = 2 - 1 = 1,$$

$$T_1(-2) = \begin{bmatrix} A(-2) \\ A^{(1)}(-2) & A(-2) \end{bmatrix}, \quad \text{rank } T_1(-2) = 2,$$

$$n_1(2) = 2 \times 2 - 2 = 2,$$

$$T_2(-2) = \begin{bmatrix} A(-2) \\ A^{(1)}(-2) & A(-2) \\ \frac{1}{2!}A^{(2)}(-2) & A^{(1)}(-2) & A(-2) \end{bmatrix}, \quad \text{rank } T_2(-2) = 4,$$

$$n_2(2) = 6 - 4 = 2.$$

Since $k(1) = n_0(1) = 1$, from Theorem 7.2.1 there is one IRED $(s + 1)^m$ with $m \geq 1$. Since $k_2(1) = n_1(1) - n_0(1) = 0$, from Theorem 7.2.2 there are no IRED $(s + 1)^m$ with $m \geq 2$. Similarly, there is $k(2) = n_0(2) = 1$ IRED $(s + 2)^m$ with $m \geq 1$, there is $n_1(2) - n_0(2) = 1$ IRED $(s + 2)^m$ with $m \geq 2$, there is $n_2(2) - n_1(2) = 0$ IRED $(s + 2)^m$ with $m \geq 3$. So far we can deduce that the Smith form of $A(s)$ is

$$S_{A(s)}^C = \text{diag}\{1, (s + 1)(s + 2)^2\}.$$

Example 7.2.2. Consider

$$A(s) = \begin{bmatrix} s^3 & 1 & 0 \\ 0 & s^3 & s^2 \\ 0 & 0 & s^3 \end{bmatrix}.$$

We easily find that $r = 3$, $\mathbf{I} = \{1\}$, that $A(s)$ has only one finite zero $\lambda_1 = 0$ and

$$A(0) = \begin{bmatrix} 0 & 1 & 0 \\ 0 & 0 & 0 \\ 0 & 0 & 0 \end{bmatrix}, \quad A^{(1)}(0) = \begin{bmatrix} 0 & 0 & 0 \\ 0 & 0 & 0 \\ 0 & 0 & 0 \end{bmatrix}, \quad \frac{1}{2!}A^{(2)}(0) = \begin{bmatrix} 0 & 0 & 0 \\ 0 & 0 & 1 \\ 0 & 0 & 0 \end{bmatrix},$$

$$\frac{1}{3!}A^{(3)}(0) = \begin{bmatrix} 1 & 0 & 0 \\ 0 & 1 & 0 \\ 0 & 0 & 1 \end{bmatrix}, \quad \frac{1}{4!}A^{(4)}(0) = \cdots = \frac{1}{7!}A^{(7)}(0) \begin{bmatrix} 0 & 0 & 0 \\ 0 & 0 & 0 \\ 0 & 0 & 0 \end{bmatrix}.$$

The Toeplitz matrices and their rank information

$$T_0(0) = A(0), \quad \text{rank } T_0(0) = 1, \quad n_0(1) = 3 - 1 = 2,$$

$$T_1(0) = \begin{bmatrix} A(0) \\ A^{(1)} & A(0) \end{bmatrix}, \quad \text{rank } T_1(0) = 2, \quad n_1(1) = 2 \times 3 - 2 = 4,$$

$$T_2(0) = \begin{bmatrix} A(0) \\ A^{(1)} & A(0) \\ \frac{1}{2!}A^{(2)}(0) & A^{(1)}(0) & A(0) \end{bmatrix}, \quad \text{rank } T_2(0) = 4,$$

$$n_2(1) = 3 \times 3 - 4 = 5,$$

$$T_3(0) = \begin{bmatrix} A(0) & & & \\ A^{(1)} & A(0) & & \\ \frac{1}{2!}A^{(2)}(0) & A^{(1)}(0) & A(0) & \\ \frac{1}{3!}A^{(3)}(0) & \frac{1}{2!}A^{(2)}(0) & A^{(1)}(0) & A(0) \end{bmatrix}, \quad \text{rank } T_3(0) = 6,$$

$$n_3(1) = 4 \times 3 - 6 = 6,$$

$$T_4(0) = \begin{bmatrix} A(0) & & & \\ A^{(1)}(0) & A(0) & & \\ \vdots & \vdots & \ddots & \\ \frac{1}{4!}A^{(4)}(0) & \frac{1}{3!}A^{(3)}(0) & \cdots & A(0) \end{bmatrix}, \quad \text{rank } T_4(0) = 8,$$

$$n_4(1) = 5 \times 3 - 8 = 7,$$

$$T_5(0) = \begin{bmatrix} A(0) & & & \\ A^{(1)}(0) & A(0) & & \\ \vdots & \vdots & \ddots & \\ \frac{1}{5!}A^{(5)}(0) & \frac{1}{4!}A^{(4)}(0) & \cdots & A(0) \end{bmatrix}, \quad \text{rank } T_5(0) = 10,$$

$$n_5(1) = 6 \times 3 - 10 = 8,$$

$$T_6(0) = \begin{bmatrix} A(0) & & & \\ A^{(1)}(0) & A(0) & & \\ \vdots & \vdots & \ddots & \\ \frac{1}{6!}A^{(6)}(0) & \frac{1}{5!}A^{(5)}(0) & \cdots & A(0) \end{bmatrix}, \quad \text{rank } T_6(0) = 12,$$

$$n_6(1) = 7 \times 3 - 12 = 9,$$

$$T_7(0) = \begin{bmatrix} A(0) & & & \\ A^{(1)}(0) & A(0) & & \\ \vdots & \vdots & \ddots & \\ \frac{1}{7!}A^{(7)}(0) & \frac{1}{6!}A^{(6)}(0) & \cdots & A(0) \end{bmatrix}, \quad \text{rank } T_7(0) = 15,$$

$$n_7(1) = 8 \times 3 - 15 = 9.$$

From Theorems 7.2.1 and 7.2.2, we deduce that:

- there are $k(1) = n_0(1) = 2$ IREDs s^m with $m \geq 1$,
- there are $k_2(1) = n_1(1) - n_0(1) = 2$ IREDs s^m with $m \geq 2$,
- there is $k_3(1) = n_2(1) - n_1(1) = 1$ IREDs s^m with $m \geq 3$,
- there is $k_4(1) = n_3(1) - n_2(1) = 1$ IREDs s^m with $m \geq 4$,
- there is $k_5(1) = n_4(1) - n_3(1) = 1$ IREDs s^m with $m \geq 5$,
- there is $k_6(1) = n_5(1) - n_4(1) = 1$ IREDs s^m with $m \geq 6$,
- there is $k_7(1) = n_6(1) - n_5(1) = 1$ IREDs s^m with $m \geq 7$,
- and finally there is $k_8(1) = n_7(1) - n_6(1) = 0$ IREDs s^m with $m \geq 8$.

Thus the Smith form of $A(s)$ is

$$S^C_{A(s)} = \text{diag}\{1, s^2, s^7\}.$$

Remark 7.2.1. For a polynomial matrix $A(s)$, consider the sequence

$$k(i) := k_1(i), k_2(i), \ldots, k_{\mu_0(i)+1}(i),$$

if $k_j(i) = k_{j+1}(i) + 1$, we can deduce that there is one IRED $(s - \lambda_i)^j$ for $A(s)$.

Remark 7.2.2. For a polynomial matrix $A(s)$, a rank search of the Toeplitz matrices $T_\mu(\lambda_i)$ gives all information about the occurrence of the IREDs in its Smith form. From Theorems 7.2.1 and 7.2.2, we know that

$$k(i) := k_1(i) \geq k_2(i) \geq \cdots \geq 0.$$

So, for any $\lambda_i \in \mathbf{I}$, there exists a minimal integer $\mu_0(i)$ such that

$$k_{\mu_0(i)+1}(i) = 0,$$

i.e.,

$$r - \text{rank } T_{\mu_0(i)}(\lambda_i) + \text{rank } T_{\mu_0(i)-1}(\lambda_i) = 0.$$

As soon as this is satisfied the search can be terminated. In the methods of Van Dooren et al. [1], Verghese and Kailath [2], and Pugh et al. [3], the Toeplitz matrices were formed from Laurent expansions. So before the rank search begins, all the coefficients of the Laurent expansions must be worked out. In the method presented here, we form the Toeplitz matrix from the derivatives of the polynomial matrix directly. As soon as the above stop criterion is satisfied, the calculation is finished. By then only a minimal number of the derivatives of the matrix has been employed in the computation. Furthermore, if a polynomial matrix $A(s)$ is of a degree n, then we definitely know that for $i \geq n + 1$, $A^{(i)}(s) = 0$. This means the proposed method can be carried out more efficiently with a minimal memory content in computation.

If

$$A(s) := A_0 + A_1 s + \cdots + A_q s^q \in \Re[s]^{p \times m},$$

where A_i, $i = 0, 1, \ldots, q$ are $p \times m$ constant matrices with

$$A_q \neq 0.$$

From Eq. (7.13), we can easily find that

1. when $\mu \leq q$

$$
T_\mu(\lambda_i) = \begin{bmatrix}
A_0 & 0 & 0 & \cdots & 0 \\
A_1 & A_0 & 0 & \cdots & 0 \\
A_2 & A_1 & A_0 & \cdots & 0 \\
\vdots & \vdots & \vdots & \ddots & \vdots \\
A_\mu & A_{\mu-1} & A_{\mu-2} & \cdots & A_0
\end{bmatrix}
$$

$$
+ \begin{bmatrix}
A_1 & 0 & 0 & \cdots & 0 \\
C_2^1 A_2 & A_1 & 0 & \cdots & 0 \\
C_3^2 A_3 & C_2^1 A_2 & A_1 & \cdots & 0 \\
\vdots & \vdots & \vdots & \ddots & \vdots \\
C_{\mu+1}^\mu A_{\mu+1} & C_\mu^{\mu-1} A_\mu & C_{\mu-1}^{\mu-2} A_{\mu-1} & \cdots & A_1
\end{bmatrix} \lambda_i
$$

$$+ \cdots$$

$$+ \begin{bmatrix} A_{q-1} & 0 & 0 & \cdots & 0 \\ C_q^1 A_q & A_{q-1} & 0 & \cdots & 0 \\ 0 & C_q^1 A_q & A_{q-1} & \cdots & 0 \\ \vdots & \ddots & \ddots & \ddots & \vdots \\ 0 & \cdots & 0 & C_q^1 A_q & A_{q-1} \end{bmatrix} \lambda_i^{q-1}$$

$$+ \begin{bmatrix} A_q & 0 & 0 & \cdots & 0 \\ 0 & A_q & 0 & \cdots & 0 \\ 0 & 0 & A_q & \cdots & 0 \\ \vdots & \ddots & \ddots & \ddots & \vdots \\ 0 & \cdots & 0 & 0 & A_q \end{bmatrix} \lambda_i^q. \tag{7.16}$$

2. when $\mu > q$

$$T_\mu(\lambda_i) = \begin{bmatrix} A_0 & 0 & \cdots & 0 & 0 & 0 & \cdots & 0 \\ A_1 & A_0 & \cdots & 0 & 0 & 0 & \cdots & 0 \\ \vdots & \vdots & \ddots & \vdots & \vdots & \vdots & \ddots & \vdots \\ A_{q-1} & A_{q-2} & \cdots & A_0 & 0 & 0 & \cdots & 0 \\ A_q & A_{q-1} & \cdots & A_1 & A_0 & 0 & \cdots & 0 \\ 0 & A_q & \cdots & A_2 & A_1 & A_0 & \cdots & 0 \\ \vdots & \vdots & \ddots & \vdots & \vdots & \vdots & \ddots & \vdots \\ 0 & 0 & \cdots & A_q & A_{q-1} & \cdots & A_1 & A_0 \end{bmatrix}$$

$$+ \begin{bmatrix} A_1 & 0 & \cdots & 0 & 0 & 0 & \cdots & 0 \\ C_2^1 A_2 & A_1 & \cdots & 0 & 0 & 0 & \cdots & 0 \\ \vdots & \vdots & \ddots & \vdots & \vdots & \vdots & \ddots & \vdots \\ C_q^{q-1} A_q & C_{q-1}^{q-2} A_{q-1} & \cdots & A_1 & 0 & 0 & \cdots & 0 \\ 0 & C_q^{q-1} A_q & \cdots & C_2^1 A_2 & A_1 & 0 & \cdots & 0 \\ 0 & 0 & \cdots & C_3^2 A_3 & C_2^1 A_2 & A_1 & \cdots & 0 \\ \vdots & \vdots & \ddots & \vdots & \vdots & \vdots & \ddots & \vdots \\ 0 & 0 & \cdots & 0 & C_q^{q-1} A_q & \cdots & C_2^1 A_2 & A_1 \end{bmatrix} \lambda_i$$

$$+ \cdots$$

$$+ \begin{bmatrix} A_{q-1} & 0 & 0 & \cdots & 0 \\ C_q^1 A_q & A_{q-1} & 0 & \cdots & 0 \\ 0 & C_q^1 A_q & A_{q-1} & \cdots & 0 \\ \vdots & \ddots & \ddots & \ddots & \vdots \\ 0 & \cdots & 0 & C_q^1 A_q & A_{q-1} \end{bmatrix} \lambda_i^{q-1}$$

$$+ \begin{bmatrix} A_q & 0 & 0 & \cdots & 0 \\ 0 & A_q & 0 & \cdots & 0 \\ 0 & 0 & A_q & \cdots & 0 \\ \vdots & \ddots & \ddots & \ddots & \vdots \\ 0 & \cdots & 0 & 0 & A_q \end{bmatrix} \lambda_i^q, \tag{7.17}$$

where the combination $C_m^n = \frac{m(m-1)\cdots(m-n+1)}{n!}$.

Remark 7.2.3. From the above, we can see that the rank information of the Toeplitz matrices is based on the coefficient matrices of $A(s)$ rather than on its power of s, the above representations of $T_\mu(\lambda_i)$ avoid the need for any differentiation in the construction of $T_\mu(\lambda_i)$. Therefore from a numerical computation point of view our method is simpler and more numerically advantageous than the aforementioned methods due to the fact that it is easier and more direct to obtain the derivatives of a polynomial matrix than to obtain its Laurent expansion.

7.3 To determine the Smith form of a polynomial matrix

When the present ideas are used to find the Smith form of a polynomial matrix, the knowledge of the finite zeros (eigenvalues) is required. If the given polynomial matrix is regular, i.e., $A(s) \in \Re[s]^{p \times m}$, $\text{rank}_\Re A(s) = r$, with $p = m = r$, then the complete knowledge about the finite zeros (Smith zeros) of $A(s)$ can be obtained from $\det A(s) = 0$. Concerning this, several numerical algorithms [6] have been proposed to find the locations of the finite zeros of a system. Also, for a general polynomial matrix $A(s)$, Bosgra and Van der Weiden [96], Hayton et al. [97], and Pugh et al. [98] have presented some procedures that reduce a general polynomial matrix to a similarly equivalent matrix pencil form. Such reduction is accomplished by a form of *equivalence* [97], which leaves invariant the finite and infinite zero structure of the transformed matrix. This reduction proceeds as follows.

Let the $p \times m$ polynomial matrix $A(s)$ correspond to the matrix polynomial defined by

$$A(s) := A_0 + A_1 s + \cdots + A_q s^q,$$

where A_i, $i = 0, 1, \ldots, q$ are $p \times m$ constant matrices with

$$A_q \neq 0.$$

Define the following matrices

$$\Pi(E) := \begin{bmatrix} A_2 & A_3 & \cdots & A_q \\ A_3 & A_4 & \cdots & 0 \\ \vdots & \vdots & & \vdots \\ A_q & 0 & \cdots & 0 \end{bmatrix}, \quad \Pi(A) := \begin{bmatrix} A_3 & A_4 & \cdots & A_q & 0 \\ A_4 & A_5 & \cdots & 0 & 0 \\ \vdots & \vdots & & \vdots & \vdots \\ A_q & 0 & \cdots & 0 & 0 \\ 0 & 0 & \cdots & 0 & 0 \end{bmatrix};$$

$$\tag{7.18}$$

$$\Pi(B) := \begin{bmatrix} A_2 \\ A_3 \\ \vdots \\ A_q \end{bmatrix}, \quad \Pi(C) := \begin{bmatrix} A_2 & A_3 & \cdots & A_q \end{bmatrix}. \tag{7.19}$$

Let $\rho(E) := \text{rank}\Pi(E)$ and let the positive sets of integers $\mho := \{i_1, i_2, \ldots, i_{\rho(E)}\}$ (resp. $\mho := \{j_1, j_2, \ldots, j_{\rho(E)}\}$) define a row (resp. column) selection also denote $\mho$ (resp. $\mho$) from $\Pi(E)$ of $\rho(E)$ linearly independent rows (resp. columns). Let P_E (resp. P_A) be that submatrix of $\Pi(E)$ (resp. columns) formed from rows of the selection $\mho$ and the columns of the selection $\mho$. Let P_C be the submatrix of $\Pi(C)$ formed from the columns of the selection $\mho$, and P_B be the submatrix of $\Pi(B)$ formed from the rows of the selection $\mho$. Bosgra and Weiden [96] and Hayton et al. [97] showed that it is possible to reduce the polynomial matrix $A(s)$ to the following matrix pencil form:

$$P_A(s) := \begin{bmatrix} P_E - sP_A & P_B s \\ -P_C s & A_1 s + A_0 \end{bmatrix} = s \begin{bmatrix} -P_A & P_B \\ -P_C & A_1 \end{bmatrix} + \begin{bmatrix} P_E & 0 \\ 0 & A_0 \end{bmatrix}. (7.20)$$

In this way we can obtain a matrix pencil $P_A(s)$ for $A(s)$, which has the same zero structure as $A(s)$. Such an equivalent matrix pencil form of the polynomial matrix has the advantage of bringing the matrix into a form suitable for computation of its finite zero structure. The above link between a polynomial matrix and matrix pencil also indicates that the method proposed in [92] can also be used in here to obtain the required Toeplitz matrix information, this detail is, however, omitted in here. One may find it is not always stable in computation of zeros in a numerical context. Relatively more numerically advantageous algorithms have, however, been developed by Moler and Stewart [99] and Ward [100] for computation of generalized eigenvalue problems of the matrix pencil $P_A(s)$.

Algorithm 7.3.1 (Determination of the Smith form of a polynomial matrix).
Step 1 Given $A(s) \in \Re[s]^{p \times m}$, if $\text{rank}_{\Re(s)}A(s) := r = m = p$, then by solving equation $\det A(s) = 0$, we obtain the finite zeros $\{\lambda_i, \ i \in \mathbf{I}\}$ of $A(s)$. Go to Step 3. If not, go to Step 2.
Step 2 Compute the corresponding matrix pencil $P_A(s)$ and its finite zeros. Go to Step 3.
Step 3 For $\lambda_i, \ i \in \mathbf{I}$, compute $T_\mu(\lambda_i)$ and $k(i)$, $k_2(i), \ldots$ if for some $\mu_0(i)$, $k_{\mu_0(i)+1}(i) = 0$, the rank search terminate for i. Recur with $i = i + 1$. Go to Step 4.
Step 4 Deduce the Smith form of $A(s)$ from $k(i), k_2(i), \ldots, k_{\mu_0(i)}(i)$.

7.4 To determine the Smith-McMillan form at infinity of a rational matrix

Verghese and Kailath [2] and Pugh et al. [3] developed a method to construct the Smith-McMillan form at infinity of the given rational matrix from its Laurent expansions about the point at infinity. We will show that the Smith-McMillan form at infinity of a rational matrix can be found in a more straightforward and much simpler way.

Let $T(s) = [t_{ij}(s)]$ be any $p \times m$ rational matrix whose entries $t_{ij}(s) = \frac{r_{ij}(s)}{p_{ij}(s)}$ are rational functions of s. We assume that $\text{rank}_{\Re(s)}T(s) = r$. If we now let $g(s)$ be the least common (monic) multiple of the $p \times m$ denominator polynomial $\{p_{ij}(s) \in \Re[s],$

$i = 1, 2, \ldots, p; \, j = 1, 2, \ldots, m\}$, then

$$
T(s) = \begin{bmatrix}
\dfrac{r_{11}^*(s)}{g(s)} & \dfrac{r_{12}^*(s)}{g(s)} & \cdots & \dfrac{r_{1m}^*(s)}{g(s)} \\[2mm]
\dfrac{r_{21}^*(s)}{g(s)} & \dfrac{r_{22}^*(s)}{g(s)} & \cdots & \dfrac{r_{2m}^*(s)}{g(s)} \\[2mm]
\vdots & \vdots & \cdots & \vdots \\[2mm]
\dfrac{r_{p1}^*(s)}{g(s)} & \dfrac{r_{p2}^*(s)}{g(s)} & \cdots & \dfrac{r_{pm}^*(s)}{g(s)}
\end{bmatrix}
$$

$$
:= \frac{1}{g(s)} T^*(s), \tag{7.21}
$$

where $T^*(s) := [r_{ij}^*(s)] \in \Re[s]^{p \times m}$ is called the *companion polynomial matrix* of $T(s)$. Obviously, $\mathrm{rank}_{\Re[s]} T^*(s) = \mathrm{rank}_{\Re[s]} T(s)$. We denote

$$
T^*(s) := T_0^* + T_1^* s + \cdots + T_{q1}^* s^{q1}. \tag{7.22}
$$

The *dual* of $T^*(s)$ was defined [18] as

$$
D_{T^*}(w) := T_0^* w^{q1} + T_1^* w^{q1-1} + \cdots + T_{q1}^* = w^{q1} T^*\left(\frac{1}{w}\right). \tag{7.23}
$$

Theorem 7.4.1. *Let* $\tau := \deg g(s)$ *and if the (local) Smith form of* $D_{T^*}(w)$ *at* $w = 0$ *is*

$$
S_{D_{T^*}}^0(w) := \left[
\begin{array}{ccccc|c}
1 & & & & & \\
 & \ddots & & & & \\
 & & 1 & & & \\
 & & & w^{p_1} & & 0_{r,m-r} \\
 & & & & \ddots & \\
 & & & & & w^{p_{k(1)}} \\
\hline
 & & 0_{p-r,r} & & & 0_{p-r,m-r}
\end{array}
\right],
$$

then the Smith-McMillan form at $s = \infty$ *of the original rational matrix is*

$$
S_{T(s)}^\infty(s) = \left[
\begin{array}{ccccc|c}
s^{q1-\tau} & & & & & \\
 & \ddots & & & & \\
 & & s^{q1-\tau} & & & \\
 & & & s^{q1-p_1-\tau} & & 0_{r,m-r} \\
 & & & & \ddots & \\
 & & & & & s^{q1-p_{k(1)}-\tau} \\
\hline
 & & 0_{p-r,r} & & & 0_{p-r,m-r}
\end{array}
\right].
$$

Proof. There exist unimodular matrices $T_L(w) \in \Re[w]^{p \times p}$, $T_R(w) \in \Re[w]^{m \times m}$ such that

$$
T_L(w) D_{T^*}(w) T_R(w) = S_{D_{T^*}}^0(w).
$$

Now using the transformation $w = \frac{1}{s}$, noticing Eq. (7.23) we have

$$T_L\left(\frac{1}{s}\right)\frac{1}{s^{q_1}}T^*(s)T_R\left(\frac{1}{s}\right) = S^0_{D_{T^*}}\left(\frac{1}{s}\right),$$

i.e.,

$$T_L\left(\frac{1}{s}\right)T^*(s)T_R\left(\frac{1}{s}\right) = s^{q_1}S^0_{D_{T^*}}\left(\frac{1}{s}\right).\tag{7.24}$$

Note that

$$T(s) = \frac{1}{g(s)}T^*(s) = \left(\frac{1}{g(s)}I_p\right)T^*(s),$$

$$\left(\frac{g(s)}{s^\tau}I_p\right)\left(\frac{1}{g(s)}I_p\right) = \frac{1}{s^\tau}I_p,$$

where I_p is the identity matrix. So from Eq. (7.24)

$$T_L\left(\frac{1}{s}\right)\left(\frac{g(s)}{s^\tau}I_p\right)\left(\frac{1}{g(s)}I_p\right)T^*(s)T_R\left(\frac{1}{s}\right) = s^{q_1-\tau}S^0_{D_{T^*}}\left(\frac{1}{s}\right),$$

i.e.,

$$T_L\left(\frac{1}{s}\right)\left(\frac{g(s)}{s^\tau}I_p\right)T(s)T_R\left(\frac{1}{s}\right) = s^{q_1-\tau}S^0_{D_{T^*}}\left(\frac{1}{s}\right).\tag{7.25}$$

Now $T_L(w), T_R(w)$ are unimodular, so $T_L\left(\frac{1}{s}\right), T_R\left(\frac{1}{s}\right)$ are biproper. Obviously,

$$\lim_{s\to\infty}\frac{g(s)}{s^\tau}I_p = I_p.$$

So $T_L\left(\frac{1}{s}\right)\left(\frac{g(s)}{s^\tau}I_p\right)$ is biproper. Thus the result follows directly from Eq. (7.25). $\qquad\square$

Remark 7.4.1. The above theorem presents a natural relationship between the local Smith form of $D_{T^*}(w)$ and the Smith-McMillan form at $s = \infty$, of the original rational matrix $T(s)$ which makes it possible to deduce $S^\infty_{T(s)}$ from $S^0_{D_{T^*}}(w)$. Due to the fact that the (local) finite zero is definitely known, the determination of $S^0_{D_{T^*}}(w)$ becomes very easy when using the procedure described in Section 7.3.

Example 7.4.1. Consider

$$T(s) = \begin{bmatrix} \frac{1}{s(s-1)} & \frac{s}{s-1} \\ 1 & \frac{s}{s-1} \end{bmatrix}.$$

The least common multiple of $\{s(s-1), s-1, 1, s-1\}$ is $g(s) = s(s-1)$. $\tau = \deg g(s) = 2$. $r = 2$. The companion polynomial matrix of $T(s)$ is

$$T^*(s) = \begin{bmatrix} 1 & s^2 \\ s(s-1) & s^2 \end{bmatrix}$$

with $q_1 = 2$. The dual of $T^*(s)$ is

$$D_{T^*}(w) = \begin{bmatrix} w^2 & 1 \\ -w+1 & 1 \end{bmatrix}.$$

To find the local Smith form of $D_{T^*}(w)$ at $w = 0$, we note that the Toeplitz matrix

$$T_0(0) = D_{T^*}(0) = \begin{bmatrix} 0 & 1 \\ 1 & 1 \end{bmatrix},$$

with rank $T_0(0) = 2$, $n_0(1) = 2 - 2 = 0$, so from Theorem 7.2.1 $k(1) = n_0(1) = 0$, there are no IRED w^m with $m \geq 1$. Thus the local Smith form of $D_{T^*}(w)$ at $w = 0$ is

$$S^0_{D_{T^*}}(w) = \text{diag}\{1, 1\}.$$

Using Theorem 7.4.1, we obtain the Smith-McMillan form at $s = \infty$ of $T_{(s)}$

$$S^\infty_{T_{(s)}}(s) = \text{diag}\{1, 1\}.$$

Example 7.4.2. Let

$$T(s) = \begin{bmatrix} s^2+1 & 1 & 0 & 0 \\ 1 & \frac{1}{s-1} & 0 & s \\ 0 & 0 & \frac{s^2}{s^2-1} & 0 \\ 0 & 0 & 0 & 0 \end{bmatrix},$$

we are to find $S^\infty_{T_{(s)}}(s)$. Obviously, $r = 3$, $g(s) = s^2 - 1$, $\tau = 2$, the companion polynomial matrix

$$T^*(s) = \begin{bmatrix} s^4-4 & s^2-1 & 0 & 0 \\ 0 & s+1 & 0 & s^3-s \\ 0 & 0 & s^2 & 0 \\ 0 & 0 & 0 & 0 \end{bmatrix}$$

with $q_1 = 4$ and its dual

$$D_{T^*}(w) = w^4 T^*\left(\frac{1}{w}\right) = \begin{bmatrix} -w^4+1 & -w^4+w^2 & 0 & 0 \\ 0 & w^4+w^3 & 0 & -w^3+w \\ 0 & 0 & w^2 & 0 \\ 0 & 0 & 0 & 0 \end{bmatrix}.$$

The Toeplitz matrices

$$T_\mu(0) = \begin{bmatrix} D_{T^*}(0) & & & \\ D_{T^*}(0)^{(1)} & D_{T^*}(0) & & \\ \vdots & \vdots & \ddots & \\ \frac{1}{\mu!}D_{T^*}(0)^{(\mu)} & \frac{1}{(\mu-1)!}D_{T^*}(0)^{(\mu-1)} & \cdots & D_{T^*}(0) \end{bmatrix}, \quad \mu = 0, 1, 2,$$

and the rank information

$$\text{rank } T_0(0) = 1, \quad n_0(1) = r - \text{rank } T_0(0) = 2;$$

$$\text{rank } T_1(0) = 3, \quad n_1(1) = 2r - \text{rank } T_1(0) = 3;$$

$$\text{rank } T_2(0) = 6, \quad n_2(1) = 3r - \text{rank } T_2(0) = 3;$$

so

$$k(1) = 2, \quad k_2(1) = 1, \quad k_3(1) = 0,$$

from Theorems 7.2.1 and 7.2.2, we deduce that the local Smith form at $w = 0$ of $D_{T^*}(w)$ is

$$S_{D_{T^*}}^0(W) = \text{diag}\{1, w, w^2, 0\}.$$

From Theorem 7.4.1, we obtain the Smith-McMillan form at $s = \infty$ of $T(s)$

$$S_{T(S)}^\infty(s) = \text{diag}\{s^2, s, 1, 0\}.$$

Algorithm 7.4.1 (Determination of the Smith-McMillan form at infinity of a rational matrix).
Step 1 Given $T(s) \in \Re(s)^{p \times m}$, calculate $g(s)$, $T^*(s)$, $D_{T^*}(w)$, τ, and q_1.
Step 2 To find $S_{D_{T^*}}^0(w)$, we construct $T_0(0)$, $T_1(0), \ldots,$ and search the rank information $n_0(1), n_1(0), \ldots$. Till for some μ_0, $n_{\mu_0}(1) = n_{\mu_0+1}(1)$. After this process is terminated, go to Step 3.
Step 3 Using Theorems 7.2.1 and 7.2.2 to deduce $S_{D_{T^*}}^0(w)$.
Step 4 By Theorem 7.4.1, we obtain $S_{T(s)}^\infty(s)$.

7.5 To determine the Smith-McMillan form of a rational matrix

In this section we discuss briefly how the procedures proposed above can be used to find the Smith-McMillan form of a rational matrix. If we can construct a right (left) coprime polynomial MFD from the given rational matrix, then the problem of finding the Smith-McMillan form of a rational matrix is reduced to that of determining the Smith forms of the polynomial matrices via the following known result.

Proposition 7.5.1 ([5]). *Let* $T(s) \in \Re(s)^{p \times m}$ *with* $\text{rank}_{\Re(s)} T(s) = r$, $r \le \min\{p, m\}$ *and let* $T(s) = R_1(s)P_1^{-1}(s) = P_2^{-1}(s)R_2(s)$ *be respectively right and left coprime polynomial MFDs of* $T(s)$, *if*

$$S_{R_1(s)}^C = S_{R_2(s)}^C = block\ diag\{\varepsilon_1(s), \ldots, \varepsilon_r(s), 0_{p-r,m-r}\} \in \Re[s]^{p \times m},$$

and

$$S_{P_1(s)}^C = block\ diag\{I_{m-r}, \Psi_r(s), \ldots, \Psi_1(s)\} \in \Re[s]^{m \times m},$$

$$S^C_{P_2(s)} = block\ diag\{I_{P-r}, \Psi_r(s), \ldots, \Psi_1(s)\} \in \Re[s]^{p \times p},$$

then

$$S^C_{T(s)} = block\ diag\left\{\frac{\varepsilon_1(s)}{\Psi_1(s)}, \frac{\varepsilon_2(s)}{\Psi_2(s)}, \ldots, \frac{\varepsilon_r(s)}{\Psi_r(s)}, 0_{p-r,m-r}\right\}.$$

The subject of computing the right and left coprime polynomials of a rational matrix is addressed in detail in Rosenbrock [5].

The following result which generalizes the well-known Wolovich-Falb "structure theorem" [71] from the proper rational matrix case to the general rational matrix case, can be applied to construct a coprime polynomial MFD from a rational matrix. Furthermore, it is of special interest by itself.

Let $T(s) \in \Re(s)^{p \times m}$ with $rank_{\Re(s)} T(s) = r$, $r \leq \min\{p, m\}$. We can write

$$T(s) = T_{sp}(s) + T_{pol}(s), \tag{7.26}$$

where $T_{sp}(s) \in \Re(s)^{p \times m}$ is strictly proper and $T_{pol}(s) \in \Re(s)^{p \times m}$ is the polynomial part of $T(s)$. If $A \in R^{n \times m}$, $B \in R^{n \times m}$, $C \in R^{p \times n}$ is a minimal realization of $T_{sp}(s)$, i.e.,

$$C(sI_n - A)^{-1} B = T_{sp}(s), \tag{7.27}$$

then it can be reduced by a nonsingular transformation Q to a *controllable companion form* $\{\hat{A}, \hat{B}, \hat{C}\} = \{QAQ^{-1}, QB, CQ^{-1}\}$. [71]. Furthermore, associated with the structured form are the (m) *controllability indices* d_i, for $i = 1, 2, \ldots, m$, which specify the dimensions of the various diagonal companion-form submatrices of $\hat{A}$, as well as the (m) ordered integers $\sigma_k = \sum_{i=1}^{k} d_i$, for $k = 1, 2, \ldots, m$, which denote the "nontrivial" row of $\hat{A}$ and $\hat{B}$. We define $\hat{A}_m$ as the $(m \times n)$ matrix consisting of the (m) ordered σ_k row of $\hat{A}$, and $\hat{B}_m$ as the $(m \times m)$ matrix consisting of the (m) ordered σ_k row of $\hat{B}$. $\hat{B}_m$ is clearly nonsingular. We denote

$$S(s) = \begin{bmatrix} 1 & 0 & \cdots & 0 \\ s & 0 & \cdots & 0 \\ \vdots & \vdots & \cdots & \vdots \\ s^{d_1-1} & 0 & \cdots & 0 \\ \hline 0 & 1 & \cdots & 0 \\ 0 & s & \cdots & 0 \\ \vdots & \vdots & \cdots & \vdots \\ 0 & s^{d_2-1} & \cdots & 0 \\ \hline & & \vdots & \\ \hline 0 & 0 & \cdots & 1 \\ 0 & 0 & \cdots & s \\ \vdots & \vdots & \cdots & \vdots \\ 0 & 0 & \cdots & s^{d_m-1} \end{bmatrix}$$

and

$$\delta(s) = \text{diag}\{s^{d_1}, s^{d_2}, \ldots, s^{d_m}\} - \hat{A}_m S(s).$$

Thus we are ready now to state the following original result, which is a generalization of Wolovich [71].

Theorem 7.5.1. *If (A, B, C) is a minimal realization of the strictly proper part $T_{sp}(s)$ of a rational matrix $T(s)$, and $(\hat{A}, \hat{B}, \hat{C})$ is the controllable companion form of (A, B, C), assuming $m \leq n$, then $T(s)$ can be expressed as the following MFD*

$$T(s) = [\hat{C}S(s) + T_{pol}(s)\hat{B}_m^{-1}\delta(s)][\hat{B}_m^{-1}\delta(s)]^{-1}, \tag{7.28}$$

where $S(s)$, $\hat{B}_m$, and $\delta(s)$ are defined above. Furthermore $R(s) := \hat{C}S(s) + T_{pol}(s)\hat{B}_m^{-1}\delta(s)$ and $P(s) := \hat{B}_m^{-1}\delta(s)$ are right coprime.
   Proof. In view of Eqs. (7.26), (7.27) and

$$\hat{C}(sI_n - \hat{A})^{-1}\hat{B} = C(sI_n - A)^{-1}B,$$

we only need to show that

$$\hat{C}(sI_n - \hat{A})^{-1}\hat{B} = \hat{C}S(s)\delta^{-1}(s)\hat{B}_m.$$

It is therefore sufficient to show that

$$(sI_n - \hat{A})^{-1}\hat{B} = S(s)[\hat{B}_m^{-1}\delta(s)]^{-1},$$

or equivalently, that

$$(sI_n - \hat{A})S(s) = \hat{B}\hat{B}_m^{-1}\delta(s). \tag{7.29}$$

Note, however, that Eq. (7.29) is an immediate consequence of the definition of $S(s)$, $\hat{B}_m$, and $\delta(s)$.
   It can be easily checked that $P(s)$ and $R(s)$ are right coprime. To this end, we assume that there exist some $s_0 \in C$ and some $q \in R^m \neq 0$ such that

$$\begin{bmatrix} P(s_0) \\ R(s_0) \end{bmatrix} q = \begin{bmatrix} \hat{B}_m^{-1}\delta(s_0) \\ \hat{C}S(s_0) + T_{pol}(s_0)\hat{B}_m^{-1}\delta(s_0) \end{bmatrix} q = 0,$$

it follows that

$$\hat{B}_m^{-1}\delta(s_0)q = 0,$$

and

$$\hat{C}S(s_0) + T_{pol}(s_0)\hat{B}_m^{-1}\delta(s_0)q = 0,$$

which result in

$$\hat{C}S(s_0)q = 0.$$

Note that $S(s_0)$ is of full column rank thus $S(s_0)q \neq 0$. (A, B, C) is a minimal realization of $T_{sp}(s)$, C is thus of full row rank, $\hat{C}$ is subsequently of full row rank, thus $\hat{C}S(s_0)q \neq 0$. Therefore the original assumption results in a contradiction, this is to say $P(s)$ and $R(s)$ are right coprime. □

Remark 7.5.1. The similar result can also be obtained for left coprime polynomial MFDs, which is based on the observable companion form of (A, B, C).

Remark 7.5.2. A related realization approach, which is based on MFDs and zero structure for nonproper system, is introduced in [101]. The realization approach suggested also provides the relevant Smith-McMillan information [5].

Algorithm 7.5.1 (Determination of the Smith-McMillan form of a rational matrix).
Step 1 Given $T(s) \in \Re(s)^{p \times m}$, compute the strictly proper part $T_{sp}(s)$ and its polynomial part $T_{pol}(s)$ of $T(s)$.

Step 2 Using the known realization techniques such as realization from Markov parameters [32, 71] to find a minimal realization (A, B, C) of $T_{sp}(s)$.

Step 3 Using the procedure proposed by Wolovich [71] reduce (A, B, C) to its controllable companion form $(\hat{A}, \hat{B}, \hat{C})$ (or the observable companion form). Obtain $\hat{A}_m, \hat{B}_m, S(s), \delta(s)$. By Theorem 7.5.1, express $T(s)$ as a right (or left) coprime polynomial MFD such that $T(s) = R(s)P(s)^{-1}$.

Step 4 Using Algorithm 7.4.1 to find $S^C_{P(x)}$ and $S^C_{R(s)}(s)$.
Step 5 By Proposition 7.5.1, we finally obtain the Smith-McMillan form of $T(s)$.

Example 7.5.1. Consider the rational matrix

$$T(s) = \begin{bmatrix} \frac{s^4-4s^3+3s^2-1}{s^4-4s^3+3s^2-s-1} & \frac{s^2}{s^4-4s^3+3s^2-s-1} \\ \frac{s^3-4s^2+1}{s^4-4s^3+3s^2-s-1} & \frac{s^5-4s^4+3s^3-4s^2+s+1}{s^4-4s^3+3s^2-s-1} \\ \frac{s^2}{s^4-4s^3+3s^2-s-1} & \frac{s^3}{s^4-4s^3+3s^2-s-1} \\ \frac{1-s}{s^4-4s^3+3s^2-s-1} & \frac{-s^2+s}{s^4-4s^3+3s^2-s-1} \end{bmatrix},$$

we are to find its Smith-McMillan form $S^C_{T(s)}$. It is easily seen that the strictly proper part and the polynomial part are

$$T_{sp}(s) = \begin{bmatrix} \frac{s}{s^4-4s^3+3s^2-s-1} & \frac{s^2}{s^4-4s^3+3s^2-s-1} \\ \frac{s^3-4s^2+1}{s^4-4s^3+3s^2-s-1} & \frac{-3s^2+2s+1}{s^4-4s^3+3s^2-s-1} \\ \frac{s^2}{s^4-4s^3+3s^2-s-1} & \frac{s^3}{s^4-4s^3+3s^2-s-1} \\ \frac{1-s}{s^4-4s^3+3s^2-s-1} & \frac{-s(s-1)}{s^4-4s^3+3s^2-s-1} \end{bmatrix}, \quad T_{pol}(s) = \begin{bmatrix} 1 & 0 \\ 0 & s \\ 0 & 0 \\ 0 & 0 \end{bmatrix}.$$

Using the Markov parameters realization technique, we find that (A, B, C) is a minimal realization of $T_{sp}(s)$ with

$$A = \begin{bmatrix} 0 & 0 & 1 & 0 \\ 3 & 0 & -3 & 1 \\ -1 & 1 & 4 & -1 \\ 1 & 0 & -1 & 0 \end{bmatrix}, \quad B = \begin{bmatrix} 0 & 0 \\ 1 & 0 \\ 0 & 1 \\ 0 & 0 \end{bmatrix}, \quad C = \begin{bmatrix} 1 & 0 & 0 & 0 \\ 0 & 1 & 0 & 0 \\ 0 & 0 & 1 & 0 \\ 0 & 0 & 0 & 1 \end{bmatrix}.$$

From the controllability matrix of (A, B, C)

$$\Phi = \begin{bmatrix} B, AB, A^2B, A^3B \end{bmatrix} = \begin{bmatrix} 0 & 0 & 0 & 1 & 1 & 4 & 4 & 13 \\ 1 & 0 & 0 & -3 & -3 & -10 & -10 & -30 \\ 0 & 1 & 1 & 4 & 4 & 13 & 13 & 41 \\ 0 & 0 & 0 & -1 & -1 & -3 & -3 & -9 \end{bmatrix},$$

we get the controllability indices $d_1 = 1$, $d_2 = 3$ and $\sigma_1 = d_1 = 1$, $\sigma_2 = d_1 + d_2 = n = 4$. The equivalence transformation matrix is

$$Q = \begin{bmatrix} 1 & 1 & 0 & -2 \\ 1 & 0 & 0 & 1 \\ 1 & 0 & 0 & 0 \\ 0 & 0 & 1 & 0 \end{bmatrix}.$$

Under this transformation (A, B, C) is changed to its controllable companion form $(\hat{A}, \hat{B}, \hat{C})$ as follows

$$\hat{A} = QAQ^{-1} = \begin{bmatrix} 0 & 1 & 0 & 0 \\ 0 & 0 & 1 & 0 \\ 0 & 0 & 0 & 1 \\ 1 & 1 & -3 & 4 \end{bmatrix}, \quad \hat{B} = QB = \begin{bmatrix} 1 & 0 \\ 0 & 0 \\ 0 & 0 \\ 0 & 1 \end{bmatrix},$$

$$\hat{C} = CQ^{-1} = \begin{bmatrix} 0 & 0 & 1 & 0 \\ 1 & 2 & -3 & 0 \\ 0 & 0 & 0 & 1 \\ 0 & 1 & -1 & 0 \end{bmatrix}.$$

Selecting the ordered σ_1 and σ_2 row of $\hat{A}, \hat{B}$ we form the matrices

$$\hat{A}_m = \begin{bmatrix} 0 & 1 & 0 & 0 \\ 1 & 1 & -3 & 4 \end{bmatrix}, \quad \hat{B}_m = \begin{bmatrix} 1 & 0 \\ 0 & 1 \end{bmatrix}.$$

Now

$$S(s) = \begin{bmatrix} 1 & 0 \\ 0 & 1 \\ 0 & s \\ 0 & s^2 \end{bmatrix}, \quad \delta(s) = \mathrm{diag}\{s^{d_1}, s^{d_2}\} - \hat{A}_m S(s) = \begin{bmatrix} s & -1 \\ -1 & s^3 - 4s^2 + 3s - 1 \end{bmatrix},$$

by Theorem 7.5.1 thus a right coprime polynomial MFD of $T(s)$ is $T(s) = R(s)P(s)^{-1}$ with

$$P(s) = \hat{B}_m^{-1} \delta(s) = \begin{bmatrix} s & -1 \\ -1 & s^3 - 4s^2 + 3s - 1 \end{bmatrix},$$

and

$$R(s) = \hat{C}S(s) + T_{pol}(s)\hat{B}_m^{-1}\delta(s) = \begin{bmatrix} s & s-1 \\ 1-s & s^4 - 4s^3 + 3s^2 - 4s + 2 \\ 0 & s^2 \\ 0 & 1-s \end{bmatrix}.$$

We now use Algorithm 7.4.1 to determine $S^C_{P(s)}(s)$ and $S^C_{R(s)}(s)$. For $P(s)$, by

$$\det p(\lambda) = \lambda^4 - 4\lambda^3 + 3\lambda^2 - \lambda - 1 = 0$$

we obtain the finite zeros of $P(s)$

$$\lambda_1 = i, \ \lambda_2 = -i, \ \lambda_3 = 2 + \sqrt{2}, \ \lambda_4 = 2 - \sqrt{2}.$$

From the fact that $\text{rank}_{\Re[s]}P(s) = 2$, $\text{rank } P(\lambda_i) = 1$, for $i = 1,2,3,4$, $k(i) = 1$, $i = 1,2,3,4$, we deduce by Theorem 7.2.1 that

$$S^C_{P(s)} = \text{diag}\{1, (s-i)(s+i)(s-2-\sqrt{2})(s-2+\sqrt{2})\}.$$

Noting that

$$\text{rank}_{\Re[s]}R(s) = \text{rank } R(\lambda), \quad \forall \lambda \in C,$$

by Theorem 7.2.1 we at once deduce that

$$S^C_{R(s)} = \text{diag}\{1, 1\}.$$

From Proposition 7.5.1, we know that the Smith-McMillan form $T(s)$ is

$$S^C_{T(s)} = \begin{bmatrix} \frac{1}{(s-i)(s+i)(s-2-\sqrt{2})(s-2+\sqrt{2})} & 0 \\ 0 & 1 \\ 0 & 0 \\ 0 & 0 \end{bmatrix}.$$

7.6 Conclusions

In this chapter a novel method of determining the finite and infinite frequency structure of a rational matrix is derived that is based on rank information obtained from certain polynomial matrices. Theorems 7.2.1 and 7.2.2 provide a good insight into the natural relationship between the rank information of these Toeplitz matrices and the occurrence of the IREDs. This relationship proves to be fundamental to the study of the finite frequency structure of a polynomial matrix and the basis for an efficient algorithm. Theorem 7.4.1 determines the linkage between the local Smith form of dual polynomial matrix and the Smith-McMillan form at infinity of the original matrix. Based on this result, the proposed technique can thus be employed to find the infinite frequency structure of a polynomial matrix by examining the finite frequency structure of the dual of its companion polynomial matrix.

As a generalization of the classical result [71], Theorem 7.5.1 shows how to construct a right (or left) coprime polynomial MFD directly from the given rational matrix. This can then be applied to extend the proposed method to find the finite frequency structure of a rational matrix via a polynomial matrix fraction description. The proposed method is neat and numerically advantageous when compared with the classical procedures based on elementary transformations with unimodular matrices.

The Toeplitz procedure that has been used in developing our technique originates from Gohberg et al. [17]. For matrix pencils a Toeplitz-based theory of computing Kronecker invariants with geometric implications has been developed through the works of [92–95]. The proposed method has the same problems as those of computing the Jordan form, as they introduce errors as well. Nevertheless, some significant progress has been made in our new methods.

Compared to the methods of Van Dooren et al. [1], Verghese and Kailath [2], and Pugh et al. [3], the present approach, which is based on analyzing the *nullity* of the Toeplitz matrices formed from the derivatives of the polynomial matrix rather than Laurent expansion, will be more straightforward and much simpler. For it is easier and more direct to obtain the derivatives of a polynomial matrix than to obtain its Laurent expansion.

The relevant Toeplitz matrices are based on the information given by the *Smith zeros* [4, 5] of the system. These are relatively easy to compute due to the fact that several numerical algorithms [6] have been proposed to find the locations of these Smith zeros. The necessary rank information of the Toeplitz matrices is based on the coefficient matrices of the original matrix polynomial rather than on the coefficient matrices of the Taylor expansion of the matrix about the relevant Smith zero. These coefficient matrices are also explicitly represented and only a minimal number of the derivatives of the matrix need to be employed in the computation. Further more, if a polynomial matrix $A(s)$ is of a degree n, then we definitely know that for $i \geq n + 1$, $A^{(i)}(s) = 0$. This means the proposed method can be carried out more efficiently with a minimal memory content in computation. Moreover, the proposed procedure will terminate after a minimal number of steps as soon as the set stop criterion is satisfied, which thus represents one more numerical advantage over Van Dooren et al. [1], Verghese and Kailath [2], and Pugh et al. [3].

The solution of a regular PMD and the set of impulsive free initial conditions

8.1 Introduction

In this chapter, we discuss the complete solution of the linear nonhomogeneous matrix differential equations (LNHMDEs). What is meant by the *complete* solution is the full solution taking into account the initial conditions of the state as well as the initial conditions of the input. The essence of the new results presented here with respect to the existing results [13, 17, 102] is that the obtained solution displays the full solution components including the impulse response and the slow response created, not only by the initial conditions of the state, but also by the initial condition of the input. Special attention is drawn to the initial conditions of the system input. Compared to the known results [13, 17, 102], in which the initial conditions of the system input are usually assumed to be zero, our approach firstly considers this issue explicitly.

The LNHMDEs or linear nonhomogeneous regular polynomial matrix descriptions (PMDs) are described by

$$A(\rho)\beta_{NH}(t) = B(\rho)\mu(t), \quad t \geq 0, \tag{8.1}$$

where $\rho := d/dt$ is the differential operator, $A(\rho) = A_{q_1}\rho^{q_1} + A_{q_1-1}\rho^{q_1-1} + \cdots + A_1\rho + A_0 \in R[\rho]^{r \times r}$, $\text{rank}_{R(\rho)}A(\rho) = r$, $A_i \in R^{r \times r}$, $i = 0, 1, 2, \ldots, q_1$, $q_1 \geq 1$, $B(\rho) = B_l\rho^l + B_{l-1}\rho^{l-1} + \cdots + B_1\rho + B_0 \in R[\rho]^{r \times m}$, $B_j \in R^{r \times m}$, $j = 0, 1, 2, \ldots, l$, $l \geq 0$, $\beta_{NH}(t) : [0, +\infty) \longrightarrow R^r$ is the *pseudo-state* of the LNHMDE, $u(t) : [0, +\infty) \longrightarrow R^m$ is an i times piecewise continuously differentiable function called the *input* of the LNHMDE. Their homogeneous cases

$$A(\rho)\beta_H(t) = 0, \quad t \geq 0$$

are called the linear homogeneous matrix differential equations (LHMDEs) or homogeneous PMDs. Both regular generalized state space systems (GSSSs), which are described by

$$E\dot{x} = Ax(t) + Bu(t),$$

where E is a singular matrix, $\text{rank}(\rho E - A) = r$, and the (regular) state space systems, which are described by

$$\dot{x} = Ax(t) + Bu(t)$$

A Generalized Framework of Linear Multivariable Control. http://dx.doi.org/10.1016/B978-0-08-101946-7.00008-1

are special cases of the above LNHMDEs (4.1). There have been many discussions about GSSSs (see, e.g., [7–11, 103, 104]). In [12], the impulsive solution to the LHMDEs was presented in a closed form. For both the LNHMDEs and LHMDEs, Vardulakis [13] developed their solutions under the assumption that both the initial conditions of the state and the input are zero. However, as we will see later, in some real cases, the initial conditions of the state might result from a random disturbance entering the system, and a feedback controller is called for. Since the precise value of the initial conditions of the state is unpredictable and the control is likely to depend on those initial conditions of the state, so this assumption is somewhat stronger than necessary.

Example 8.1.1. This example is to display the fact that the initial conditions of $u(t)$ are not necessarily equal to zero in some situations.

To the singular perturbation model [105]

$$\left[\begin{array}{c} \dot{x}(t) \\ \varepsilon\dot{z}(t) \end{array}\right] = \left[\begin{array}{cc} A_{11} & A_{12} \\ A_{21} & A_{22} \end{array}\right]\left[\begin{array}{c} x(t) \\ z(t) \end{array}\right] + \left[\begin{array}{c} B_1 \\ B_2 \end{array}\right]u(t), \quad \left[\begin{array}{c} x(0) \\ z(0) \end{array}\right] = \left[\begin{array}{c} x^0 \\ z^0 \end{array}\right], \quad t \geq 0$$

$$y = \left[\begin{array}{cc} M_1, M_2 \end{array}\right]\left[\begin{array}{c} x(t) \\ z(t) \end{array}\right],$$

where $x(t) \in R^n$, $z(t) \in R^m$, $u(t) \in R^r$, and $y(t) \in R^p$. Kokotovic et al. [105] considered the so-called near-optimal regulator problem, i.e., to find a control $u(t) \in R^r$, $t \geq 0$ so as to regulate the state $\left[\begin{array}{c} x(t) \\ z(t) \end{array}\right]$ to the original by way of minimizing the quadratic performance index

$$J = \frac{1}{2}\int_0^\infty (y^T(t)y(t) + u^T(t)Ru(t))dt, \quad R > 0,$$

where the $r \times r$ weighting matrix R penalizes excessive values of control $u(t)$. The solution of the above problem is the optimal linear feedback control law

$$u_{opt}(t) = -R^{-1}B^T K(\varepsilon)\left[\begin{array}{c} x(t) \\ z(t) \end{array}\right] = G(\varepsilon)\left[\begin{array}{c} x(t) \\ z(t) \end{array}\right], \quad t \geq 0,$$

where $K(\varepsilon)$ satisfies the algebraic matrix Riccati equation. When the original system satisfies some conditions, $K(\varepsilon)$ exist as a positive-definite matrix (see [105, page 111, Theorem 4.1]).

Let us investigate a special case with $B_1 = I_n$, $B_2 = I_m$, and note that the matrix $K(\varepsilon)$ is positive-definite, and thus is nonsingular. In this case, the matrix $G(\varepsilon)$ is of full column rank. Therefore if

$$\left[\begin{array}{c} x(0) \\ z(0) \end{array}\right] \neq 0,$$

then

$$\mu_{opt}(0) = G(\varepsilon) \begin{bmatrix} x(0) \\ z(0) \end{bmatrix} \neq 0.$$

In this chapter, we will present a solution to Eq. (8.1) that displays the impulse response and the slow response created not only by the initial conditions of $\beta_{NH}(t)$, but also by the initial condition of $u(t)$. Also reformulations to the solution of the LNHMDEs in terms of the *regular derivatives* of $u(t)$ are given. By defining the *slow state* (*smooth state*) and the *fast state* (*impulsive state*) of them, it is shown that the system behaviors of the LNHMDEs can be decomposed into the *slow response* (*smooth response*) and the *fast response* (*impulsive response*) completely. This approach is applied conveniently to discuss the impulse free initial conditions of Eq. (8.1).

So far there have been many discussions about the impulse free condition either of the GSSSs (see, e.g., [14, 15]) or of the LNHMDEs and the LNHMDEs (see, e.g., [13, 16]). However, the common concern in the known results is the impulse created by the appropriate initial conditions of the state alone. For LNHMDEs, a different analysis concerning this issue is carried out that considers both the impulse created by the initial conditions of $u(t)$ and that created by the initial conditions of $\beta_{NH}(t)$.

8.2 Preliminary results

Any rational matrix $A(s)$ is equivalent [87] at $s = \infty$ to its Smith-McMillan form, having the form:

$$S_{A(s)}^{\infty}(s) = \text{block diag}\left[s^{q_1}, s^{q_2}, \dots, s^{q_k}, \frac{1}{s^{\hat{q}_{k+1}}}, \dots, \frac{1}{s^{\hat{q}_r}}, 0_{p-r,m-r} \right], \tag{8.2}$$

where $1 \leq k \leq r$ and $q_1 \geq q_2 \geq \cdots \geq q_k \geq 0$, $\hat{q}_r \geq \hat{q}_{r-1} \geq \cdots \geq \hat{q}_{k+1} \geq 0$. Any polynomial matrix $A(s) = A_0 + A_1 s + \cdots + A_{q_1} s^{q_1} \in R^{r \times r}[s]$ with $\text{rank}_{\Re[s]} A(s) = r$ can be transformed by unimodular transformation to its Smith form

$$S_{A(s)}^{C}(s) = \text{diag}[1, 1, \dots, 1, f_k(s), f_{k+1}(s), \dots, f_r(s)],$$

where $1 \leq k \leq r, f_j(s)/f_{j+1}(s), j = k, k+1, \dots, r-1$. We denote $x_{j,0}^i, x_{j,1}^i, \dots, x_{j,\sigma_{ij}-1}^i \in R^r (x_{j,0}^i \neq 0)$, $i \in I$, $j = k, k+1, \dots, r$ as the Jordan Chain of lengths σ_{ij} corresponding to the eigenvalue λ_i of $A(s)$ and consider matrices

$$C_i = [x_{k,0}^i, \dots, x_{k,\sigma_{i,k}-1}^i] \cdots [x_{r,0}^i, \dots, x_{r,\sigma_{i,r}-1}^i] \in R^{r \times m_i}, \tag{8.3}$$

where $m_i = \sum_{j=k}^{r} \sigma_{ij}, i = 1, 2, \dots$ and

$$J_i = \text{block diag}[J_{i,k}, J_{i,k+1}, \dots, J_{i,r}] \in R^{m_i \times m_i}, \quad i = 1, 2, \dots, \tag{8.4}$$

where $J_{i,j} \in R^{\sigma_{ij} \times \sigma_{ij}}, i \in I, j = k, k+1, \dots, r$, is the Jordan block matrix.

Definition 8.2.1. Finite Jordan pair of $A(s)$ is defined as (C, J), $C = [C_1, C_2, \ldots, C_l] \in R^{r \times n}$, $J = \text{block diag}[J_1, J_2, \ldots, J_l] \in R^{n \times n}$, with $n := m_1 + m_2 + \cdots + m_l = \deg \det(A(s))$.

Definition 8.2.2 ([13]). The pair: $C_\infty \in R^{r \times v}$, $J_\infty = \text{block diag}[J_{\infty 1}, J_{\infty 2}, \ldots, J_{\infty \xi}] \in R^{v \times v}$, where

$$
J_{\infty i} = \begin{bmatrix}
0 & 1 & 0 & \cdots & 0 \\
0 & 0 & 1 & \cdots & 0 \\
\cdot & \cdot & \cdot & \cdots & \cdot \\
0 & 0 & 0 & \cdots & 1 \\
0 & 0 & 0 & \cdots & 0
\end{bmatrix} \in R^{v_i \times v_i}, \quad i = 1, 2, \ldots, \xi
$$

$v = \sum_{i=1}^{\xi} v_i, v_i, \xi \in N$, is called an *infinite Jordan pair* of $A(s)$ if it is a (finite) Jordan pair of the "dual" polynomial matrix: $\tilde{A}(w) = w^{q_1} A(1/w) = A_0 w^{q_1} + A_1 w^{q_1 - 1} + \cdots + A_{q_1} \in R[w]^{r \times r}$ corresponding to its zero at $w = 0$.

Theorem 8.2.1 ([13]). *For a regular matrix polynomial $A(s)$, its inverse matrix $A^{-1}(s)$ can be written*

$$
A^{-1}(s) = \begin{bmatrix} C, C_\infty \end{bmatrix} \begin{bmatrix} sI_n - J & 0 \\ 0 & sJ_\infty - I_\mu \end{bmatrix}^{-1} \begin{bmatrix} B \\ B_\infty \end{bmatrix}, \tag{8.5}
$$

where $n = \deg \det A(s)$, $\mu = \sum_{j=k+1}^{r} (\hat{q}_j + 1) = \sum_{j=k+1}^{r} \hat{q}_j + (r - k)$ where $\hat{q}_j, j = k + 1, \ldots, r$ are the orders of the zeros at $s = \infty$ of $A(s)$.

Proposition 8.2.1 ([106]). *If $u(t)^{(i)}$ denotes the distributional derivative of $u(t)$, $u(t)^{[i]}$ denotes the regular derivative of $u(t)$, $\delta(t)$ denotes the impulsive function, then we have the identity*

$$
u(t)^{(i)} = u(t)^{[i]} + \delta(t)u(0)^{[i-1]} + \delta(t)^{(1)}u(0)^{[i-2]} + \cdots + \delta(t)^{(i-1)}u(0),
$$
$$
i = 1, 2, \ldots \tag{8.6}
$$

8.3 A solution for the LNHMDEs

To the LNHMDEs

$$
A(\rho)\beta_{NH}(t) = B(\rho)u(t), \quad t \geq 0. \tag{8.7}
$$

Assuming that the initial values of $u(t)$ and its $(l - 1)$-derivatives at $t = 0$ are $u(0)$, $u^{(1)}(0) \cdots u^{(l-1)}(0)$ and that the initial values of $\beta_{NH}(t)$ and its $(q_1 - 1)$-derivatives at $t = 0$ are $\beta_{NH}(0), \beta_{NH}^{(1)}(0), \ldots, \beta_{NH}^{(q_1-1)}(0)$. Taking the Laplace transformation of Eq. (8.7), we obtain

$$
A(s)\hat{\beta}_{NH}(s) - \hat{\alpha}_\beta(s) = B(s)\hat{u}(s) - \hat{\alpha}_u(s), \tag{8.8}
$$

where $\hat{\beta}_{NH}(s) := \int_0^{+\infty} \beta_{NH}(t)e^{-st}dt$, $\hat{u}(s) := \int_0^{+\infty} u(t)e^{-st}dt$. Because of the Laplace-transformation rule

$$
L\left\{ \frac{d^i}{dt^i} \beta_{NH}(t) \right\} = s^i \hat{\beta}_{NH}(s) - s^{i-1}\beta_{NH}(0) - \cdots - s\beta_{NH}^{i-2}(0) - \beta_{NH}^{i-1}(0); \quad i = 0, 1, \ldots
$$

$$L\left\{\frac{d^j}{dt^j}u(t)\right\} = s^j\hat{u}(s) - s^{j-1}u(0) - \cdots - su^{j-2}(0) - u^{j-1}(0); \quad j = 0, 1, \ldots$$

$\hat{\alpha}_\beta(s), \hat{\alpha}_u(s)$ can be written [107] as follows

$$\hat{\alpha}_\beta(s) = \begin{bmatrix} s^{q_1-1}I_r, s^{q_1-2}I_r, \ldots, sI_r, I_r \end{bmatrix} \begin{bmatrix} A_{q_1} & 0 & \cdots & 0 \\ A_{q_1-1} & A_{q_1} & \cdots & 0 \\ \vdots & \vdots & \ddots & \vdots \\ A_1 & A_2 & \cdots & A_{q_1} \end{bmatrix} \begin{bmatrix} \beta_{NH}(0) \\ \beta_{NH}^{(1)}(0) \\ \vdots \\ \beta_{NH}^{(q_1-1)}(0) \end{bmatrix}$$

$$\hat{\alpha}_u(s) = \begin{bmatrix} s^{l-1}I_r, s^{l-2}I_r, \ldots, sI_r, I_r \end{bmatrix} \begin{bmatrix} B_l & 0 & \cdots & 0 \\ B_{l-1} & B_l & \cdots & 0 \\ \vdots & \vdots & \ddots & \vdots \\ B_1 & B_2 & \cdots & B_l \end{bmatrix} \begin{bmatrix} u(0) \\ u^{(1)}(0) \\ \vdots \\ u^{(l-1)}(0) \end{bmatrix}.$$

We denote $A^{-1}(s) = H_{pol}(s) + H_{sp}(s)$, where $H_{pol}(s)$ is the polynomial part of $A^{-1}(s)$ and $H_{sp}(s)$ is the strictly proper part of $A^{-1}(s)$. To find a minimal realization of $H_{sp}(s)(C, J, B)$ and a minimal realization of $H_{pol}(s)(C_\infty, J_\infty, B_\infty)$ such that (C, J) and (C_∞, J_∞) are a finite Jordan pair and an infinite Jordan pair of $A(s)$, respectively, and

$$A^{-1}(s) = C_\infty(sJ_\infty - I_\mu)^{-1}B_\infty + C(sI_n - J)^{-1}B.$$

We obtain from [13]

$$A^{-1}(s)\hat{\alpha}_\beta(s) = [C, C_\infty] \begin{bmatrix} sI_n - J & 0_{n,\mu} \\ \hline 0_{\mu,n} & sJ_\infty - I_\mu \end{bmatrix}^{-1} \begin{bmatrix} x_{s\beta}(0) \\ J_\infty x_{f\beta}(0) \end{bmatrix},$$

where

$$\begin{bmatrix} x_{s\beta}(0) \\ J_\infty x_{f\beta}(0) \end{bmatrix} = \begin{bmatrix} J^{q_1-1}B, \ldots, B & 0_{n,q_1\mu} \\ \hline 0_{\mu,q_1n} & J_\infty B_\infty, \ldots, J_\infty^{q_1}B_\infty \end{bmatrix}$$

$$\begin{bmatrix} A_{q_1} & 0 & \cdots & 0 \\ A_{q_1-1} & A_{q_1} & \cdots & 0 \\ \vdots & \vdots & \ddots & \vdots \\ A_1 & A_2 & \cdots & A_{q_1} \\ \hline A_0 & A_1 & \cdots & A_{q_1-1} \\ 0 & A_0 & \cdots & A_{q_1-2} \\ \vdots & \vdots & \ddots & \vdots \\ 0 & 0 & \cdots & A_0 \end{bmatrix} \begin{bmatrix} \beta_{NH}(0) \\ \beta_{NH}^{(1)}(0) \\ \vdots \\ \beta_{NH}^{(q_1-1)}(0) \end{bmatrix} \in R^{(n+\mu)\times 1}.$$

Similarly,

$$A^{-1}(S)\hat{\alpha}_\mu(s) = [C, C_\infty] \begin{bmatrix} sI_n - J & 0_{n,\mu} \\ \hline 0_{\mu,n} & sJ_\infty - I_\mu \end{bmatrix}^{-1} \begin{bmatrix} x_{su}(0) \\ J_\infty x_{fu}(0) \end{bmatrix},$$

where

$$
\begin{bmatrix} x_{su}(0) \\ J_\infty x_{fu}(0) \end{bmatrix} = \left[\begin{array}{c|c} J^{l-1}B, \ldots, B & 0_{n,l\mu} \\ \hline 0_{\mu,ln} & J_\infty B_\infty, \ldots, J_\infty^l B_\infty \end{array} \right]
$$

$$
\left[\begin{array}{cccc} B_l & 0 & \cdots & 0 \\ B_{l-1} & B_l & \cdots & 0 \\ \vdots & \vdots & \ddots & \vdots \\ B_1 & B_2 & \cdots & B_l \\ \hline B_0 & B_1 & \cdots & B_{l-1} \\ 0 & B_0 & \cdots & B_{l-2} \\ \vdots & \vdots & \ddots & \vdots \\ 0 & 0 & \cdots & B_0 \end{array} \right] \begin{bmatrix} u(0) \\ u^{(1)}(0) \\ \vdots \\ u^{(l-1)}(0) \end{bmatrix} \in R^{(n+\mu)\times 1}.
$$

From [13] we also obtain

$$
A^{-1}(s)B(s)\hat{u}(s) = [C, C_\infty]\Psi\Phi \begin{bmatrix} I_m \\ sI_m \\ \vdots \\ s^{\hat{q}_r+1}I_m \end{bmatrix} \hat{u}(s) + C[sI_n - J]^{-1}\Omega\hat{u}(s),
$$

where

$$
\Psi := \left[\begin{array}{c|c} J^{l-1}B, J^{l-2}B, \ldots, B & 0_{n,(\hat{q}_r+1)r} \\ \hline 0_{\mu,lr} & B_\infty, J_\infty B_\infty, \ldots, J_\infty^{\hat{q}_r} B_\infty \end{array} \right] \in R^{(n+\mu)\times(\hat{q}_r+l+1)r}
$$

$$
\Phi := \left[\begin{array}{cccccccc} B_l & 0 & \cdots & 0 & 0 & 0 & \cdots & 0 \\ B_{l-1} & B_l & \cdots & 0 & 0 & 0 & \cdots & 0 \\ \vdots & \vdots & \ddots & \vdots & \vdots & \vdots & \ddots & \vdots \\ B_1 & B_2 & \cdots & B_l & 0 & 0 & \cdots & 0 \\ \hline B_0 & B_1 & \cdots & B_{l-1} & B_l & 0 & \cdots & 0 \\ 0 & B_0 & \cdots & B_{l-2} & B_{l-1} & B_l & \cdots & 0 \\ \vdots & \vdots & \ddots & \vdots & \vdots & \vdots & \ddots & \vdots \\ 0 & 0 & \cdots & B_0 & B_1 & \cdots & B_{l-1} & B_l \end{array} \right] \in R^{r(\hat{q}_r+l+1)\times(\hat{q}_r+l+1)m}
$$

$$
\Omega := J^l BB_l + J^{l-1}BB_{l-1} + \cdots + JBB_l + BB_0 \in R^{n\times m}. \tag{8.9}
$$

Now Eq. (8.8) can be written as

$$
\hat{\beta}_{NH}(s) = A^{-1}(s)\hat{\alpha}_\beta(s) + A^{-1}(s)B(s)u(s) - A^{-1}(s)\hat{\alpha}_u(s). \tag{8.10}
$$

By taking the inverse Laplace transformation of the above, we finally obtain the following theorem that represents the required complete solution of the LNHMDE (8.7) created by the nonzero initial conditions on both the control $u(t)$ and $\beta_{NH}(t)$.

Theorem 8.3.1. *The solution of the LNHMDE (8.7) corresponding to nonzero initial conditions both on the pseudo-state $\beta_{NH}(t)$ and the input $u(t)$ is*

$$\beta_{NH}(t) = Ce^{Jt}x_{s\beta}(0) - Ce^{Jt}x_{su}(0) - C_\infty \sum_{i=1}^{\hat{q}_r} \delta(t)^{(i-1)} J_\infty^{i-1}(J_\infty x_{f\beta}(0))$$

$$+ C_\infty \sum_{i=1}^{\hat{q}_r} \delta(t)^{(i-1)} J_\infty^{i-1}(J_\infty x_{fu}(0)) + \int_0^t Ce^{J(t-\tau)}\Omega u(\tau)d\tau$$

$$+ [C, C_\infty]\Psi\Phi \begin{bmatrix} u(t) \\ u(t)^1 \\ \vdots \\ u(t)^{(\hat{q}_r+l)} \end{bmatrix}, \quad t \geq 0 \qquad (8.11)$$

Remark 8.3.1. The solution (8.11) is an extension of that given by Vardulakis [13] to the case where the initial conditions of the pseudo-state and the input are not zero. Also the solution and impulsive behavior of PMDs of free linear multivariable systems given in [102] can be obtained as special cases from our result here simply by letting $B(\rho) = 0$.

Remark 8.3.2. From the above result, it is clearly seen that the nonzero initial conditions of $\beta(t)$ and $u(t)$ both contribute to the corresponding *slow (smooth) zero input response* by means of the terms

$$Ce^{Jt}x_{s\beta}(0), \quad -Ce^{Jt}x_{su}(0)$$

and to the fast (impulse) response through the terms

$$-C_\infty \sum_{i=1}^{\hat{q}_r} \delta(t)^{(i-1)} J_\infty^{i-1}(J_\infty x_{f\beta}(0)), \quad C_\infty \sum_{i=1}^{\hat{q}_r} \delta(t)^{(i-1)} J_\infty^{i-1}(J_\infty x_{fu}(0)).$$

8.4 The smooth and impulsive solution components and impulsive free initial conditions: C_∞ is of full row rank

The main aims of this section are to analyze the *fast* and *slow* components in the solution of the LNHMDEs and then to characterize the *impulsive behavior* of the system. To this end, one is suggested to reformulate the solution that is given by Theorem 8.3.1 in terms of the *regular derivative* of $u(t)$. This is the subject of the following result.

Consider the following notations:

$$x_s(0) := x_{s\beta}(0) - x_{su}(0), \quad J_\infty x_f(0) := J_\infty x_{f\beta}(0) - J_\beta x_{fu}(0),$$

$$\widehat{U(t)} := \begin{bmatrix} u(t) \\ u^{[1]}(t) \\ u^{[2]}(t) \\ \vdots \\ u^{[\hat{q}_r+l]}(t) \end{bmatrix} \in R^{(\hat{q}_r+l+1)m}, \quad \widehat{\delta(t)} := \begin{bmatrix} \delta(t) \\ \delta^{(1)}(t) \\ \vdots \\ \delta^{\hat{q}_r+l+1}(t) \end{bmatrix} \in R^{(\hat{q}_r+l)},$$

$$\mathcal{U}(0) := \begin{bmatrix} 0 & 0 & \cdots & 0 & 0 \\ u(0) & 0 & \cdots & 0 & 0 \\ u^{[1]}(0) & u(0) & \cdots & 0 & 0 \\ \vdots & \vdots & \ddots & \vdots & \vdots \\ u^{[\hat{q}_r+l-1]}(0) & u^{[\hat{q}_r+l-2]}(0) & \cdots & u^{[1]}(0) & u(0) \end{bmatrix} \in R^{(\hat{q}_r+l+1)m \times (\hat{q}_r+l)},$$

$$\widehat{J_\infty x_f(0)} := [J_\infty x_f(0), J_\infty^2 x_f(0), \ldots, J_\infty^{\hat{q}_r} x_f(0)] \in R^{\mu \times \hat{q}_r},$$

$$\hat{\widehat{J_\infty x_f(0)}} := [\widehat{J_\infty x_f(0)}, 0_{\mu \times l}] \in R^{\mu \times (\hat{q}_r+l)}, \tag{8.12}$$

$$\Psi_1 := [J^{l-1}B, J^{l-2}B, \ldots, B] \in R^{n \times lr}, \quad \Psi_2 := [B_\infty, J_\infty B_\infty, \ldots, J_\infty^{\hat{q}_r} B_\infty] \in R^{\mu \times (\hat{q}_r+1)r},$$

$$\Phi_1(B) := \begin{bmatrix} B_l & 0 & \cdots & 0 & 0 & 0 & \cdots & 0 \\ B_{l-1} & B_l & \cdots & 0 & 0 & 0 & \cdots & 0 \\ \vdots & \vdots & \ddots & \vdots & \vdots & \vdots & \ddots & \vdots \\ B_1 & B_2 & \cdots & B_l & 0 & 0 & \cdots & 0 \end{bmatrix} \in R^{lr \times (\hat{q}_r+l+1)m},$$

$$\Phi_2(B) := \begin{bmatrix} B_0 & B_1 & \cdots & B_{l-1} & B_l & 0 & \cdots & 0 \\ 0 & B_0 & \cdots & B_{l-2} & B_{l-1} & B_l & \cdots & 0 \\ \vdots & \vdots & \ddots & \vdots & \vdots & \vdots & \ddots & \vdots \\ 0 & 0 & \cdots & B_0 & B_1 & \cdots & B_{l-1} & B_l \end{bmatrix} \in R^{(\hat{q}_r+1)r \times (\hat{q}_r+l+1)m}.$$

Theorem 8.4.1. *With the above notations and assume that C_∞ has full row rank and that its {1}-inverse is $C_\infty^{(1)}$. Denote*

$$\beta_{s(NH)}(t) := e^{Jt} x_s(0) + \int_0^t e^{J(t-\tau)} \Omega u(\tau) d\tau$$

$$\beta_{f(NH)}(t) := (\Psi_2 \Phi_2(B) + C_\infty^{(1)} C \Psi_1 \Phi_1(B)) \widehat{U(t)} + [(\Psi_2 \Phi_2(B)$$
$$+ C_\infty^{(1)} C \Psi_1 \Phi_1(B)) \mathcal{U}(0) - \hat{\widehat{J_\infty x_f(0)}}] \widehat{\delta(t)},$$

then

$$\beta_{NH}(t) = [C, C_\infty] \begin{bmatrix} \beta_{s(NH)}(t) \\ \beta_{f(NH)}(t) \end{bmatrix}. \tag{8.13}$$

Proof. From Eqs. (10.9), (8.12), we know

$$\Psi = \left[\begin{array}{c|c} \Psi_1 & 0_{n,(\hat{q}_r+1)r} \\ \hline 0_{\mu,lr} & \Psi_2 \end{array} \right], \quad \Phi = \left[\begin{array}{c} \Phi_1(B) \\ \Phi_2(B) \end{array} \right], \tag{8.14}$$

from Eq. (8.6), we have

$$\left[\begin{array}{c} u(t) \\ u^{(1)}(t) \\ \vdots \\ u^{(\hat{q}_r+l)}(t) \end{array} \right] = \widehat{U(t)} + \mathcal{U}(0)\widehat{\delta(t)}, \tag{8.15}$$

if C_∞ has full row rank, we have $C_\infty C_\infty^{(1)} = I_{r \times r}$. Substituting the above into Eq. (10.11), we obtain

$$\begin{aligned} \beta_{NH}(t) &= Ce^{Jt}x_s(0) + C\int_0^t e^{J(t-\tau)}\Omega u(\tau)d\tau \\ &\quad - C_\infty \widehat{J_\infty x_f(0)}\widehat{\delta(t)} + (C\Psi_1\Phi_1(B) + C_\infty\Psi_2\Phi_2(B))(\widehat{U(t)} + \mathcal{U}(0)\widehat{\delta(t)}) \\ &= Ce^{Jt}x_s(0) + C\int_0^t e^{J(t-\tau)}\Omega u(\tau)d\tau \\ &\quad + C_\infty(\Psi_2\Phi_2(B) + C_\infty^{(1)}C\Psi_1\Phi_1(B))\widehat{U(t)} \\ &\quad + C_\infty[(\Psi_2\Phi_2(B) + C_\infty^{(1)}C\Psi_1\Phi_1(B))\mathcal{U}(0) - \widehat{J_\infty x_f(0)}]\widehat{\delta(t)} \\ &= [C, C_\infty] \left[\begin{array}{c} \beta_{s(NH)}(t) \\ \beta_{f(NH)}(t) \end{array} \right]. \end{aligned}$$

$\square$

Remark 8.4.1. The assumption that C_∞ has full row rank will be relaxed later.

Remark 8.4.2. The above theorem is interesting not least for its clear designation of the *fast and slow* components of the system solution structure. It provides further interest from the point of view of characterizing those initial conditions that give rise to completely smooth solutions (the so-called impulse free initial conditions) of regular PMDs. By treating it thus, the impulsive behavior of the system can be interpreted in a natural and clear way as the following theorem reveals.

Now we shall generalize the concept of the state to the LNHMDEs as [13] has done for the LHMDEs and the GSSS.

Definition 8.4.1. We define the vector

$$x_{NH}(t) := \left[\begin{array}{c} \beta_{s(NH)}(t) \\ \beta_{f(NH)}(t) \end{array} \right] \in R^{(n+\mu) \times 1}, \tag{8.16}$$

where

$$\beta_{s(NH)}(t) := e^{Jt}x_s(0) + \int_0^t e^{J(t-\tau)}\Omega u(\tau)d\tau,$$

$$\beta_{f(NH)}(t) := (\Psi_2\Phi_2(B) + C_\infty^{(1)}C\Psi_1\Phi_1(B))\widehat{U(t)} + [(\Psi_2\Phi_2(B)$$
$$+ C_\infty^{(1)}C\Psi_1\Phi_1(B))\mathcal{U}(0) - \widehat{J_\infty x_f(0)}]\widehat{\delta(t)},$$

as the *state* of the LNHMDE. $\beta_{s(NH)}(t)$ is called the *slow state or smooth state*, $\beta_{f(NH)}(t)$ is called the *fast state or impulsive state*.

For the LHMDEs

$$A(\rho)\beta_H(t) = 0,$$

its solution is

$$\beta_H(t) = [C, C_\infty]\begin{bmatrix} x_s(t) \\ x_f(t) \end{bmatrix},$$

where

$$x_s(t) = e^{Jt}x_s(0),$$
$$x_f(t) = -\sum_{i=1}^{\hat{q}_r} \delta(t)^{(i-1)} J_\infty^i x_f(0).$$

It is clear that the *impulsive behavior* of $\beta_H(t)$ at $t = 0$ only depends on $x_f(0)$, which is only related to the initial conditions of $\beta_H(t)$ and its derivatives at $t = 0$. Thus the set of impulse free initial conditions for LHMDE $A(\rho)\beta_H(t) = 0, t \geq 0$ is [13–16]

$$H_{I(H)} = \left\{ x(0) = \begin{bmatrix} x_s(0) \\ x_f(0) \end{bmatrix} : x_s(0) \in R^n, x_f(0) \in KerJ_\infty \right\} = R^n \oplus KerJ_\infty.$$

However, for the LNHMDEs, this issue becomes much more complicated, for in this case not only the initial conditions of $\beta_{NH}(t)$, but also the initial conditions of $u(t)$ influence the solution structure. The following result provides an answer to this difficulty.

Theorem 8.4.2. Assume that C_∞ has full row rank. The set of the impulsive free initial conditions *for the LNHMDE (8.7) is*

$$H_{I(NH)} = \left\{ x_{NH}(0) = \begin{bmatrix} \beta_{s(NH)}(0) \\ \beta_{f(NH)}(0) \end{bmatrix} : \beta_{s(NH)}(0) = x_s(0) \in R^n, \right.$$
$$\left. \beta_{f(NH)}(0) = (\Psi_2\Phi_2(B) + C_\infty^{(1)}C\Psi_1\Phi_1(B))\widehat{U(0)} \right\} \tag{8.17}$$

Proof. For the LNHMDE (8.7), under the notations in Theorems 8.3.1 and 8.4.1, we partition

$$(\Psi_2\Phi_2(B) + C_\infty^{(1)}C\Psi_1\Phi_1(B))\mathcal{U}(0) := [\Pi_1(\mathcal{U}(0)), \Pi_2(\mathcal{U}(0))],$$

where

$$\Pi_1(\mathcal{U}(0)) \in R^{\mu \times \hat{q}_r}, \quad \Pi_2(\mathcal{U}(0)) \in R^{\mu \times l}.$$

Observing Eqs. (8.12), (8.13), we can see that $\beta_{NH}(t)$ is impulse free at $t = 0$ iff

$$(\Psi_2\Phi_2(B) + C_\infty^{(1)}C\Psi_1\Phi_1(B))\mathcal{U}(0) - \widehat{J_\infty x_f}(0) = 0, \tag{8.18}$$

which hold true iff

$$\Pi_1(\mathcal{U}(0)) - \widehat{J_\infty x_f}(0) = 0,$$

i.e.,

$$x_f(0) \in Ker J_\infty + Im(\Pi_1(\mathcal{U}(0))),$$

and

$$\Pi_2(\mathcal{U}(0)) = 0.$$

The set of the impulsive free initial conditions is thus derived from Theorem 8.4.1. $\square$

Theorem 8.4.3. *If initial conditions $\beta_H^{(i)}(0)$, $i = 0, 1, \ldots, q_1 - 1$ of the LHMDE are compatible with $A(\rho)\beta_H(t) = 0$, $t \geq 0, u(0) = u^{(i)}(0)$, $i = 1, 2, \ldots, l - 1$ and $\mathcal{U}(0) \in Ker(\Psi_1\Phi_1(B)) \oplus Ker(\Psi_2\Phi_2(B))$, then the state of the LNHMDE $A(\rho)\beta_{NH}(t) = B(\rho)u(t)$, $t \geq 0$ is "impulse free".*

Proof. We have

$$\mathcal{U}(0) \in Ker(\Psi_2\Phi_2(B)) \oplus Ker(\Psi_1\Phi_1(B))$$
$$\subseteq Ker(\Psi_2\Phi_2(B)) + Ker(C_\infty^{(1)}C\Psi_1\Phi_1(B))$$
$$= Ker(\Psi_2\Phi_2(B)) + C_\infty^{(1)}C\Psi_1\Phi_1(B).$$

The initial conditions of the LHMDEs are compatible, from [13] it is known $x_{f\beta}(0) = 0$. From $u(0) = u^{(1)}(0), i = 1, 2, \ldots, l - 1$, we know $x_{fu}(0) = 0$. Subsequently from Eq. (8.12)), $\widehat{J_\infty x_f}(0) = 0$. So

$$\beta_{f(NH)}(t) = (\Psi_2\Phi_2(B)) + (C_\infty^{(1)}C\Psi_1\Phi_1(B))\widehat{U}(t),$$

the state of $A(\rho)\beta_{NH}(t) = B(\rho)u(t), t \geq 0$ is *impulse free*. $\square$

8.5 The smooth and impulsive solution components and impulsive free initial conditions: C_∞ is not of full row rank

This section is devoted to analyze the solution components and impulsive free initial conditions of the LNHMDEs without assuming that C_∞ is of full row rank.

If C_∞ is not of full row rank, say its row rank is $r_1 < r$, then there exists a transformation called "row compression," which reduces C_∞ to the form

$$PC_\infty = \begin{bmatrix} C_\infty^* \\ 0 \end{bmatrix}, \tag{8.19}$$

where C_∞^* is of full row rank r_1, it thus satisfies

$$C_\infty^*(C_\infty^*)^{(1)} = I_{r_1}. \tag{8.20}$$

If one denotes

$$P\beta_{NH}(t) := \begin{bmatrix} \beta_{NH1}(t) \\ \beta_{NH2}(t) \end{bmatrix}, \quad PC := \begin{bmatrix} C_1 \\ C_2 \end{bmatrix}, \tag{8.21}$$

where

$$\beta_{NH1}(t) \in R^{r_1 \times l}, \quad \beta_{NH2}(t) \in R^{(r-r_1) \times 1}$$

$$C_1 \in R^{r_1 \times n}, \quad C_2 \in R^{(r-r_1) \times n}$$

then one has the following result, which designates the *fast* and the *slow* components of the system solution structure.

Theorem 8.5.1. *With the above notations, further denoting*

$$\beta_{s(NH)1}(t) = \beta_{s(NH)2}(t) := e^{Jt}\chi_s(0) + \int_0^t e^{J(t-\tau)}\Omega u(\tau)d\tau \tag{8.22}$$

and

$$\beta_{f(NH)1}(t) := (\Psi_2\Phi_2(B) + (C_\infty^*)^{(1)}C_1\Psi_1\Phi_1(B)\widehat{U(t)}$$
$$+ \left[(\Psi_2\Phi_2(B) + (C_\infty^*)^{(1)}C_1\Psi_1\Phi_1(B))u(0) - \widehat{J_\infty x_f}(0)\right]\widehat{\delta(t)}, \tag{8.23}$$

$$\beta_{f(NH)2}(t) := \Psi_1\Phi_1(B)\widehat{U(t)} + \Psi_1\Phi_1(B)\mathcal{U}(0)\widehat{\delta(t)}, \tag{8.24}$$

the solution for the LNHMDE (8.7) is

$$\beta_{NH}(t) = P^{-1}\begin{bmatrix} C_1 & 0 & C_\infty^* & 0 \\ 0 & C_2 & 0 & C_2 \end{bmatrix}\begin{bmatrix} \beta_{s(NH)1}(t) \\ \beta_{s(NH)2}(t) \\ \hline \beta_{f(NH)1}(t) \\ \beta_{f(NH)2}(t) \end{bmatrix}. \tag{8.25}$$

Proof. From Theorem 8.3.1, one knows that the solution of the LNHMDE (8.7) corresponding to nonzero initial conditions both on the pseudo-state $\beta_{NH}(t)$ and the input $u(t)$ is

$$\beta_{NH}(t) = Ce^{Jt}x_{s\beta}(0) - Ce^{Jt}x_{su}(0)$$
$$- C_\infty\sum_{i=1}^{\widehat{q_r}}\delta(t)^{(i-1)}J_\infty^{i-1}(J_\infty x_{f\beta}(0)) + C_\infty\sum_{i=1}^{\widehat{q_r}}\delta(t)^{(i-1)}J_\infty^{i-1}(J_\infty x_{fu}(0))$$
$$+ \int_0^t Ce^{J(t-\tau)}\Omega u(\tau)d\tau + [C, C_\infty]\Psi\Phi\begin{bmatrix} u(t) \\ u(t)^{(1)} \\ \vdots \\ u(t)^{(\widehat{q_r}+l)} \end{bmatrix}, \quad t \geq 0. \tag{8.26}$$

From the Proof of Theorem 8.4.1, one further obtains

$$\beta_{NH}(t) = Ce^{Jt}x_s(0) + C \int_0^t e^{J(t-\tau)}\Omega u(\tau)d\tau$$
$$- C_\infty \widehat{J_\infty} x_f(0)\widehat{\delta(t)} + (C\Psi_1\Phi_1(B) + C_\infty\Psi_2\Phi_2(B))(\widehat{U(t)} + \mathcal{U}(0)\widehat{\delta(t)}).$$

Under the transformation P, the above $\beta_{NH}(t)$ is seen to be transformed into

$$P\beta_{NH}(t) := \begin{bmatrix} \beta_{NH1}(t) \\ \beta_{NH2}(t) \end{bmatrix}, \tag{8.27}$$

where

$$\beta_{NH1}(t) = C_1 e^{Jt}x_s(0) + C_1 \int_0^t e^{J(t-\tau)}\Omega u(\tau)d\tau$$
$$- C_\infty^* \widehat{J_\infty} x_f(0)\widehat{\delta(t)} + (C_1\Psi_1\Phi_1(B) + C_\infty^*\Psi_2\Phi_2(B))(\widehat{U(t)} + \mathcal{U}(0)\widehat{\delta(t)}),$$

by the virtue of that C_∞^* is of full row rank and by using Theorem 8.4.1, the above formulation can subsequently be written into

$$\beta_{NH1}(t) = \begin{bmatrix} C_1, C_\infty^* \end{bmatrix} \begin{bmatrix} \beta_{s(NH)1}(t) \\ \beta_{f(NH)1}(t) \end{bmatrix}, \tag{8.28}$$

where the components $\beta_{s(NH)1}(t)$ and $\beta_{f(NH)1}(t)$ are given by Eqs. (8.22), (8.23), respectively

$$\beta_{NH2}(t) = C_2 e^{Jt}x_s(0) + C_2 \int_0^t e^{J(t-\tau)}\Omega u(\tau)d\tau$$
$$+ C_2\Psi_1\Phi_1(B)\widehat{U(t)} + C_2\Psi_1\Phi_1(B)\mathcal{U}(0)\widehat{\delta(t)}$$
$$= [C_2, C_2] \begin{bmatrix} \beta_{s(NH)2}(t) \\ \beta_{f(NH)2}(t) \end{bmatrix}, \tag{8.29}$$

where the components $\beta_{s(NH)2}(t)$ and $\beta_{f(NH)2}(t)$ are given by Eqs. (8.22), (8.24), respectively. By combining Eqs. (8.28), (8.29) and taking the inverse transformation P^{-1} of Eq. (8.27) one finally establishes the required results. $\qquad\square$

Remark 8.5.1. The above theorem serves to designate the fast and slow components in the solutions of LNHMDEs in a general setting; it also enables us to analyze the impulse free initial conditions of the systems in the following manner without assuming that C_∞ is of full row rank.

Theorem 8.5.2. *The set of impulsive free initial conditions for the LNHMDE (8.7) is*

$$H_{I(NH)} = \left\{ \begin{bmatrix} \beta_{s(NH)1}(0) \\ \beta_{s(NH)2}(0) \\ \hline \beta_{f(NH)1}0 \\ \beta_{f(NH)2}(0) \end{bmatrix} : \beta_{s(NH)1}(0) = \beta_{s(NH)2}(0) = \chi_s(0) \in R^n, \right.$$

$$\beta_{f(NH)1}(0) = (\Psi_2\Phi_2(B) + (C_\infty^*)^{(1)}C_1\Psi_1\Phi_1(B))\widehat{U(0)},$$

$$\left. \beta_{f(NH)2}(0) = \Psi_1\Phi_1(B)\widehat{U(0)} \right\} \tag{8.30}$$

Proof. From Theorem 8.5.1, it is clearly seen that β_{NH} is impulsive free at $t = 0$ if and only if $\beta_{f(NH)1}(t)$ and $\beta_{f(NH)2}(t)$ are both impulse free at $t = 0$.

If one partitions

$$(\Psi_2\Phi_2(B) + (C_\infty^*)^{(1)}C_1\Psi_1\Phi_1(B))\mathcal{U}(0) := \left[\Pi_1^*(\mathcal{U}(0)), \ \Pi_2^*(\mathcal{U}(0))\right]$$

where

$$\Pi_1^*(\mathcal{U}(0)) \in R^{\times\widehat{q_r}}, \quad \Pi_2^*(\mathcal{U}(0)) \in R^{\mu\times l}.$$

Observing Eqs. (8.12), (8.23), one can see that $\beta_{f(NH)1}(t)$ is impulse free at $t = 0$ iff

$$(\Psi_2\Phi_2(B) + (C_\infty^*)^{(1)}C_1\Psi_1\Phi_1(B))\mathcal{U}(0) - \widehat{J_\infty x_f}(0) = 0 \tag{8.31}$$

which hold true iff

$$\Pi_1^*(\mathcal{U}(0)) - \widehat{J_\infty x_f}(0) = 0,$$

i.e.,

$$x_f(0) \in KerJ_\infty + Im(\Pi_1^*(\mathcal{U}(0))),$$

and

$$\Pi_2^*(\mathcal{U}(0)) = 0.$$

$\beta_{f(NH)2}(t)$ is impulse free at $t = 0$, if and only if

$$\mathcal{U}(0) \in Ker\Psi_1\Phi_1(B).$$

The set of the impulsive free initial conditions is thus derived from Theorem 8.5.1. $\qquad\square$

8.6 Illustrative example

This example is to display the fact that the initial conditions of $\beta_{NH}(t)$ and $u(t)$ must satisfy some conditions in order that the system is impulsive free and the fact that the proposed approach is efficient in analyzing the impulsive property of the system.

Consider the following LNHMDE:

$$\begin{bmatrix} \rho+1 & \rho^2 \\ 0 & 1 \end{bmatrix}\begin{bmatrix} \beta_{1(NH)}(t) \\ \beta_{2(NH)}(t) \end{bmatrix} = \begin{bmatrix} \rho^2+1 & \rho & 1 \\ 0 & \rho+1 & 1 \end{bmatrix}\begin{bmatrix} u_1(t) \\ u_2(t) \\ u_3(t) \end{bmatrix}, \quad t \geq 0. \tag{8.32}$$

with the initial values

$$\beta_{NH}(0) := \begin{bmatrix} \beta_{1(NH)}(0) \\ \beta_{2(NH)}(0) \end{bmatrix}, \quad \beta_{NH}{}^{(1)}(0) := \begin{bmatrix} \beta_{1(NH)}^{(1)}(0) \\ \beta_{2(NH)}^{(1)}(0) \end{bmatrix}$$

and

$$u(0) = \begin{bmatrix} u_1(0) \\ u_2(0) \\ u_3(0) \end{bmatrix}, \quad u^{(1)}(0) = \begin{bmatrix} u_1^{(1)}(0) \\ u_2^{(1)}(0) \\ u_3^{(1)}(0) \end{bmatrix}$$

we have

$$A_2 = \begin{bmatrix} 0 & 1 \\ 0 & 0 \end{bmatrix}, \quad A_1 = \begin{bmatrix} 1 & 0 \\ 0 & 0 \end{bmatrix}, \quad A_0 = \begin{bmatrix} 1 & 0 \\ 0 & 1 \end{bmatrix},$$

$$B_2 = \begin{bmatrix} 1 & 0 & 0 \\ 0 & 0 & 0 \end{bmatrix}, \quad B_1 = \begin{bmatrix} 0 & 1 & 0 \\ 0 & 1 & 0 \end{bmatrix}, \quad B_0 = \begin{bmatrix} 1 & 0 & 1 \\ 0 & 1 & 1 \end{bmatrix},$$

and the corresponding polynomial matrix (in the s-domain)

$$A(s) = \begin{bmatrix} s+1 & s^2 \\ 0 & 1 \end{bmatrix}, \quad S_{A(s)}^{\infty}(s) = \begin{bmatrix} s^2 & 0 \\ 0 & 1/s \end{bmatrix}.$$

So, $r = 2$, $n = 1$, $\mu = \hat{q}_2 + 1 = 1 + 1 = 2$, $q_1 = 2$, $\hat{q}_2 = 1$, $m = 3$, $l = 2$, we can find the finite and infinite Jordan triples (see [13])

$$C = \begin{bmatrix} 1 \\ 0 \end{bmatrix}, \quad J = [-1], \quad B = [1, -1],$$

$$C_\infty = \begin{bmatrix} -1 & 1 \\ 0 & 1 \end{bmatrix}, \quad J_\infty = \begin{bmatrix} 0 & 1 \\ 0 & 0 \end{bmatrix}, \quad B_\infty = \begin{bmatrix} 0 & 0 \\ 0 & 1 \end{bmatrix}.$$

Now let

$$\widehat{U(t)} = \begin{bmatrix} u(t) \\ u(t)^{[1]} \\ u(t)^{[2]} \\ \vdots \\ u(t)^{[\hat{q}_r+l]} \end{bmatrix} = \begin{bmatrix} u_1(t) \\ u_2(t) \\ u_3(t) \\ u_1(t)^{[1]} \\ u_2(t)^{[1]} \\ u_3(t)^{[1]} \\ u_1(t)^{[2]} \\ u_2(t)^{[2]} \\ u_3(t)^{[2]} \\ u_1(t)^{[3]} \\ u_2(t)^{[3]} \\ u_3(t)^{[3]} \end{bmatrix} \in R^{12},$$

$$\mathcal{U}(0) = \begin{bmatrix} 0 & 0 & 0 \\ 0 & 0 & 0 \\ 0 & 0 & 0 \\ u_1(0) & 0 & 0 \\ u_2(0) & 0 & 0 \\ u_3(0) & 0 & 0 \\ u_1(0)^{[1]} & u_1(0) & 0 \\ u_2(0)^{[1]} & u_2(0) & 0 \\ u_3(0)^{[1]} & u_3(0) & 0 \\ u_1(0)^{[2]} & u_1(0)^{[1]} & u_1(0) \\ u_2(0)^{[2]} & u_2(0)^{[1]} & u_2(0) \\ u_3(0)^{[2]} & u_3(0)^{[1]} & u_3(0) \end{bmatrix} \in R^{12 \times 3},$$

$$\widehat{\delta(t)} = \begin{bmatrix} \delta(t) \\ \delta(t)^{(1)} \\ \delta(t)^{(2)} \end{bmatrix} \in R^3.$$

By some simple calculation, we obtain

$$C_\infty^{(1)} = \begin{bmatrix} -1 & 1 \\ 0 & 1 \end{bmatrix}, \quad \Psi_1 = [JB, B] = [-1, 1, 1, -1],$$

$$\Psi_2 = [B_\infty, J_\infty B_\infty] = \begin{bmatrix} 0 & 0 & 0 & 1 \\ 0 & 1 & 0 & 0 \end{bmatrix},$$

$$\Phi_1(B) = \begin{bmatrix} B_2 & 0 & 0 & 0 \\ B_1 & B_2 & 0 & 0 \end{bmatrix} = \left[\begin{array}{ccc|ccc|ccc|ccc} 1 & 0 & 0 & 0 & 0 & 0 & 0 & 0 & 0 & 0 & 0 & 0 \\ 0 & 0 & 0 & 0 & 0 & 0 & 0 & 0 & 0 & 0 & 0 & 0 \\ \hline 0 & 1 & 0 & 1 & 0 & 0 & 0 & 0 & 0 & 0 & 0 & 0 \\ 0 & 1 & 0 & 0 & 0 & 0 & 0 & 0 & 0 & 0 & 0 & 0 \end{array} \right],$$

$$\Phi_2(B) = \begin{bmatrix} B_0 & B_1 & B_2 & 0 \\ 0 & B_0 & B_1 & B_2 \end{bmatrix} = \left[\begin{array}{ccc|ccc|ccc|ccc} 1 & 0 & 1 & 0 & 1 & 0 & 1 & 0 & 0 & 0 & 0 & 0 \\ 0 & 1 & 1 & 0 & 1 & 0 & 0 & 0 & 0 & 0 & 0 & 0 \\ \hline 0 & 0 & 0 & 1 & 0 & 1 & 0 & 1 & 0 & 1 & 0 & 0 \\ 0 & 0 & 0 & 0 & 1 & 1 & 0 & 1 & 0 & 0 & 0 & 0 \end{array} \right],$$

$$\Omega = \left[J^2 BB_2 + JBB_1 + BB_0 \right] = [2, -1, 0].$$

We have

$$x_{s\beta}(0) = [JB, B] \begin{bmatrix} A_2 & 0 \\ A_1 & A_2 \end{bmatrix} \begin{bmatrix} \beta_{NH}(0) \\ \beta_{NH}^{(1)}(0) \end{bmatrix}$$

$$= \beta_{1(NH)}(0) - \beta_{2(NH)}(0) + \beta_{2(NH)}^{(1)}(0),$$

$$x_{f\beta}(0) = [B_\infty, J_\infty B_\infty] \begin{bmatrix} A_0 & A_1 \\ 0 & A_0 \end{bmatrix} \begin{bmatrix} \beta_{NH}(0) \\ \beta_{NH}^{(1)}(0) \end{bmatrix}$$

$$= \begin{bmatrix} \beta_{2(NH)}^{(1)}(0) \\ \beta_{2(NH)}(0) \end{bmatrix},$$

$$x_{su}(0) = [JB, B] \begin{bmatrix} B_2 & 0 \\ B_1 & B_2 \end{bmatrix} \begin{bmatrix} u(0) \\ u^{(1)}(0) \end{bmatrix}$$

$$= u_1^{(1)}(0) - u_1(0),$$

$$x_{fu}(0) = [B_\infty, J_\infty B_\infty] \begin{bmatrix} B_0 & B_1 \\ 0 & B_0 \end{bmatrix} \begin{bmatrix} u(0) \\ u^{(1)}(0) \end{bmatrix}$$

$$= \begin{bmatrix} u_2^{(1)}(0) + u_3^{(1)}(0) \\ u_2(0) + u_3(0) + u_2^{(1)}(0) \end{bmatrix}.$$

$$x_s(0) = x_{s\beta}(0) - x_{su}(0)$$

$$= \beta_{1(NH)}(0) - \beta_{2(NH)}(0) + \beta_{2(NH)}^{(1)}(0) - u_1^{(1)}(0) + u_1(0),$$

$$x_f(0) = \chi_{f\beta}(0) - \chi_{fu}(0)$$

$$= \begin{bmatrix} \beta_{2(NH)}^{(1)}(0) - u_2^{(1)}(0) - u_3^{(1)}(0) \\ \beta_{2(NH)}(0) - u_2(0) - u_3(0) - u_2^{(1)}(0) \end{bmatrix}.$$

$$C_\infty^{(1)} C\Psi_1\Phi_1(B) = \begin{bmatrix} 1 & 0 & 0 & -1 & 0 & 0 & 0 & 0 & 0 & 0 & 0 & 0 \\ 0 & 0 & 0 & 0 & 0 & 0 & 0 & 0 & 0 & 0 & 0 & 0 \end{bmatrix},$$

$$\Psi_2\Phi_2(B) = \begin{bmatrix} 0 & 0 & 0 & 0 & 1 & 1 & 0 & 1 & 0 & 0 & 0 & 0 \\ 0 & 1 & 1 & 0 & 1 & 0 & 0 & 0 & 0 & 0 & 0 & 0 \end{bmatrix},$$

$$\Psi_2\Phi_2(B) + C_\infty^{(1)} C\Psi_1\Phi_1(B) = \begin{bmatrix} 1 & 0 & 0 & -1 & 1 & 1 & 0 & 1 & 0 & 0 & 0 & 0 \\ 0 & 1 & 1 & 0 & 1 & 0 & 0 & 0 & 0 & 0 & 0 & 0 \end{bmatrix},$$

$$(\Psi_2\Phi_2(B) + C_\infty^{(1)} C\Psi_1\Phi_1(B))\widehat{U(t)} = \begin{bmatrix} u_1(t) - u_1(t)^{[1]} + u_2(t)^{[1]} + u_3(t)^{[1]} + u_2(t)^{[2]} \\ u_2(t) + u_3(t) + u_2(t)^{[1]} \end{bmatrix},$$

$$(\Psi_2\Phi_2(B) + C_\infty^{(1)} C\Psi_1\Phi_1(B))\mathcal{U}(0) = \begin{bmatrix} -u_1(0) + u_2(0) + u_3(0) + u_2(0)^{[1]} & u_2(0) & 0 \\ u_2(0) & 0 & 0 \end{bmatrix},$$

$$\Pi_1(\mathcal{U}(0)) = \begin{bmatrix} -u_1(0) + u_2(0) + u_3(0) + u_2(0)^{[1]} \\ u_2(0) \end{bmatrix},$$

$$\Pi_2(\mathcal{U}(0)) = \begin{bmatrix} u_2(0) & 0 \\ 0 & 0 \end{bmatrix},$$

$$\widehat{J_\infty\chi_f}(0) = \begin{bmatrix} \beta_{2(NH)}(0) - u_2(0) - u_3(0) - u_2^{(1)}(0) \\ 0 \end{bmatrix},$$

$$\widehat{J_\infty\chi_f}(0) = \begin{bmatrix} \beta_{2(NH)}(0) - u_2(0) - u_3(0) - u_2^{(1)}(0) & 0 & 0 \\ 0 & 0 & 0 \end{bmatrix},$$

$$\left[(\Psi_2 \Phi_2(B) + C_\infty^{(1)} C \Psi_1 \Phi_1(B)) \mathcal{U}(0) - \widehat{J_\infty \chi_f}(0) \right] \widehat{\delta(t)}$$

$$= \begin{bmatrix} -u_1(0) + 2u_2(0) + 2u_3(0) + u_2(0)^{[1]} + u_2^{(1)}(0) - \beta_{2(NH)}(0) \begin{vmatrix} u_2(0) & 0 \\ 0 & 0 \end{vmatrix} \\ u_2(0) \end{bmatrix} \begin{bmatrix} \delta(t) \\ \delta(t)^{(1)} \\ \delta(t)^{(2)} \end{bmatrix},$$

so from Theorem 8.4.1, we have

$$\beta_{s(NH)}(t) = e^{Jt} x_s(0) + \int_0^t e^{J(t-\tau)} \Omega u(\tau) d\tau$$

$$= e^{-t}(\beta_{1(NH)}(0) - \beta_{2(NH)}(0) + \beta_{2(NH)}^{(1)}(0) + u_1(0) - u_1^{(1)}(0))$$

$$+ e^{-t} \int_0^t (2u_1(\tau) - u_2(\tau)) e^\tau d\tau,$$

$$\beta_{f(NH)}(t) = (\Psi_2 \Phi_2(B) + C_\infty^{(1)} C \Psi_1 \Phi_1(B)) \widehat{U(t)}$$

$$+ \left[(\Psi_2 \Phi_2(B) + C_\infty^{(1)} C \Psi_1 \Phi_1(B)) \mathcal{U}(0) - \widehat{J_\infty \chi_f}(0) \right] \widehat{\delta(t)},$$

$$= \begin{bmatrix} u_1(t) - u_1(t)^{[1]} + u_2(t)^{[1]} + u_3(t)^{[1]} + u_2(t)^{[2]} \\ u_2(t) + u_3(t) + u_2(t)^{[1]} \end{bmatrix}$$

$$+ \begin{bmatrix} -u_1(0) + 2u_2(0) + 2u_3(0) + u_2(0)^{[1]} + u_2^{(1)}(0) - \beta_{2(NH)}(0) \begin{vmatrix} u_2(0) & 0 \\ 0 & 0 \end{vmatrix} \\ u_2(0) \end{bmatrix} \begin{bmatrix} \delta(t) \\ \delta(t)^{(1)} \\ \delta(t)^{(2)} \end{bmatrix}.$$

It can easily be seen that the state $\beta_{NH}(t)$ is *impulse free* at $t = 0$ if and only if

$$\Pi_2(\mathcal{U}(0)) = \begin{bmatrix} u_2(0) & 0 \\ 0 & 0 \end{bmatrix} = 0, \quad \rightarrow u_2(0) = 0,$$

and

$$\Pi_1(\mathcal{U}(0)) - \widehat{J_\infty \chi_f}(0) = \begin{bmatrix} -u_1(0) + 2u_2(0) + 2u_3(0) + u_2(0)^{[1]} + u_2^{(1)}(0) - \beta_{2(NH)}(0) \\ u_2(0) \end{bmatrix}$$

$$= 0$$

$$\rightarrow \beta_{2(NH)}(0) = -u_1(0) + 2u_3(0) + u_2(0)^{[1]} + u_2^{(1)}(0).$$

The set of the impulse free initial conditions for the LNHMDE (8.32) is

$$H_{I(NH)} = \left\{ \chi_{NH}(0) = \begin{bmatrix} \beta_{s(NH)}(0) \\ \beta_{f(NH)}(0) \end{bmatrix} : \beta_{s(NH)}(0) = \beta_{1(NH)}(0) - \beta_{2(NH)}(0) \right.$$

$$+ \beta_{2(NH)}^{(1)}(0) + u_1(0) - u_1^{(1)}(0),$$

$$\left. \beta_{f(NH)}(0) = \begin{bmatrix} u_1(0) - u_1(0)^{[1]} + u_3(0)^{[1]} + u_2(0)^{[1]} + u_2(0)^{[2]} \\ u_3(0) + u_2(0)^{[1]} \end{bmatrix} \right\}.$$

$$(8.33)$$

8.7 Conclusions

In this chapter the complete solution of the LNHMDEs has been investigated. Special interest is paid to the nonzero initial conditions of the state as well as the nonzero initial conditions of the input, both of which might influence the solution structure of the system substantially. This is reflected completely in the solution components and in the set of impulse free initial conditions of the system. With respect to the complete solution of the LNHMDE the initial conditions of the system input are firstly considered. The solution obtained is an extension to that given by Gohberg et al. [17], Vardulakis [13], and Vardulakis et al. [102]. The solution to the LNHMDE can be decomposed into the *smooth* (*slow*) response and the *impulse* (*fast*) response. The notions of *state, slow state* (*smooth state*) and the *fast state* (*impulse state*) were generalized to the LNHMDE case. One example has been presented to show how to decompose the solution components and how to analyze the impulsive behavior of the system using this approach.

A refined resolvent decomposition of a regular polynomial matrix and application to the solution of regular PMDs

9

9.1 Introduction

Regular *polynomial matrix descriptions* (PMDs) are described by

$$\Sigma' : \begin{cases} T(\rho)\beta(t) = E(\rho)u(t) \\ y(t) = C(\rho)\beta(t) + D(\rho)u(t), \quad t \geq 0 \end{cases}$$

where $\rho := d/dt$ is the differential operator,

$$T(\rho) \in R[\rho]^{l \times l}, \quad \text{rank}_{\Re[\rho]} T(\rho) = l,$$

$$E(\rho) \in R[\rho]^{l \times m'}, \quad C(\rho) \in R[\rho]^{p \times l},$$

$$D(\rho) \in R[\rho]^{p \times m'},$$

$\beta(t) : [0, +\infty) \to R^l$ is the pseudo-state of Σ', $u(t) : [0, +\infty) \to R^{m'}$ is the control input and $y(t)$ the output of Σ'. Σ' can be written in the form:

$$\underbrace{\begin{bmatrix} T(\rho) & E(\rho) & 0 \\ -C(\rho) & D(\rho) & I_p \\ 0 & -I_{m'} & 0 \end{bmatrix}}_{A(\rho)} \underbrace{\begin{bmatrix} \beta(t) \\ -u(t) \\ y(t) \end{bmatrix}}_{\xi(t)} = \underbrace{\begin{bmatrix} 0 \\ 0 \\ I_{m'} \end{bmatrix}}_{B} u(t)$$

$$y(t) = \underbrace{[0, 0, I_p]}_{\Psi} \underbrace{\begin{bmatrix} \beta(t) \\ -u(t) \\ y(t) \end{bmatrix}}_{\xi(t)}, \quad t \geq 0.$$

In order to investigate the solution of the above PMD, we thus focus our interests on the following form of equation

$$\Sigma : A(\rho)\xi(t) = Bu(t), \quad t \geq 0, \tag{9.1}$$

A Generalized Framework of Linear Multivariable Control. http://dx.doi.org/10.1016/B978-0-08-101946-7.00009-3

where $A(\rho) \in R[\rho]^{r \times r}$, with $\text{rank}_{\Re[\rho]} A(\rho) = r$, $\xi(t) \in R^r, B \in R^{r \times m}, u(t) \in R^m$. The homogeneous case of Σ is

$$\Sigma_0 : A(\rho)\xi(t) = 0, \quad t \geq 0. \tag{9.2}$$

Both regular generalized state space systems which are described by

$$E\dot{x} = Ax(t) + Bu(t),$$

where E is a singular matrix, $\text{rank}(\rho E - A) = r$, and the (regular) state space systems which are described by

$$\dot{x} = Ax(t) + Bu(t)$$

are special cases of the regular PMDs (Eq. 9.1).

The solution of the above PMDs can be proposed on the basis of any resolvent decomposition of $A(s)$ given by

$$A^{-1}(s) = C(sI_n - J)^{-1}Z + C_\infty(sJ_\infty - I_\varepsilon)^{-1}Z_\infty, \tag{9.3}$$

where (C, J) is the finite Jordan pair of $A(s)$, and (C_∞, J_∞) is the infinite Jordan pair [13, 17] of $A(s)$. Such resolvent decompositions are not unique, but play an important role in formulating the solution of regular PMDs. The difficulties in the problem of obtaining the solution of the PMD are specific to the particular resolvent decomposition used. From a computation point of view, when the matrices in Eq. (9.3) are of minimal dimensions, it is obviously easiest to obtain the inverse matrix of $A(s)$.

There is fundamental interest (notably [13, 17, 102]) in the solution of the regular PMD and its homogeneous case. For Eq. (9.1) Gohberg et al. [17] proposed a particular resolvent decomposition, and the solution of it was formulated according to this decomposition. However, the impulsive property of the system at $t = 0$ was not properly considered in this work. One advantage of this resolvent Decomposition, however, is that it is constructive, due to the fact that the matrices Z, Z_∞ in Eq. (9.3) are formulated in terms of the finite and infinite Jordan pairs. On the other hand in this resolvent decomposition, there appears to be certain redundant information, in that the dimensions of the infinite Jordan pair are much larger than is usually necessary, this in turn brings some inconvenience in computing the inverse matrix of $A(s)$.

In fact some of the *infinite elementary divisors* [18] that correspond to the infinite poles of $A(s)$ actually contribute nothing to the solution, as will be seen subsequently. Thus a resolvent decomposition can be proposed in a simpler form that has the further advantage of giving a more precise insight into the system structure as well as bringing some convenience in actual computation. Vardulakis [13] obtained a general solution of Eq. (9.1) under the assumption that the initial conditions of $\xi(t)$ are zero. The main idea of this approach is to find minimal realizations of the strictly proper part and the polynomial part of $A^{-1}(s)$ and then to obtain the required resolvent decomposition. Although this procedure does give good insight into the system structure, the realization approach is not so straightforward by itself and it is consequently more difficult to apply in actual computation. Furthermore, in this procedure, no explicit formula for Z and Z_∞ is available.

In [102] the solution and the impulsive behavior of autonomous linear multivariable systems whose pseudo-state obeys the linear matrix differential equation (9.2) are examined. The impulsive solution part of the general nonhomogeneous PMD (Eq. 9.1) is, however, not displayed in the solution of Eq. (9.2) given by Vardulakis et al. [102]. Apparently, differences arise between the solution of Gohberg et al. [17] and that of Vardulakis [13] due to the fact that these two solutions are expressed through two different resolvent decompositions. Although it is found that the redundant information contained in the solution of Gohberg et al. [17] decouples in the meaning described in this book, an overly large resolvent decomposition definitely brings some inconvenience to actual computation.

The main purpose of this chapter is to present a resolvent decomposition that is a refinement of both results obtained by Gohberg et al. [17] and Vardulakis [13]. It is formulated in terms of the notions of the finite Jordan pairs, infinite Jordan pairs, and the generalized infinite Jordan pairs that were defined by Gohberg et al. [17] and Vardulakis [13]. We clarify the issue of the *infinite Jordan pair* noted by Gohberg et al. [17] and Vardulakis [13]. This refined resolvent decomposition captures the essential feature of the system structure, and the redundant information that is included in the resolvent decomposition of Gohberg et al. [17] is deleted through certain transformations. The resulting resolvent decomposition thus inherits the advantages of both the results of Gohberg et al. [17] and that of Vardulakis [13].

Based on this proposed resolvent decomposition, a complete solution of a PMD follows that reflects the detailed structure of the *zero state response* and the *zero input response* of the system. The complete impulsive properties of the system are also displayed in our solution. Such impulsive properties are not properly displayed in the solution of Gohberg et al. [17]. Although for the homogeneous case it is considered in Vardulakis [13], Vardulakis et al. [102], such a complete treatment of the impulsive properties of the system for the general nonhomogeneous regular PMD is not available in Vardulakis [13] and Vardulakis et al. [102].

Compared with the complete solution of regular PMDs given in Chapter 8, where the solution is proposed based on any resolvent decomposition, the resolvent decomposition obtained in this chapter is of minimal dimension, the solution constructed subsequently is specific to this refined resolvent decomposition.

9.2 Infinite Jordan pairs

In this section, we analyze two definitions of *infinite Jordan pairs*. The first was given by Gohberg et al. [17] and the other was given by Vardulakis [13]. The essential differences between them are elucidated and these discussions enable us to establish a refined resolvent decomposition.

Let $A(s) = A_0 + A_1 s + \cdots + A_{q_1} s^{q_1} \in R^{r \times r}[s]$, $\text{rank}_{\Re[s]} A(s) = r$. $S^{\infty}_{A(s)}$ denotes the Smith-McMillan form of $A(s)$ at $s = \infty$,

$$
S^{\infty}_{A(s)}(s) = \text{diag}\left[\underbrace{s^{q_1}, s^{q_2}, \ldots, s^{q_k}}_{k}, \underbrace{1, \ldots, 1}_{r-h-k}, \underbrace{{}^{1}/_{s^{\hat{q}_{r-h+1}}}, \ldots, {}^{1}/_{s^{\hat{q}_r}}}_{h} \right],
$$

where

$$q_1 \geq q_2 \geq \cdots \geq q_k > 0$$

are the orders of the infinite poles at $s = \infty$ of $A(s)$, and

$$\widehat{q}_r \geq \widehat{q}_{r-1} \geq \cdots \geq \widehat{q}_{r-h+1} > 0$$

are the orders of the infinite zeros at $s = \infty$ of $A(s)$. The *dual polynomial* of $A(s)$ is defined as

$$D_{A(s)}(w) := w^{q_1} A\left(\frac{1}{w}\right) = A_0 w^{q_1} + A_1 w^{q_1-1} + \cdots + A_{q_1}.$$

The *local Smith form* of $D_{A(s)}(w)$ at $w = 0$ is

$$S^0_{D_{A(s)}}(w) = w^{q_1} S^\infty_{A(s)}\left(\frac{1}{w}\right)$$

$$= \operatorname{diag}\left[1, w^{q_1-q_2}, \ldots, w^{q_1-q_k}, w^{q_1}, \ldots, w^{q_1}, w^{q_1+\widehat{q}_{r-h+1}}, \ldots, w^{q_1+\widehat{q}_r}\right].$$

It is seen that the polynomial matrices have in general three kinds of IEDs, or equivalently $D_{A(s)}(w)$ has three kinds of finite zero at $w = 0$. The first kind of IEDs with orders $q_1 - q_j$, $j = 2, 3, \ldots, k$ correspond to the poles of $A(s)$ at $s = \infty$. The second kind of IEDs comprise those corresponding to the zeros of $A(s)$ at $s = \infty$ with orders $q_1 + \widehat{q}_j$, $j = r - h + 1, \ldots, r$. The third kind with orders q_1 is not dynamically important [18].

Let $\tilde{U}_L(w) \in R[s]^{r \times r}$, $\tilde{U}_R(w) \in R[s]^{r \times r}$ be unimodular matrices reducing $D_{A(s)}(w)$ to its local Smith form $S^0_{D_{A(s)}}(w)$, i.e.,

$$\tilde{U}_L(w) D_{A(s)}(w) \tilde{U}_R(w) = S^0_{D_{A(s)}}(w).$$

Let $u_j(w) \in R[w]^{r \times 1}$, $v_j(w) \in R[w]^{r \times 1}$, $j \doteq 1, 2, \ldots, r$ be the columns of $\tilde{U}_R(w)$ and $\tilde{U}_L^{-1}(w)$, then

$$D_{A(s)}(w) u_j(w) = v_j(w) w^{\tau(j)},$$

where

$$\tau(j) := \begin{cases} q_1 - q_j, & j = 1, 2, \ldots, k \\ q_1 + \widehat{q}_j, & j = k+1, \ldots, r \end{cases}$$

Let

$$\beta_{jq} := \frac{1}{q!} u_j^{(q)}(0), \quad j = 2, 3, \ldots, r, \quad q = 0, 1, 2, \ldots, \tau(j) - 1,$$

then for $j = 2, 3, \ldots, r$, the vectors

$$\beta_{j0}, \beta_{j1}, \ldots, \beta_{j(\tau(j)-1)} \in R^r$$

form a Jordan chain corresponding to the zero $w = 0$ of the IEDs $w^{\tau(j)}$. Now we can compare the two different definitions of the infinite Jordan pair.

Definition 9.2.1 (Gohberg et al. [17]). The *generalized infinite Jordan pair* (C'_∞, J'_∞) of $A(s)$ are defined as

$$C'_\infty := \left[C'_2, C'_3, \ldots, C'_k, C'_{k+1}, \ldots, C'_r \right] \in R^{r \times v},$$

$$J'_\infty := \text{block diag} \left[J'_2, \ldots, J'_r \right] \in R^{v \times v},$$

where

$$C'_j = \left[\beta_{j0}, \beta_{j1}, \ldots, \beta_{j(\tau(j)-1)} \right] \in R^{r \times \tau(j)}, \quad j = 2, 3, \ldots, r,$$

$$J'_j = \begin{bmatrix} 0 & 1 & 0 & \cdots & 0 \\ 0 & 0 & 1 & \cdots & 0 \\ \cdot & \cdot & \cdot & \cdots & \cdot \\ 0 & 0 & 0 & \cdots & 1 \\ 0 & 0 & 0 & \cdots & 0 \end{bmatrix} \in R^{\tau(j) \times \tau(j)}, \quad j = 2, 3, \ldots, r.$$

We observe that

1. J'_∞ is nilpotent with an index of nil-potency $q_1 + \hat{q}_r$, because the largest Jordan block is $(q_1 + \hat{q}_r) \times (q_1 + \hat{q}_r)$.
2. $v = rq_1 - \sum_{i=1}^{k} q_i$ (the total number of the infinite poles of $A(s)$) $+ \sum_{i=1}^{r} \hat{q}_{k+i}$ (the total number of the infinite zeros of $A(s)$).

Definition 9.2.2 (Vardulakis [13]). The *infinite Jordan pair* (C_∞, J_∞) of $A(s)$ is defined as follows

$$C_\infty := [C_{k+1}, \ldots, C_r] \in R^{r \times \mu},$$

$$J_\infty := \text{block diag} [J_{k+1}, \ldots, J_r] \in R^{\mu \times \mu},$$

where

$$C_j = \left[\beta_{j0}, \beta_{j1}, \ldots, \beta_{j\hat{q}_j} \right] \in R^{r \times (\hat{q}_j+1)}, \quad j = k+1, \ldots, r,$$

$$J_j = \begin{bmatrix} 0 & 1 & 0 & \cdots & 0 \\ 0 & 0 & 1 & \cdots & 0 \\ \cdot & \cdot & \cdot & \cdots & \cdot \\ 0 & 0 & 0 & \cdots & 1 \\ 0 & 0 & 0 & \cdots & 0 \end{bmatrix} \in R^{(\hat{q}_j+1) \times (\hat{q}_j+1)}, \quad j = k+1, \ldots, r.$$

We observe that

1. J_∞ is nilpotent with an index of nil-potency $\hat{q}_r + 1$, because the largest Jordan block is $(\hat{q}_r + 1) \times (\hat{q}_r + 1)$.
2. $\mu = \sum_{j=k+1}^{r} (\hat{q}_j + 1) = \sum_{j=k+1}^{r} \hat{q}_j + r - k$.

The above two interpretations of infinite Jordan pairs display an essential difference. For example, the roles that the infinite poles and the infinite zeros play in the system behaviors are considered in totally different manners. It is obvious that $v \geq \mu$, and the dimensions of the generalized infinite Jordan pair of Definition 9.2.1 are unnecessarily larger than that of the infinite Jordan pair of Definition 9.2.2 due to some redundant information being included. This can be seen clearly from the following example.

Example 9.2.1. Consider the polynomial matrix

$$A(s) = \begin{bmatrix} 1 & s^3 & 0 \\ 0 & 1 & s \\ 0 & 0 & 1 \end{bmatrix} \in R[s]^{3 \times 3},$$

its Smith-McMillan form is

$$S_{A(s)}^\infty = \begin{bmatrix} s^3 & 0 & 0 \\ 0 & s & 0 \\ 0 & 0 & \frac{1}{s^4} \end{bmatrix}, i.e., q_1 = 3, q_2 = 1, \widehat{q}_3 = 4.$$

Thus $A(s)$ has two infinite poles at degrees 1 and 3 and an infinite zero at degree 4. The dual polynomial matrix to $A(s)$ is

$$D_{A(s)}(w) = w^{q_1} A\left(\frac{1}{w}\right) = \begin{bmatrix} w^3 & 1 & 0 \\ 0 & w^3 & w^2 \\ 0 & 0 & w^3 \end{bmatrix} \in R[w]^{3 \times 3}$$

and

$$\tilde{U}_L(w) D_{A(s)}(w) \tilde{U}_R(w) = S_{D_{A(s)}}^0(w) = w^{q_1} S_{A(s)}^\infty\left(\frac{1}{w}\right) = \begin{bmatrix} 1 & 0 & 0 \\ 0 & w^2 & 0 \\ 0 & 0 & w^7 \end{bmatrix}$$

is the local Smith form of $D_{A(S)}(w)$ at $w = 0$, where

$$\tilde{U}_L(w) = \begin{bmatrix} 1 & 0 & 0 \\ -w^3 & 1 & 0 \\ w^4 & -w & 1 \end{bmatrix}, \quad \tilde{U}_R(w) = \begin{bmatrix} 0 & 0 & 2 \\ 1 & 0 & -w^3 \\ 0 & 1 & w^4 \end{bmatrix}$$

$$:= [u_1(w), u_2(w), u_3(w)].$$

So we get three kinds of infinite elementary divisor:

- w^2 corresponding to the infinite pole of $A(s)$,
- w^7 corresponding to the infinite zero of $A(s)$ and the IED $1 = w^0$.

From

$$\beta_{2q} = \frac{1}{q!} u_2^{(q)}(0), \quad q = 0, 1,$$

$$\beta_{3q} = \frac{1}{q!} u_3^{(q)}(0), \quad q = 0, 1, 2, 3, 4, 5, 6,$$

we have

$$\beta_{20} = \begin{bmatrix} 0 \\ 0 \\ 1 \end{bmatrix}, \quad \beta_{21} = \begin{bmatrix} 0 \\ 0 \\ 0 \end{bmatrix},$$

$$\beta_{30} = \begin{bmatrix} 1 \\ 0 \\ 0 \end{bmatrix}, \quad \beta_{31} = \begin{bmatrix} 0 \\ 0 \\ 0 \end{bmatrix}, \quad \beta_{32} = \begin{bmatrix} 0 \\ 0 \\ 0 \end{bmatrix},$$

$$\beta_{33} = \begin{bmatrix} 0 \\ -1 \\ 0 \end{bmatrix}, \quad \beta_{34} = \begin{bmatrix} 0 \\ 0 \\ 1 \end{bmatrix}, \quad \beta_{35} = \begin{bmatrix} 0 \\ 0 \\ 0 \end{bmatrix}, \quad \beta_{36} = \begin{bmatrix} 0 \\ 0 \\ 0 \end{bmatrix}.$$

According to Definition 9.2.1, the generalized infinite Jordan pair in the sense of Gohberg et al. [17] is (C'_∞, J'_∞):

$$C'_\infty = \begin{bmatrix} C'_2, C'_3 \end{bmatrix} = [\beta_{20}, \beta_{21}, \beta_{30}, \beta_{31}, \beta_{32}, \beta_{33}, \beta_{34}, \beta_{35}, \beta_{36}]$$

$$= \begin{bmatrix} 0 & 0 & 1 & 0 & 0 & 0 & 0 & 0 & 0 \\ 0 & 0 & 0 & 0 & 0 & -1 & 0 & 0 & 0 \\ 1 & 0 & 0 & 0 & 0 & 0 & 1 & 0 & 0 \end{bmatrix} \in R^{r \times v} \quad (r = 3, v = 9),$$

$$J'_\infty = \begin{bmatrix} 0 & 1 & 0 & 0 & 0 & 0 & 0 & 0 & 0 \\ 0 & 0 & 0 & 0 & 0 & 0 & 0 & 0 & 0 \\ 0 & 0 & 0 & 1 & 0 & 0 & 0 & 0 & 0 \\ 0 & 0 & 0 & 0 & 1 & 0 & 0 & 0 & 0 \\ 0 & 0 & 0 & 0 & 0 & 1 & 0 & 0 & 0 \\ 0 & 0 & 0 & 0 & 0 & 0 & 1 & 0 & 0 \\ 0 & 0 & 0 & 0 & 0 & 0 & 0 & 1 & 0 \\ 0 & 0 & 0 & 0 & 0 & 0 & 0 & 0 & 1 \\ 0 & 0 & 0 & 0 & 0 & 0 & 0 & 0 & 0 \end{bmatrix} \in R^{9 \times 9},$$

While the infinite Jordan pair in the sense of Vardulakis [13] is (C_∞, J_∞) with:

$$C_\infty = [C_3] = [\beta_{30}, \beta_{31}, \beta_{32}, \beta_{33}, \beta_{34}]$$

$$= \begin{bmatrix} 1 & 0 & 0 & 0 & 0 \\ 0 & 0 & 0 & -1 & 0 \\ 0 & 0 & 0 & 0 & 1 \end{bmatrix} \in R^{r \times \mu} \quad (r = 3, \mu = \widehat{q}_3 + 1 = 5),$$

$$J_\infty = \begin{bmatrix} 0 & 1 & 0 & 0 & 0 \\ 0 & 0 & 1 & 0 & 0 \\ 0 & 0 & 0 & 1 & 0 \\ 0 & 0 & 0 & 0 & 1 \\ 0 & 0 & 0 & 0 & 0 \end{bmatrix} \in R^{5 \times 5}.$$

In the above example it is noted that the dimensions of the generalized infinite Jordan pair, which are 3×9 and 9×9, are unnecessarily larger than that of the infinite Jordan pair, which are 3×5 and 5×5. Such a tendency will become more evident as the number of the infinite poles of $A(s)$, the dimension of $A(s)$ and the degree of $A(s)$ are increased. However, the following observation arising from the above two definitions of infinite Jordan pairs suggests the possibility of deleting the redundant information that is contained in the generalized infinite Jordan pair.

Theorem 9.2.1. *If (C'_∞, J'_∞) is a generalized infinite Jordan pair, there exists an elementary matrix P such that*

$$C'_\infty P = [C_\infty, C_0],$$

$$P^{-1} J'_\infty P = \left[\begin{array}{c|c} J_\infty & J' \\ \hline 0 & J_0 \end{array}\right],$$

where $C_0 \in R^{r \times (v-\mu)}$, $J' \in R^{\mu \times (v-\mu)}$, $J_0 \in R^{(v-\mu) \times (v-\mu)}$, $0 \in R^{(v-\mu) \times \mu}$ and (C_∞, J_∞) is an infinite Jordan pair. Furthermore, J_0 is nilpotent with an index of nil-potency $q_1 - 1$.

Proof. Note that C_∞ is a submatrix of C'_∞. By a series of interchanging of two columns in C'_∞, C'_∞ can be brought to the form $[C_\infty, C_0]$. We denote this elementary matrix as P, then $P^{-1} = P^T$, $C'_\infty P = [C_\infty, C_0]$, where $C_0 \in R^{r \times (v-\mu)}$. Similarly, by interchanging the corresponding columns and the corresponding rows of J'_∞, J'_∞ can be brought to a block matrix, i.e.,

$$P^{-1} J'_\infty P = \left[\begin{array}{c|c} J_\infty & J' \\ \hline 0 & J_0 \end{array}\right],$$

where

$$J' = [a(\sigma_1, \sigma_2)]_{(\sigma_1, \sigma_2)} \in R^{\mu \times (v-\mu)},$$

$$a(\sigma_1, \sigma_2) = \begin{cases} 1, & \sigma_1 = \sum_{\sigma=k+1}^{j} (\widehat{q}_\sigma + 1), \sigma_2 = \sum_{j=2}^{k} \tau(j) + (\widehat{q}_j + 2), \\ & j = k+1, \ldots, r \\ 0, & \text{other} \end{cases}$$

$$J_0 = \text{block diag}\left[J_2', \ldots, J_k', J_{k+1}', \ldots, J_r'\right],$$

$$J_j'(j = 2, 3, \ldots, k) \text{ are Jordan matrix of } (q_1 - q_j) \times (q_1 - q_j) (j = 2, \ldots, k),$$

$$J_i'(i = k+1, \ldots, r) \text{ are Jordan matrix of } (q_1 - 1) \times (q_1 - 1).$$

It can easily be seen that J_0 is nilpotent with an index of nil-potency $q_1 - 1$. □

Remark 9.2.1. The above elementary matrix P has the effect of deleting the redundant information in two ways. First, it deletes the redundant information in those blocks in the infinite Jordan pair of Gohberg et al. [17] that corresponds to the infinite zeros and brings them to the correct sizes. Second, it deletes all blocks in C'_∞, J'_∞ that correspond to the infinite poles and all blocks that are not dynamically important. In this way the resulting resolvent decomposition more precisely reflects the relevant system structure, since the deleted blocks in the infinite Jordan pairs display no dynamics in the state solution. This will be seen explicitly when the effect of P on Z has been determined in a subsequent result.

Example 9.2.2. Recall Example 9.2.1, we have

$$
\underbrace{\begin{bmatrix}
0 & 0 & 1 & 0 & 0 & 0 & 0 & 0 & 0 \\
0 & 0 & 0 & 0 & 0 & -1 & 0 & 0 & 0 \\
1 & 0 & 0 & 0 & 0 & 0 & 1 & 0 & 0
\end{bmatrix}}_{C'_\infty}
\underbrace{\begin{bmatrix}
0 & 0 & 0 & 0 & 0 & 1 & 0 & 0 & 0 \\
0 & 0 & 0 & 0 & 0 & 0 & 1 & 0 & 0 \\
1 & 0 & 0 & 0 & 0 & 0 & 0 & 0 & 0 \\
0 & 1 & 0 & 0 & 0 & 0 & 0 & 0 & 0 \\
0 & 0 & 1 & 0 & 0 & 0 & 0 & 0 & 0 \\
0 & 0 & 0 & 1 & 0 & 0 & 0 & 0 & 0 \\
0 & 0 & 0 & 0 & 1 & 0 & 0 & 0 & 0 \\
0 & 0 & 0 & 0 & 0 & 0 & 0 & 1 & 0 \\
0 & 0 & 0 & 0 & 0 & 0 & 0 & 0 & 1
\end{bmatrix}}_{P}
$$

$$
= \left[\begin{array}{ccccc|cccc}
1 & 0 & 0 & 0 & 0 & 0 & 0 & 0 & 0 \\
0 & 0 & 0 & -1 & 0 & 0 & 0 & 0 & 0 \\
0 & 0 & 0 & 0 & 1 & 1 & 0 & 0 & 0
\end{array}\right],
$$

with $\underbrace{}_{C_\infty}$ and $\underbrace{}_{C_0}$

$$
\underbrace{\begin{bmatrix}
0 & 0 & 1 & 0 & 0 & 0 & 0 & 0 & 0 \\
0 & 0 & 0 & 1 & 0 & 0 & 0 & 0 & 0 \\
0 & 0 & 0 & 0 & 1 & 0 & 0 & 0 & 0 \\
0 & 0 & 0 & 0 & 0 & 1 & 0 & 0 & 0 \\
0 & 0 & 0 & 0 & 0 & 0 & 1 & 0 & 0 \\
1 & 0 & 0 & 0 & 0 & 0 & 0 & 0 & 0 \\
0 & 1 & 0 & 0 & 0 & 0 & 0 & 0 & 0 \\
0 & 0 & 0 & 0 & 0 & 0 & 0 & 1 & 0 \\
0 & 0 & 0 & 0 & 0 & 0 & 0 & 0 & 1
\end{bmatrix}}_{P^{-1}}
\underbrace{\begin{bmatrix}
0 & 1 & 0 & 0 & 0 & 0 & 0 & 0 & 0 \\
0 & 0 & 0 & 0 & 0 & 0 & 0 & 0 & 0 \\
0 & 0 & 0 & 1 & 0 & 0 & 0 & 0 & 0 \\
0 & 0 & 0 & 0 & 1 & 0 & 0 & 0 & 0 \\
0 & 0 & 0 & 0 & 0 & 1 & 0 & 0 & 0 \\
0 & 0 & 0 & 0 & 0 & 0 & 1 & 0 & 0 \\
0 & 0 & 0 & 0 & 0 & 0 & 0 & 1 & 0 \\
0 & 0 & 0 & 0 & 0 & 0 & 0 & 0 & 1 \\
0 & 0 & 0 & 0 & 0 & 0 & 0 & 0 & 0
\end{bmatrix}}_{J'_\infty}
$$

$$
\times
\underbrace{\begin{bmatrix}
0 & 0 & 0 & 0 & 0 & 1 & 0 & 0 & 0 \\
0 & 0 & 0 & 0 & 0 & 0 & 1 & 0 & 0 \\
1 & 0 & 0 & 0 & 0 & 0 & 0 & 0 & 0 \\
0 & 1 & 0 & 0 & 0 & 0 & 0 & 0 & 0 \\
0 & 0 & 1 & 0 & 0 & 0 & 0 & 0 & 0 \\
0 & 0 & 0 & 1 & 0 & 0 & 0 & 0 & 0 \\
0 & 0 & 0 & 0 & 1 & 0 & 0 & 0 & 0 \\
0 & 0 & 0 & 0 & 0 & 0 & 0 & 1 & 0 \\
0 & 0 & 0 & 0 & 0 & 0 & 0 & 0 & 1
\end{bmatrix}}_{P}
=
\left[\begin{array}{ccccc|cccc}
0 & 1 & 0 & 0 & 0 & 0 & 0 & 0 & 0 \\
0 & 0 & 1 & 0 & 0 & 0 & 0 & 0 & 0 \\
0 & 0 & 0 & 1 & 0 & 0 & 0 & 0 & 0 \\
0 & 0 & 0 & 0 & 1 & 0 & 0 & 0 & 0 \\
0 & 0 & 0 & 0 & 0 & 0 & 0 & 1 & 0 \\
\hline
0 & 0 & 0 & 0 & 0 & 0 & 1 & 0 & 0 \\
0 & 0 & 0 & 0 & 0 & 0 & 0 & 0 & 0 \\
0 & 0 & 0 & 0 & 0 & 0 & 0 & 0 & 1 \\
0 & 0 & 0 & 0 & 0 & 0 & 0 & 0 & 0
\end{array}\right]
$$

$$
:= \left[\begin{array}{c|c}
J_\infty & J' \\
\hline
0 & J_0
\end{array}\right].
$$

It is noted that here J_0 is nilpotent with index of nil-potency 2.

The key point in here is to find the above elementary matrix P to delete the redundant information. Based on the information carried by the Smith-McMillan form at $s = \infty$ of $A(s)$, P results from a series of elementary operations from an identity matrix, which is very easy to obtain. Once the redundant information is deleted from the generalized infinite Jordan pair, the generalized infinite Jordan pair and the elementary matrix P are not involved in computation in the solution procedure.

9.3 The solution of regular PMDs

We consider now the homogeneous regular PMD:

$$A(\rho)\xi(t) = 0, \quad t \geq 0. \tag{9.4}$$

Let $S^{\infty}_{A(s)}(s)$ be the Smith-McMillan form at $s = \infty$ of $A(s) = L_-[A(\rho)] = A_0 + A_1 s + \cdots + A_{q_1} s^{q_1}$:

$$S^{\infty}_{A(s)}(s) = \mathrm{diag}\left[s^{q_1}, s^{q_2}, \ldots, s^{q_k}, \frac{1}{s^{\widehat{q}_{k+1}}}, \ldots, \frac{1}{s^{\widehat{q}_r}}\right] \in R^{r \times r}(s)$$

and if $A(s)$ has at least one zero at $s = \infty$ then the Laurent expansion of $A^{-1}(s)$ can be written as follows

$$A^{-1}(s) = H_{\widehat{q}_r} s^{\widehat{q}_r} + H_{\widehat{q}_r-1} s^{\widehat{q}_r-1} + \cdots + H_1 s + H_0 + H_{-1} s^{-1} + H_{-2} s^{-2} + \cdots$$
$$= H_{pol}(s) + H_{sp}(s),$$

where $H_{pol}(s) \in R^{r \times r}[s]$ is the polynomial part of $A^{-1}(s)$ and $H_{sp}(s) \in R^{r \times r}(s)$ is the strictly proper part of $A^{-1}(s)$. Let a resolvent decomposition of $A(s)$ be given by

$$A^{-1}(s) = C_{\infty}(sJ_{\infty} - I_{\mu})^{-1}Z_2 + C(sI_n - J)^{-1}Z_1,$$

where $n := \deg \det(A(s))$. It should be noted that (C, J, Z_1) is a realization of $H_{sp}(s)$, and $(C_{\infty}, J_{\infty}, Z_2)$ is a realization of $H_{pol}(s)$. Considering the Laplace transformation of Eq. (9.4), we obtain

$$\hat{\xi}(s) = A^{-1}(s)\hat{\alpha}(s) \in R^{r \times 1}, \tag{9.5}$$

where $\hat{\alpha}(s) \in R^{r \times 1}[s]$ is the initial condition vector associated with the initial values of $\xi(t)$ and its $(q_1 - 1)$-derivatives at $t = 0-$, i.e., $\xi(0-), \xi^{(1)}(0-), \ldots, \xi^{(q_1-1)}(0-)$ given by Fossard [108], Pugh [107]

$$\hat{\alpha}(s) = \left[s^{q_1-1}I_r, s^{q_1-2}I_r, \ldots, sI_r, I_r\right] \begin{bmatrix} A_{q_1} & 0 & \cdots & 0 \\ A_{q_1-1} & A_{q_1} & \cdots & 0 \\ \vdots & \vdots & \ddots & \vdots \\ A_1 & A_2 & \cdots & A_{q_1} \end{bmatrix} \begin{bmatrix} \xi(0-) \\ \xi^{(1)}(0-) \\ \vdots \\ \xi^{(q_1-1)}(0-) \end{bmatrix}. \tag{9.6}$$

We obtain

$$\widehat{\xi}(s) = [C, C_{\infty}] \left[\begin{array}{c|c} sI_n - J & 0_{n,\mu} \\ \hline 0_{\mu,n} & sJ_{\infty} - I_{\mu} \end{array} \right]^{-1} \left[\begin{array}{c} x_s(0-) \\ J_{\infty}x_f(0-) \end{array} \right], \tag{9.7}$$

where the vector

$$\left[\begin{array}{c} x_s(0-) \\ J_\infty x_f(0-) \end{array}\right] = \left[\begin{array}{c|c} J^{q_1-1}Z_1,\ldots,Z_1 & 0_{n,q_1\mu} \\ \hline 0_{\mu,q_1n} & J_\infty Z_2,\ldots,J_\infty^{q_1}Z_2 \end{array}\right]$$

$$\left[\begin{array}{cccc} A_{q_1} & 0 & \cdots & 0 \\ A_{q_1-1} & A_{q_1} & \cdots & 0 \\ \vdots & \vdots & \ddots & \vdots \\ A_1 & A_2 & \cdots & A_{q_1} \\ \hline A_0 & A_1 & \cdots & A_{q_1-1} \\ 0 & A_0 & \cdots & A_{q_1-2} \\ \vdots & \vdots & \ddots & \vdots \\ 0 & 0 & \cdots & A_0 \end{array}\right] \left[\begin{array}{c} \xi(0-) \\ \xi^{(1)}(0-) \\ \vdots \\ \xi^{(q_1-1)}(0-) \end{array}\right] \in R^{(n+\mu)\times 1}. \tag{9.8}$$

Now by taking the inverse L_- Laplace transformation of Eq. (9.7), we have the following result.

Lemma 9.3.1 (Vardulakis [13]). *The general solution of homogeneous PMD (Eq. 9.4) is*

$$\xi(t) = Ce^{Jt}x_s(0-) - C_\infty \sum_{i=1}^{\widehat{q}_r} \delta(t)^{(i-1)} J_\infty^{i-1}(J_\infty x_f(0-)),$$

where $\delta(t)$ is the Dirac impulse function.

The next result constructs a resolvent decomposition for a regular polynomial matrix, which follows from Gohberg et al. [17].

Lemma 9.3.2. *Let $A(s) = A_0 + A_1 s + \cdots + A_{q_1}s^{q_1} \in R^{r\times r}[s]$, $\text{rank}_{\Re[s]}A(s) = r$. If (C,J) is the finite Jordan pair of $A(s)$, and (C'_∞, J'_∞) is the generalized infinite Jordan pair of $A(s)$ in the sense of Gohberg et al. [17], $n := \deg \det(A(s))$. Put*

$$V = \left[A_{q_1}CJ^{q_1-1}, -\sum_{i=0}^{q_1-1} A_i C'_\infty (J'_\infty)^{q_1-1-i}\right],$$

$$S = \left[\begin{array}{cc} C & C'_\infty(J'_\infty)^{q_1-2} \\ CJ & C'_\infty(J'_\infty)^{q_1-3} \\ \vdots & \vdots \\ CJ^{q_1-2} & C'_\infty \end{array}\right]$$

and

$$Z' := \left[\begin{array}{c} Z_1 \\ Z'_2 \end{array}\right] = \left[\begin{array}{c|c} I_n & 0 \\ \hline 0 & (J'_\infty)^{q_1-1} \end{array}\right]\left[\begin{array}{c} S \\ V \end{array}\right]^{-1}[0,\ldots,0,I]^T, \tag{9.9}$$

then

$$A^{-1}(s) = [C, C'_\infty]\left[\begin{array}{c|c} I_n s - J & 0 \\ \hline 0 & J'_\infty s - I_v \end{array}\right]^{-1}\left[\begin{array}{c} Z_1 \\ Z'_2 \end{array}\right]. \tag{9.10}$$

An interesting observation from Eq. (9.9) is the following.

Proposition 9.3.1. *If we partition*

$$Z' := \begin{bmatrix} Z_1 \\ Z_2' \end{bmatrix} := \begin{bmatrix} Z_1 \\ Z_{21}' \\ Z_{22}' \end{bmatrix},$$

where $Z_1 \in R^{n \times r}$, $Z_{21}' \in R^{\mu \times r}$, $Z_{22}' \in R^{(\nu - \mu) \times r}$, *then*

$$\left[\begin{array}{c|c} I_n & 0 \\ \hline 0 & P^{-1} \end{array} \right] \begin{bmatrix} Z_1 \\ \hline Z_{21}' \\ Z_{22}' \end{bmatrix} = \begin{bmatrix} Z_1 \\ Z_2 \\ 0 \end{bmatrix},$$

where $Z_2 \in R^{\mu \times r}$.

Proof. Let

$$\begin{bmatrix} \Pi_1 \\ \Pi_2 \\ \Pi_3 \end{bmatrix} := \begin{bmatrix} S \\ V \end{bmatrix}^{-1} [0, \ldots, 0, I]^T,$$

where $\Pi_1 \in R^{n \times 1}$, $\Pi_2 \in R^{\mu \times 1}$, $\Pi_3 \in R^{(\nu - \mu) \times 1}$. From Theorem 9.2.1, there exists an elementary matrix $P \in R^{\nu \times \nu}$, such that

$$J_\infty' = P \left[\begin{array}{c|c} J_\infty & J' \\ \hline 0 & J_0 \end{array} \right] P^{-1},$$

so

$$(J_\infty')^{q_1 - 1} = P \left[\begin{array}{c|c} (J_\infty)^{q_1 - 1} & Q \\ \hline 0 & (J_0)^{q_1 - 1} \end{array} \right] P^{-1},$$

where $Q = J_\infty^{q_1 - 2} J' + J_\infty^{q_1 - 3} J' J_0 + \cdots + J' J_0^{q_1 - 2}$.

$$Z' = \left[\begin{array}{c|c} I_n & 0 \\ \hline 0 & P \end{array} \right] \left[\begin{array}{c|cc} I_n & 0 & 0 \\ \hline 0 & J_\infty^{q_1 - 1} & Q \\ 0 & 0 & J_0^{q_1 - 1} \end{array} \right] \left[\begin{array}{c|c} I_n & 0 \\ \hline 0 & P^{-1} \end{array} \right] \begin{bmatrix} \Pi_1 \\ \Pi_2 \\ \Pi_3 \end{bmatrix} := \begin{bmatrix} Z_1 \\ Z_{21}' \\ Z_{22}' \end{bmatrix}.$$

Noticing J_0 is nilpotent with an index of nil-potency $q_1 - 1$, i.e., $J_0^{q_1 - 1} = 0$, it follows that:

$$\left[\begin{array}{c|c} I_n & 0 \\ \hline 0 & P^{-1} \end{array} \right] \begin{bmatrix} Z_1 \\ Z_{21}' \\ Z_{22}' \end{bmatrix} = \begin{bmatrix} Z_1 \\ Z_2 \\ 0 \end{bmatrix}.$$

□

The above observation is interesting, not least for the way in which the redundant information contained in Z' is deleted. More importantly than this, however, is that such a mechanism of decoupling makes the computation of P attractive and facilitates the proposed refined resolvent decomposition.

The following result proposes a new approach to constructing a resolvent decomposition in terms of the finite Jordan pair, the infinite Jordan pair and the generalized infinite Jordan pair, which is fundamental for us to formulate the solution of the PMD. As one of our main results, we are ready to state it as follows.

Theorem 9.3.1. *If (C, J) is the finite Jordan pair of $A(s)$ and (C_∞, J_∞) is the infinite Jordan pair of $A(s)$ in the sense of Vardulakis [13], $Z' := \begin{bmatrix} Z_1 \\ Z_2' \end{bmatrix} := \begin{bmatrix} Z_1 \\ Z_{21}' \\ Z_{22}' \end{bmatrix}$ is given by Lemma 9.3.2, let $Z_2 \in R^{\mu \times r}$ be given by*

$$\begin{bmatrix} Z_2 \\ 0 \end{bmatrix} = P^{-1} \begin{bmatrix} Z_{21}' \\ Z_{22}' \end{bmatrix},$$

then

$$A^{-1}(s) = \begin{bmatrix} C, C_\infty \end{bmatrix} \left[\begin{array}{c|c} I_n s - J & 0 \\ \hline 0 & J_\infty s - I_\mu \end{array} \right]^{-1} \begin{bmatrix} Z_1 \\ Z_2 \end{bmatrix}. \tag{9.11}$$

Proof. This follows readily from Eq. (9.10) and Theorem 9.2.1 on noting that

$$C_\infty'(J_\infty's - I_v)^{-1}Z_2' = \begin{bmatrix} C_\infty, C_0 \end{bmatrix} P^{-1} P \left[\begin{array}{c|c} sJ_\infty - I_\mu & sJ' \\ \hline 0 & sJ_0 - I_{v-\mu} \end{array} \right]^{-1} P^{-1} P \begin{bmatrix} Z_2 \\ 0 \end{bmatrix}$$

$$= C_\infty(J_\infty s - I_\mu)^{-1}Z_2.$$

□

Remark 9.3.1. The resolvent decomposition proposed by Vardulakis [13] reflects the system solution structure precisely; however, the realization approach is not so straightforward by itself. On the other hand in the resolvent decomposition proposed by Gohberg et al. [17] (Lemma 9.3.2), Z_1, Z_2' were formulated explicitly. However, there is some redundant information as described in Theorem 9.2.1 and it does not give such a precise insight into the system structure. The resolvent decomposition proposed here is a refinement of these two results and inherits the advantages of both results. Due to the fact that the redundant information is deleted, the proposed refined resolvent decomposition has minimal dimensions, the proposed approach has an obvious advantage, i.e., computation of the inverse matrix of $A(s)$ can be carried out more efficiently.

Remark 9.3.2. The proposed resolvent decomposition has a dimension that is exactly the sum of the dimension of the minimal realization of the strictly proper part of $A^{-1}(s)$ and the dimension of the minimal realization of the polynomial part of $A^{-1}(s)$. Such a resolvent decomposition has thus minimal dimension. For a proof of this statement, one can refer to Vardulakis [13].

For the nonhomogeneous regular PMD

$$A(\rho)\xi(t) = Bu(t), \quad t \geq 0 \tag{9.12}$$

the following theorem gives the complete solution.

Theorem 9.3.2. Let $A(s) = A_0 + A_1 s + \cdots + A_{q_1} s^{q_1} \in R^{r \times r}[s]$ be regular, and let $S_{A(s)}^{\infty}(s) = \text{diag}[s^{q_1}, s^{q_2}, \ldots, s^{q_k}, 1/s^{\hat{q}_{k+1}}, \ldots, 1/s^{\hat{q}_r}]$ be the Smith-McMillan form of $A(s)$ at $s = \infty$. If (C, J) is the finite Jordan pair of $A(s)$ and (C_{∞}, J_{∞}) is the infinite Jordan pair of $A(s)$ (in the sense of Vardulakis [13]), (C, J, Z_1) and $(C_{\infty}, J_{\infty}, Z_2)$ satisfy the resolvent decomposition (Eq. 9.11), then the complete solution of regular PMD (Eq. 9.12) is

$$
\begin{aligned}
\xi(t) = {} & Ce^{Jt}x_s(0-) + C \int_{0-}^{t} e^{J(t-\tau)} Z_1 Bu(\tau) d\tau \\
& - C_{\infty} \sum_{i=1}^{\hat{q}_r} \delta(t)^{(i-1)} J_{\infty}^{i-1} (J_{\infty} x_f(0-)) - C_{\infty} \sum_{i=0}^{\hat{q}_r} J_{\infty}^{i} Z_2 Bu^{(i)}(t),
\end{aligned}
\tag{9.13}
$$

where $x_s(0-), J_{\infty} x_f(0-)$ are initial values given by Eq. (9.8).

Proof. From Lemma 9.3.1, one knows that

$$
\xi(t) = Ce^{Jt}x_s(0-) - C_{\infty} \sum_{i=1}^{\hat{q}_r} \delta(t)^{(i-1)} J_{\infty}^{i-1} (J_{\infty} x_f(0-))
$$

generates the general solution of the homogeneous PMD, so we only need to check that the formula (9.13) does indeed produce a solution of the nonhomogeneous PMD (Eq. 9.12). To this end, when $t > 0$, we have

$$
\begin{aligned}
\xi^{(j)}(t) = {} & CJ^j e^{Jt} x_s(0-) + \sum_{k=0}^{j-1} CJ^{j-1-k} Z_1 Bu^{(k)}(t) \\
& + C \int_{0-}^{t} J^j e^{J(t-\tau)} Z_1 Bu(\tau) d\tau - C_{\infty} \sum_{i=0}^{\hat{q}_r} J_{\infty}^{i} Z_2 Bu^{(i+j)}(t),
\end{aligned}
$$

so

$$
\begin{aligned}
A(\rho)\xi(t) = {} & \sum_{j=0}^{q_1} A_j \xi^{(j)}(t) \\
= {} & \sum_{j=0}^{q_1} A_j CJ^j e^{Jt} x_s(0-) + \sum_{j=0}^{q_1} \sum_{k=0}^{j-1} A_j CJ^{j-1-k} Z_1 Bu^{(k)}(t) \\
& + \int_{0-}^{t} \left[\sum_{j=0}^{q_1} A_j CJ^j \right] e^{J(t-\tau)} Z_1 Bu(\tau) d\tau - \sum_{j=0}^{\hat{q}_r} \sum_{j=0}^{q_1} A_j C_{\infty} J_{\infty}^{i} Z_2 Bu^{(i+j)}(t).
\end{aligned}
$$

Using the property $\sum_{j=0}^{q_1} A_j C J^j = 0$ and rearranging the terms, we have

$$A(\rho)\xi(t) = \sum_{j=0}^{q_1} A_j \xi^{(j)}(t)$$

$$= \sum_{k=0}^{q_1+\hat{q}_r} \left\{ \sum_{j=k+1}^{q_1} A_j C J^{j-1-k} Z_1 - \sum_{j=0}^{k} A_j C_\infty J_\infty^{k-j} Z_2 \right\} B u^{(k)}(t).$$

Noticing

$$A^{-1}(s) = [C, C_\infty] \left[\begin{array}{c|c} (I_n s - J)^{-1} & 0 \\ \hline 0 & (J_\infty s - I_\mu)^{-1} \end{array} \right] \begin{bmatrix} Z_1 \\ Z_2 \end{bmatrix}$$

$$= C_\infty (J_\infty s - I)^{-1} Z_2 + C(Is - J)^{-1} Z_1$$

$$= -C_\infty J_\infty^{\hat{q}_r} Z_2 s^{\hat{q}_r} - \cdots - C_\infty J_\infty Z_2 s - C_\infty Z_2$$

$$+ C Z_1 s^{-1} + C J Z_1 s^{-2} + \cdots,$$

from

$$A(s)A^{-1}(s) = I,$$

it follows that

$$\sum_{j=k+1}^{q_1} A_j C J^{j-1-k} Z_1 - \sum_{j=0}^{k} A_j C_\infty J_\infty^{k-j} Z_2 = \begin{cases} I, & k = 0 \\ 0, & k > 0. \end{cases}$$

Hence

$$A(\rho)\xi(t) = Bu(t),$$

which finishes the proof.   □

Remark 9.3.3. The above solution displays precisely the *zero state response* by means of

$$C \int_{0-}^{t} e^{J(t-\tau)} Z_1 Bu(\tau) d\tau - C_\infty \sum_{i=0}^{\hat{q}_r} J_\infty^i Z_2 Bu^{(i)}(t)$$

and the *zero input response* through the term

$$C e^{Jt} x_s(0-) - C_\infty \sum_{i=1}^{\hat{q}_r} \delta(t)^{(i-1)} J_\infty^{(i-1)} (J_\infty x_f(0-)).$$

Also the impulsive properties of the system are displayed in this solution due to the fact that the initial conditions of the pseudo-state are considered. Such an impulsive part to the general nonhomogeneous PMDs solution is, however, unavailable from the solution of Gohberg et al. [17] and Vardulakis [13].

   Remark 9.3.4. According to Gohberg et al. [17], the solution of PMD (Eq. 9.12) is following (without the impulse solution part):

$$\xi(t) = Ce^{Jt}x_s(0-) + C \int_{0-}^{t} e^{J(t-\tau)}Z_1 Bu(\tau)d\tau - C'_{\infty} \sum_{i=0}^{\hat{q}_r+q_1-1} (J'_{\infty})^i Z'_2 Bu^{(i)}(t),$$

where $(C'_{\infty}, J'_{\infty})$ is the infinite Jordan pair in the sense of Gohberg et al. [17]. Z_1, Z'_2 are given by Eq. (9.9). By using Theorem 9.2.1, Proposition 9.3.1 and noticing that J_{∞} is nilpotent with an index of nil-potency $\hat{q}_r + 1$, one finds that

$$\sum_{i=0}^{\hat{q}_r+q_1-1} C'_{\infty}(J'_{\infty})^i Z'_2 Bu^{(i)}(t) = \sum_{i=0}^{\hat{q}_r+q_1-1} [C_{\infty}, C_0]P^{-1}P \begin{bmatrix} J_{\infty} & J' \\ \hline 0 & J_0 \end{bmatrix}^i P^{-1}P \begin{bmatrix} Z_2 \\ 0 \end{bmatrix} Bu^{(i)}(t)$$

$$= \sum_{i=0}^{\hat{q}_r} C_{\infty}J_{\infty}^i Z_2 Bu^{(i)}(t).$$

It is thus clearly seen that the redundant information included in the solution of Gohberg et al. [17] due to using the generalized infinite Jordan pairs does not appear in our solution any more. Also this helps to clarify the mechanism of decoupling in the solution of Gohberg et al. [17].

   By comparing our solution with the solution of Gohberg et al. [17] and the solution of Vardulakis [13], it can be seen that, besides the fact that it displays the impulsive solution part, our solution can be carried out more efficiently in actual computation. The main reason for this is that the refined resolvent decomposition does not contain any redundant information, since the infinite Jordan pairs used in our solution are of minimal dimensions, and the matrices Z_1 and Z_2 are formulated explicitly. The refined resolvent decomposition facilitates computation of the inverse matrix of $A(s)$ because the dimension of the matrices used are minimal. Rather than calculating the inverse matrix $A^{-1}(s)$ with an overly large dimension, it is suggested first to calculate the easily available elementary matrix P to delete all redundant information and then to obtain the inverse matrix with the minimal dimension. Once the refined resolvent decomposition is obtained, the generalized infinite Jordan pair and the elementary matrix P are not necessary for the calculation of the solution of the regular PMD. This thus presents another advantage of this method, which is algorithmically attractive when applied in actual computation.

9.4   Algorithm and examples

As mentioned before, the refined resolvent decomposition plays a key role in our approach, the difficulty in obtaining the solution of the regular PMDs depends on the calculation of this refined resolvent decomposition. To help clarify the construction of this refined resolvent decomposition, for the implementation of Theorem 9.3.1, the following algorithm is now given, which is useful for symbolic computational packages Maple.

Algorithm 9.4.1 (Computation of the refined resolvent decomposition of $A(s) \in R[s]^{r \times r}$).

Step 1 Let $A(s) = \sum_{i=0}^{q_1} A_i s^i$, calculate the Smith form and the finite Jordan pair (C, J).

Step 2 Find the dual polynomial $D_{A(s)}(w)$ of $A(s)$, calculate the local Smith form of $D_{A(s)}(w)$ at $w = 0$ $S_{D_{A(s)}}^0(w)$ and the unimodular matrices $\tilde{U}_L(w)$, $\tilde{U}_R(w)$ such that

$$\tilde{U}_L(w)D_{A(s)}(w)\tilde{U}_R(w) = S_{D_{A(s)}}^0(w).$$

Step 3 Differentiate the matrix $\tilde{U}_R(w)$ and then construct the generalized infinite Jordan pair (C_∞', J_∞') and the infinite Jordan pair (C_∞, J_∞).

Step 4 Find the elementary matrix P by using Theorem 9.2.1.

Step 5 Calculate V, S and Z' by Lemma 9.3.2.

Step 6 Partition $Z' = \begin{bmatrix} Z_1 \\ Z_2' \end{bmatrix} = \begin{bmatrix} Z_1 \\ Z_{21}' \\ Z_{22}' \end{bmatrix}$, from $\begin{bmatrix} Z_2 \\ 0 \end{bmatrix} = P^{-1} \begin{bmatrix} Z_{21}' \\ Z_{22}' \end{bmatrix}$ calculate Z_2.

Step 7 Finally, obtain the refined resolvent decomposition

$$A^{-1}(s) = C(sI_n - J)^{-1}Z_1 + C_\infty(sJ_\infty - I_\mu)^{-1}Z_2.$$

Example 9.4.1. Let

$$A(s) = \begin{bmatrix} 1 & s^3 & 0 \\ 0 & 1 & s \\ 0 & 0 & s^2 \end{bmatrix} := A_0 + A_1 s + A_2 s^2 + A_3 s^3,$$

with

$$A_0 = \begin{bmatrix} 1 & 0 & 0 \\ 0 & 1 & 0 \\ 0 & 0 & 0 \end{bmatrix}, \quad A_1 = \begin{bmatrix} 0 & 0 & 0 \\ 0 & 0 & 1 \\ 0 & 0 & 0 \end{bmatrix},$$

$$A_2 = \begin{bmatrix} 0 & 0 & 0 \\ 0 & 0 & 0 \\ 0 & 0 & 1 \end{bmatrix}, \quad A_3 = \begin{bmatrix} 0 & 1 & 0 \\ 0 & 0 & 0 \\ 0 & 0 & 0 \end{bmatrix},$$

$r = 3$, $n = \deg \det(A(s)) = 2$. The finite Jordan pairs of $A(s)$ is (C, J):

$$C = \begin{bmatrix} 0 & 0 \\ 0 & -1 \\ 1 & 1 \end{bmatrix}, \quad J = \begin{bmatrix} 0 & 1 \\ 0 & 0 \end{bmatrix}.$$

The Smith-McMillan form of $A(s)$ at $s = \infty$ is

$$S_{A(s)}^\infty(s) = \text{diag}[s^3, s^2, 1/s^3],$$

so $A(s)$ has two poles at $s = \infty$ with orders $q_1 = 3$ and $q_2 = 2$ respectively and one zero at $s = \infty$ with orders $\hat{q}_3 = 3$. The dual polynomial matrix to $A(s)$ is

$$D_{A(s)}(w) = w^{q_1} A\left(\frac{1}{w}\right) = \begin{bmatrix} w^3 & 1 & 0 \\ 0 & w^3 & w^2 \\ 0 & 0 & w \end{bmatrix} \in R[w]^{3\times 3}$$

and

$$\tilde{U}_L(w) D_{A(s)}(w) \tilde{U}_R(w) = S^0_{D_{A(s)}(w)} = w^{q_1} S^\infty_{A(s)}\left(\frac{1}{w}\right) = \begin{bmatrix} 1 & 0 & 0 \\ 0 & w & 0 \\ 0 & 0 & w^6 \end{bmatrix}$$

is the local Smith form of $D_{As}(w)$ at $w = 0$, where

$$\tilde{U}_L(w) = \begin{bmatrix} 1 & 0 & 0 \\ 0 & 0 & 1 \\ -w^3 & 1 & -w \end{bmatrix}, \quad \tilde{U}_R(w) \begin{bmatrix} 0 & 0 & -1 \\ 1 & 0 & w^3 \\ 0 & 1 & 0 \end{bmatrix}.$$

One easily obtains the *generalized infinite Jordan pair*(C'_∞, J'_∞) (in the sense of Gohberg et al. [17]) of $A(s)$ as following

$$C'_\infty = \left[\begin{array}{c|cccc|cc} 0 & -1 & 0 & 0 & 0 & 0 & 0 \\ 0 & 0 & 0 & 0 & 1 & 0 & 0 \\ 1 & 0 & 0 & 0 & 0 & 0 & 0 \end{array} \right], \tag{9.14}$$

$$J'_\infty = \left[\begin{array}{c|cccccc} 0 & 0 & 0 & 0 & 0 & 0 & 0 \\ \hline 0 & 0 & 1 & 0 & 0 & 0 & 0 \\ 0 & 0 & 0 & 1 & 0 & 0 & 0 \\ 0 & 0 & 0 & 0 & 1 & 0 & 0 \\ 0 & 0 & 0 & 0 & 0 & 1 & 0 \\ 0 & 0 & 0 & 0 & 0 & 0 & 1 \\ 0 & 0 & 0 & 0 & 0 & 0 & 0 \end{array} \right], \tag{9.15}$$

and the *infinite Jordan pair*(C_∞, J_∞) (in the sense of Vardulakis [13]) of $A(s)$ as

$$C_\infty = \begin{bmatrix} -1 & 0 & 0 & 0 \\ 0 & 0 & 0 & 1 \\ 0 & 0 & 0 & 0 \end{bmatrix}, \quad J_\infty = \begin{bmatrix} 0 & 1 & 0 & 0 \\ 0 & 0 & 1 & 0 \\ 0 & 0 & 0 & 1 \\ 0 & 0 & 0 & 0 \end{bmatrix}.$$

It is easy to find P,

$$P = \begin{bmatrix} 0 & 0 & 0 & 0 & 1 & 0 & 0 \\ 1 & 0 & 0 & 0 & 0 & 0 & 0 \\ 0 & 1 & 0 & 0 & 0 & 0 & 0 \\ 0 & 0 & 1 & 0 & 0 & 0 & 0 \\ 0 & 0 & 0 & 1 & 0 & 0 & 0 \\ 0 & 0 & 0 & 0 & 0 & 1 & 0 \\ 0 & 0 & 0 & 0 & 0 & 0 & 1 \end{bmatrix},$$

such that

$$\underbrace{\begin{bmatrix} 0 & -1 & 0 & 0 & 0 & 0 & 0 \\ 0 & 0 & 0 & 0 & 1 & 0 & 0 \\ 1 & 0 & 0 & 0 & 0 & 0 & 0 \end{bmatrix}}_{C'_\infty} \underbrace{\begin{bmatrix} 0 & 0 & 0 & 0 & 1 & 0 & 0 \\ 1 & 0 & 0 & 0 & 0 & 0 & 0 \\ 0 & 1 & 0 & 0 & 0 & 0 & 0 \\ 0 & 0 & 1 & 0 & 0 & 0 & 0 \\ 0 & 0 & 0 & 1 & 0 & 0 & 0 \\ 0 & 0 & 0 & 0 & 0 & 1 & 0 \\ 0 & 0 & 0 & 0 & 0 & 0 & 1 \end{bmatrix}}_{P}$$

$$= \begin{bmatrix} -1 & 0 & 0 & 0 & 0 & 0 & 0 \\ 0 & 0 & 0 & 1 & 0 & 0 & 0 \\ 0 & 0 & 0 & 0 & 1 & 0 & 0 \end{bmatrix}$$

$$\underbrace{\phantom{\begin{matrix}-1&0&0&0\end{matrix}}}_{\widetilde{C}_\infty} \quad \underbrace{\phantom{\begin{matrix}0&0&0\end{matrix}}}_{\widetilde{C}_0}$$

$$\underbrace{\begin{bmatrix} 0 & 1 & 0 & 0 & 0 & 0 & 0 \\ 0 & 0 & 1 & 0 & 0 & 0 & 0 \\ 0 & 0 & 0 & 1 & 0 & 0 & 0 \\ 0 & 0 & 0 & 0 & 1 & 0 & 0 \\ 1 & 0 & 0 & 0 & 0 & 0 & 0 \\ 0 & 0 & 0 & 0 & 0 & 1 & 0 \\ 0 & 0 & 0 & 0 & 0 & 0 & 1 \end{bmatrix}}_{P^{-1}} \underbrace{\begin{bmatrix} 0 & 0 & 0 & 0 & 0 & 0 & 0 \\ 0 & 0 & 1 & 0 & 0 & 0 & 0 \\ 0 & 0 & 0 & 1 & 0 & 0 & 0 \\ 0 & 0 & 0 & 0 & 1 & 0 & 0 \\ 0 & 0 & 0 & 0 & 0 & 1 & 0 \\ 0 & 0 & 0 & 0 & 0 & 0 & 1 \\ 0 & 0 & 0 & 0 & 0 & 0 & 0 \end{bmatrix}}_{J'_\infty} \underbrace{\begin{bmatrix} 0 & 0 & 0 & 0 & 1 & 0 & 0 \\ 1 & 0 & 0 & 0 & 0 & 0 & 0 \\ 0 & 1 & 0 & 0 & 0 & 0 & 0 \\ 0 & 0 & 1 & 0 & 0 & 0 & 0 \\ 0 & 0 & 0 & 1 & 0 & 0 & 0 \\ 0 & 0 & 0 & 0 & 0 & 1 & 0 \\ 0 & 0 & 0 & 0 & 0 & 0 & 1 \end{bmatrix}}_{P}$$

$$= \begin{bmatrix} 0 & 1 & 0 & 0 & 0 & 0 & 0 \\ 0 & 0 & 1 & 0 & 0 & 0 & 0 \\ 0 & 0 & 0 & 1 & 0 & 0 & 0 \\ 0 & 0 & 0 & 0 & 0 & 1 & 0 \\ \hline 0 & 0 & 0 & 0 & 0 & 0 & 0 \\ 0 & 0 & 0 & 0 & 0 & 0 & 1 \\ 0 & 0 & 0 & 0 & 0 & 0 & 0 \end{bmatrix} := \begin{bmatrix} J_\infty & J' \\ \hline 0 & J_0 \end{bmatrix}.$$

According to Lemma 9.3.2, one has

$$V = [A_3 CJ^2, \ -(A_0 C'_\infty (J'_\infty)^2 + A_1 C'_\infty J'_\infty + A_2 C'_\infty)]$$

$$= \begin{bmatrix} 0 & 0 & 0 & 0 & 0 & 1 & 0 & 0 & 0 \\ 0 & 0 & 0 & 0 & 0 & 0 & 0 & 0 & -1 \\ 0 & 0 & -1 & 0 & 0 & 0 & 0 & 0 & 0 \end{bmatrix},$$

$$S = \begin{bmatrix} C & C'_\infty J'_\infty \\ CJ & C'_\infty \end{bmatrix} = \begin{bmatrix} 0 & 0 & 0 & 0 & -1 & 0 & 0 & 0 & 0 \\ 0 & -1 & 0 & 0 & 0 & 0 & 0 & 1 & 0 \\ 1 & 1 & 0 & 0 & 0 & 0 & 0 & 0 & 0 \\ 0 & 0 & 0 & -1 & 0 & 0 & 0 & 0 & 0 \\ 0 & 0 & 0 & 0 & 0 & 0 & 1 & 0 & 0 \\ 0 & 1 & 1 & 0 & 0 & 0 & 0 & 0 & 0 \end{bmatrix},$$

so

$$
Z' = \begin{bmatrix} Z_1 \\ Z'_2 \end{bmatrix} = \begin{bmatrix} I_n & 0 \\ \hline 0 & (J'_\infty)^{q_1-1} \end{bmatrix} \begin{bmatrix} S \\ V \end{bmatrix}^{-1} [0,\dots,0,I]^T
$$

$$
= \begin{bmatrix} 1 & 0 & 0 & 0 & 0 & 0 & 0 & 0 & 0 \\ 0 & 1 & 0 & 0 & 0 & 0 & 0 & 0 & 0 \\ \hline 0 & 0 & 0 & 0 & 0 & 0 & 0 & 0 & 0 \\ 0 & 0 & 0 & 0 & 0 & 1 & 0 & 0 & 0 \\ 0 & 0 & 0 & 0 & 0 & 0 & 1 & 0 & 0 \\ 0 & 0 & 0 & 0 & 0 & 0 & 0 & 1 & 0 \\ 0 & 0 & 0 & 0 & 0 & 0 & 0 & 0 & 1 \\ 0 & 0 & 0 & 0 & 0 & 0 & 0 & 0 & 0 \\ 0 & 0 & 0 & 0 & 0 & 0 & 0 & 0 & 0 \end{bmatrix} \begin{bmatrix} 0 & 0 & 0 & 0 & -1 & 0 & 0 & 0 & 0 \\ 0 & -1 & 0 & 0 & 0 & 0 & 0 & 1 & 0 \\ 1 & 1 & 0 & 0 & 0 & 0 & 0 & 0 & 0 \\ 0 & 0 & 0 & -1 & 0 & 0 & 0 & 0 & 0 \\ 0 & 0 & 0 & 0 & 0 & 0 & 1 & 0 & 0 \\ 0 & 1 & 1 & 0 & 0 & 0 & 0 & 0 & 0 \\ \hline 0 & 0 & 0 & 0 & 0 & 1 & 0 & 0 & 0 \\ 0 & 0 & 0 & 0 & 0 & 0 & 0 & 0 & -1 \\ 0 & 0 & -1 & 0 & 0 & 0 & 0 & 0 & 0 \end{bmatrix}^{-1}
$$

$$
\times \begin{bmatrix} 0 & 0 & 0 \\ 0 & 0 & 0 \\ 0 & 0 & 0 \\ \hline 0 & 0 & 0 \\ 0 & 0 & 0 \\ 0 & 0 & 0 \\ \hline 1 & 0 & 0 \\ 0 & 1 & 0 \\ 0 & 0 & 1 \end{bmatrix} = \begin{bmatrix} 0 & 0 & -1 \\ 0 & 0 & 1 \\ 0 & 0 & 0 \\ \hline 1 & 0 & 0 \\ 0 & 0 & 0 \\ 0 & 0 & 1 \\ \hline 0 & -1 & 0 \\ 0 & 0 & 0 \\ 0 & 0 & 0 \end{bmatrix}. \tag{9.16}
$$

We obtain

$$
Z_1 = \begin{bmatrix} 0 & 0 & -1 \\ 0 & 0 & 1 \end{bmatrix}, \quad Z'_2 = \begin{bmatrix} Z'_{21} \\ Z'_{22} \end{bmatrix} = \begin{bmatrix} 0 & 0 & 0 \\ 1 & 0 & 0 \\ 0 & 0 & 0 \\ 0 & 0 & 1 \\ \hline 0 & -1 & 0 \\ 0 & 0 & 0 \\ 0 & 0 & 0 \end{bmatrix},
$$

and from

$$
P^{-1}Z'_2 = \begin{bmatrix} 0 & 1 & 0 & 0 & 0 & 0 & 0 \\ 0 & 0 & 1 & 0 & 0 & 0 & 0 \\ 0 & 0 & 0 & 1 & 0 & 0 & 0 \\ 0 & 0 & 0 & 0 & 1 & 0 & 0 \\ 1 & 0 & 0 & 0 & 0 & 0 & 0 \\ 0 & 0 & 0 & 0 & 0 & 1 & 0 \\ 0 & 0 & 0 & 0 & 0 & 0 & 1 \end{bmatrix} \begin{bmatrix} 0 & 0 & 0 \\ 1 & 0 & 0 \\ 0 & 0 & 0 \\ 0 & 0 & 1 \\ 0 & -1 & 0 \\ 0 & 0 & 0 \\ 0 & 0 & 0 \end{bmatrix} = \begin{bmatrix} 1 & 0 & 0 \\ 0 & 0 & 0 \\ 0 & 0 & 1 \\ 0 & -1 & 0 \\ 0 & 0 & 0 \\ 0 & 0 & 0 \\ 0 & 0 & 0 \end{bmatrix},
$$

we obtain

$$
Z_2 = \begin{bmatrix} 1 & 0 & 0 \\ 0 & 0 & 0 \\ 0 & 0 & 1 \\ 0 & -1 & 0 \end{bmatrix}.
$$

Hence, by Theorem 9.3.1 we have constructed a resolvent form for the regular polynomial matrix $A(s)$ as following

$$A^{-1}(s) = [C, C_\infty] \left[\begin{array}{c|c} I_n s - J & 0 \\ \hline 0 & J_\infty s - I_\mu \end{array} \right]^{-1} \left[\begin{array}{c} Z_1 \\ Z_2 \end{array} \right]$$

$$= \left[\begin{array}{ccc|ccc} 0 & 0 & -1 & 0 & 0 & 0 \\ 0 & -1 & 0 & 0 & 0 & 1 \\ 1 & 1 & 0 & 0 & 0 & 0 \end{array} \right] \left[\begin{array}{cc|cccc} s & -1 & 0 & 0 & 0 & 0 \\ 0 & s & 0 & 0 & 0 & 0 \\ \hline 0 & 0 & -1 & s & 0 & 0 \\ 0 & 0 & 0 & -1 & s & 0 \\ 0 & 0 & 0 & 0 & -1 & s \\ 0 & 0 & 0 & 0 & 0 & -1 \end{array} \right]^{-1} \left[\begin{array}{ccc} 0 & 0 & -1 \\ 0 & 0 & 1 \\ 1 & 0 & 0 \\ 0 & 0 & 0 \\ 0 & 0 & 1 \\ 0 & -1 & 0 \end{array} \right].$$

Note that in the above refined resolvent decomposition the matrices C_∞, J_∞ and Z_2 have the minimal dimensions of 3×4, 4×4, and 4×3. Compared to the resolvent decomposition of Gehberg et al. [17] given by

$$A^{-1}(s) = [C, C'_\infty] \left[\begin{array}{c|c} I_n s - J & 0 \\ \hline 0 & J'_\infty s - I_\nu \end{array} \right]^{-1} \left[\begin{array}{c} Z_1 \\ Z'_2 \end{array} \right],$$

where $C'_\infty \in \mathfrak{R}^{3 \times 7}$, $J'_\infty \in \mathfrak{R}^{7 \times 7}$, and $Z'_2 \in \mathfrak{R}^{7 \times 3}$ are given by Eqs. (9.14), (9.15), (9.16), respectively, due to the fact that the redundant information has been deleted, the above refined resolvent decomposition is obviously in a much more concise form which will definitely bring some convenience when applied to the solution of the regular PMDs.

Example 9.4.2. Consider the following regular polynomial matrix

$$A(s) = \sum_{i=0}^{7} A_i s^i$$

$$= \left[-\frac{1}{4}s^7 + \frac{5}{8}s^6 - \frac{3}{8}s^5 - \frac{1}{4}s^4 + \frac{3}{8}s^2 - \frac{1}{8}s - 1, s^2 - \frac{1}{2}s, \right.$$

$$\left. \frac{1}{4}s^6 - \frac{3}{8}s^5 + \frac{1}{4}s^3 - \frac{3}{4}s^2 - \frac{1}{8}s + \frac{1}{4}, -\frac{1}{4}s^5 + \frac{3}{8}s^4 - \frac{1}{4}s^2 + \frac{1}{4}s - \frac{1}{8} \right]$$

$$\left[\frac{1}{4}s^2 - \frac{1}{2}s + \frac{1}{2}s^3 + \frac{1}{4}s^6 - \frac{1}{2}s^5, -2s - 1, \right.$$

$$\left. -\frac{1}{4}s^5 + \frac{1}{4}s^4 + \frac{1}{4}s^3 - \frac{1}{4}s^2 + \frac{3}{2}s + \frac{3}{2}, \frac{1}{4}s^4 - \frac{1}{4}s^3 - \frac{1}{4}s^2 + \frac{1}{4}s \right]$$

$$[0, s + 1, -s - 1, 0]$$

$$\left[-\frac{1}{2}s^5 + s^4 - s^2 + \frac{1}{2}s, 0, \frac{1}{2}s^4 - \frac{1}{2}s^3 - \frac{1}{2}s^2 + \frac{1}{2}s, \right.$$

$$\left. -\frac{1}{2}s^3 + \frac{1}{2}s^2 + \frac{1}{2}s - \frac{1}{2} \right],$$

$r = 4, q_1 = 7, n = \deg \det(A(s)) = 5$. The finite Jordan pairs of $A(s)$ is (C, J):

$$
C = \begin{bmatrix} 0 & 0 & 0 & 0 & 0 \\ 1 & 0 & 0 & 0 & 0 \\ 1 & 0 & 0 & 1 & 0 \\ 1 & 1 & 0 & 1 & 1 \end{bmatrix}, \quad
J = \begin{bmatrix} 1 & 0 & 0 & 0 & 0 \\ 0 & 1 & 1 & 0 & 0 \\ 0 & 0 & 1 & 0 & 0 \\ 0 & 0 & 0 & -1 & 0 \\ 0 & 0 & 0 & 0 & -1 \end{bmatrix}.
$$

The dual polynomial matrix to $A(s)$ is

$$
D_{A(s)}(w) = w^{q_1} A\left(\frac{1}{w}\right)
$$

$$
= \left[-\frac{1}{4} + \frac{5}{8}w - \frac{3}{8}w^2 - \frac{1}{4}w^3 + \frac{3}{8}w^5 - \frac{1}{8}w^6 - w^7, w^5 - \frac{1}{2}w^6, \right.
$$

$$
\frac{1}{4}w - \frac{3}{8}w^2 + \frac{1}{4}w^4 - \frac{3}{4}w^5 - \frac{1}{8}w^6 + \frac{1}{4}w^7,
$$

$$
\left. -\frac{1}{4}w^2 + \frac{3}{8}w^3 - \frac{1}{4}w^5 + \frac{1}{4}w^6 - \frac{1}{8}w^7 \right]
$$

$$
\left[\frac{1}{4}w^5 - \frac{1}{2}w^6 + \frac{1}{2}w^4 + \frac{1}{4}w - \frac{1}{2}w^2, -2w^6 - w^7, -\frac{1}{4}w^2 + \frac{1}{4}w^3 \right.
$$

$$
\left. +\frac{1}{4}w^4 - \frac{1}{4}w^5 + \frac{3}{2}w^6 + \frac{3}{2}w^7, \frac{1}{4}w^3 - \frac{1}{4}w^4 - \frac{1}{4}w^5 + \frac{1}{4}w^6 \right]
$$

$$
\left[0, w^6 + w^7, -w^6 - w^7, 0 \right]
$$

$$
\left[-\frac{1}{2}w^2 + w^3 - w^5 + \frac{1}{2}w^6, 0, \frac{1}{2}w^3 - \frac{1}{2}w^4 - \frac{1}{2}w^5 + \frac{1}{2}w^6, \right.
$$

$$
\left. -\frac{1}{2}w^4 + \frac{1}{2}w^5 + \frac{1}{2}w^6 - \frac{1}{2}w^7 \right]
$$

and

$$
\tilde{U}_L(w)\tilde{D}_{A(s)}(w)\tilde{U}_R(w) = S^0_{D_{A(s)}}(w) = w^{q_1} S^\infty_{A(s)}\left(\frac{1}{w}\right) = \text{diag}\{1, w^6, w^8, w^9\}
$$

is the local Smith form of $D_{A(s)}(w)$ at $w = 0$. The Smith-McMillan form of $A(s)$ at $s = \infty$ is

$$
S^\infty_{A(s)}(s) = s^{q_1} D_{A(s)}\left(\frac{1}{s}\right) = \text{diag}\left\{s^7, s, \frac{1}{s}, \frac{1}{s^2}\right\}
$$

so $A(s)$ has two poles at $s = \infty$ with orders $q_1 = 7$ and $q_2 = 1$ respectively and two zeros at $s = \infty$ with orders $\hat{q}_3 = 1$ and $\hat{q}_4 = 2$ respectively

$$
v = rq_1 - (q_1 + q_2) + (\hat{q}_3 + \hat{q}_4) = 23
$$
$$
\mu = (\hat{q}_3 + 1) + (\hat{q}_4 + 1) = 5.
$$

Use the local Smith form of $D_{A(s)}$ at $w = 0$ to find the generalized infinite Jordan pair (C'_∞, J'_∞) (in the sense of Gohberg et al. [17]) of $A(s)$:

$$
C'_\infty = \left[0, 2, 0, 0, 0, 1, 0, 2, 0, 0, 0, 0, -2, \frac{-449}{32}, 0, 1, -1, -1, -1, -1, -1, 1, \frac{385}{32} \right.
$$
$$
\left[\frac{9}{4}, \frac{-3}{2}, \frac{-385}{128}, \frac{705}{32}, \frac{-3605}{64}, \frac{1485}{16}, 2, -1, \frac{-385}{128}, \frac{689}{32}, \frac{-3621}{64}, \frac{1493}{16}, \right.
$$
$$
\frac{-8623}{128}, \frac{-1215}{32}, 1, -1, 0, \frac{417}{128}, \frac{-3173}{128}, \frac{8683}{128}, \frac{-9731}{128}, \frac{-1089}{64}, \frac{5075}{64} \right]
$$
$$
[2, -1, 0, 0, 0, 0, 2, -1, 0, 0, 0, 0, 0, 0, 1, -1, 0, 0, 0, 0, 0, 0, 0]
$$
$$
[1, 0, 0, 0, 0, 0, 1, 0, 0, 0, 0, 0, 0, 0, 1, 0, 0, 0, 0, 0, 0, 0, 0]
$$

(9.17)

$$
J'_\infty = [0, 1, 0, 0, 0, 0, 0, 0, 0, 0, 0, 0, 0, 0, 0, 0, 0, 0, 0, 0, 0, 0, 0]
$$
$$
[0, 0, 1, 0, 0, 0, 0, 0, 0, 0, 0, 0, 0, 0, 0, 0, 0, 0, 0, 0, 0, 0, 0]
$$
$$
[0, 0, 0, 1, 0, 0, 0, 0, 0, 0, 0, 0, 0, 0, 0, 0, 0, 0, 0, 0, 0, 0, 0]
$$
$$
[0, 0, 0, 0, 1, 0, 0, 0, 0, 0, 0, 0, 0, 0, 0, 0, 0, 0, 0, 0, 0, 0, 0]
$$
$$
[0, 0, 0, 0, 0, 1, 0, 0, 0, 0, 0, 0, 0, 0, 0, 0, 0, 0, 0, 0, 0, 0, 0]
$$
$$
[0, 0, 0, 0, 0, 0, 0, 0, 0, 0, 0, 0, 0, 0, 0, 0, 0, 0, 0, 0, 0, 0, 0]
$$
$$
[0, 0, 0, 0, 0, 0, 0, 1, 0, 0, 0, 0, 0, 0, 0, 0, 0, 0, 0, 0, 0, 0, 0]
$$
$$
[0, 0, 0, 0, 0, 0, 0, 0, 1, 0, 0, 0, 0, 0, 0, 0, 0, 0, 0, 0, 0, 0, 0]
$$
$$
[0, 0, 0, 0, 0, 0, 0, 0, 0, 1, 0, 0, 0, 0, 0, 0, 0, 0, 0, 0, 0, 0, 0]
$$
$$
[0, 0, 0, 0, 0, 0, 0, 0, 0, 0, 1, 0, 0, 0, 0, 0, 0, 0, 0, 0, 0, 0, 0]
$$
$$
[0, 0, 0, 0, 0, 0, 0, 0, 0, 0, 0, 1, 0, 0, 0, 0, 0, 0, 0, 0, 0, 0, 0]
$$
$$
[0, 0, 0, 0, 0, 0, 0, 0, 0, 0, 0, 0, 1, 0, 0, 0, 0, 0, 0, 0, 0, 0, 0]
$$
$$
[0, 0, 0, 0, 0, 0, 0, 0, 0, 0, 0, 0, 0, 1, 0, 0, 0, 0, 0, 0, 0, 0, 0]
$$

(9.18)

$$
[0, 0, 0, 0, 0, 0, 0, 0, 0, 0, 0, 0, 0, 0, 1, 0, 0, 0, 0, 0, 0, 0, 0]
$$
$$
[0, 0, 0, 0, 0, 0, 0, 0, 0, 0, 0, 0, 0, 0, 0, 1, 0, 0, 0, 0, 0, 0, 0]
$$
$$
[0, 0, 0, 0, 0, 0, 0, 0, 0, 0, 0, 0, 0, 0, 0, 0, 1, 0, 0, 0, 0, 0, 0]
$$
$$
[0, 0, 0, 0, 0, 0, 0, 0, 0, 0, 0, 0, 0, 0, 0, 0, 0, 1, 0, 0, 0, 0, 0]
$$
$$
[0, 0, 0, 0, 0, 0, 0, 0, 0, 0, 0, 0, 0, 0, 0, 0, 0, 0, 1, 0, 0, 0, 0]
$$
$$
[0, 0, 0, 0, 0, 0, 0, 0, 0, 0, 0, 0, 0, 0, 0, 0, 0, 0, 0, 1, 0, 0, 0]
$$
$$
[0, 0, 0, 0, 0, 0, 0, 0, 0, 0, 0, 0, 0, 0, 0, 0, 0, 0, 0, 0, 1, 0, 0]
$$
$$
[0, 0, 0, 0, 0, 0, 0, 0, 0, 0, 0, 0, 0, 0, 0, 0, 0, 0, 0, 0, 0, 1, 0]
$$
$$
[0, 0, 0, 0, 0, 0, 0, 0, 0, 0, 0, 0, 0, 0, 0, 0, 0, 0, 0, 0, 0, 0, 1]
$$
$$
[0, 0, 0, 0, 0, 0, 0, 0, 0, 0, 0, 0, 0, 0, 0, 0, 0, 0, 0, 0, 0, 0, 0]
$$

and the infinite Jordan pair (C_∞, J_∞) (in the sense of Vardulakis [13]) of $A(s)$

$$C_\infty = \begin{bmatrix} 0 & 2 & 0 & 1 & -1 \\ 2 & -1 & 1 & -1 & 0 \\ 2 & -1 & 1 & -1 & 0 \\ 1 & 0 & 1 & 0 & 0 \end{bmatrix}, \quad J_\infty = \begin{bmatrix} 0 & 1 & 0 & 0 & 0 \\ 0 & 0 & 0 & 0 & 0 \\ 0 & 0 & 0 & 1 & 0 \\ 0 & 0 & 0 & 0 & 1 \\ 0 & 0 & 0 & 0 & 0 \end{bmatrix}.$$

It should be noted that the generalized infinite Jordan pair (C'_∞, J'_∞), which has dimensions of 4×23 and 23×23, is much unnecessarily larger than the infinite Jordan pair (C_∞, J_∞) with dimensions of only 4×5 and 5×5. The following matrix $P \in \Re^{23 \times 23}$ is found to delete the redundant information inherited in the generalized infinite Jordan pair.

$$
\begin{aligned}
P = \; &[0,0,0,0,0,0,1,0,0,0,0,0,0,0,0,0,0,0,0,0,0,0,0] \\
&[0,0,0,0,0,0,0,1,0,0,0,0,0,0,0,0,0,0,0,0,0,0,0] \\
&[0,0,0,0,0,0,0,0,0,0,0,0,0,0,1,0,0,0,0,0,0,0,0] \\
&[0,0,0,0,0,0,0,0,0,0,0,0,0,0,0,1,0,0,0,0,0,0,0] \\
&[0,0,0,0,0,0,0,0,0,0,0,0,0,0,0,0,1,0,0,0,0,0,0] \\
&[0,0,0,0,0,1,0,0,0,0,0,0,0,0,0,0,0,0,0,0,0,0,0] \\
&[1,0,0,0,0,0,0,0,0,0,0,0,0,0,0,0,0,0,0,0,0,0,0] \\
&[0,1,0,0,0,0,0,0,0,0,0,0,0,0,0,0,0,0,0,0,0,0,0] \\
&[0,0,0,0,0,0,0,0,1,0,0,0,0,0,0,0,0,0,0,0,0,0,0] \\
&[0,0,0,0,0,0,0,0,0,1,0,0,0,0,0,0,0,0,0,0,0,0,0] \\
&[0,0,0,0,0,0,0,0,0,0,1,0,0,0,0,0,0,0,0,0,0,0,0] \\
&[0,0,0,0,0,0,0,0,0,0,0,1,0,0,0,0,0,0,0,0,0,0,0] \\
&[0,0,0,0,0,0,0,0,0,0,0,0,1,0,0,0,0,0,0,0,0,0,0] \\
&[0,0,0,0,0,0,0,0,0,0,0,0,0,1,0,0,0,0,0,0,0,0,0] \\
&[0,0,1,0,0,0,0,0,0,0,0,0,0,0,0,0,0,0,0,0,0,0,0] \\
&[0,0,0,1,0,0,0,0,0,0,0,0,0,0,0,0,0,0,0,0,0,0,0] \\
&[0,0,0,0,1,0,0,0,0,0,0,0,0,0,0,0,0,0,0,0,0,0,0] \\
&[0,0,0,0,0,0,0,0,0,0,0,0,0,0,0,0,0,1,0,0,0,0,0] \\
&[0,0,0,0,0,0,0,0,0,0,0,0,0,0,0,0,0,0,1,0,0,0,0] \\
&[0,0,0,0,0,0,0,0,0,0,0,0,0,0,0,0,0,0,0,1,0,0,0] \\
&[0,0,0,0,0,0,0,0,0,0,0,0,0,0,0,0,0,0,0,0,1,0,0] \\
&[0,0,0,0,0,0,0,0,0,0,0,0,0,0,0,0,0,0,0,0,0,1,0] \\
&[0,0,0,0,0,0,0,0,0,0,0,0,0,0,0,0,0,0,0,0,0,0,1]
\end{aligned}
$$

satisfies Theorem 9.2.1. According to Lemma 9.3.2, one has

$$V = [A_6 C J^6, -(A_0 C'_\infty (J'_\infty)^6 + \cdots + A_1 C'_\infty J'_\infty + A_2 C'_\infty)]$$
$$= \left[0,0,0,0,0,0, \frac{3}{4}, 0, 0, \frac{-1}{4}, 1, 0, \frac{3}{4}, 0, 0, 0, \frac{1}{2}, \frac{321}{128}, \frac{-625}{32}, \right.$$
$$\left. 0, \frac{1}{2}, \frac{-1}{4}, \frac{-1}{4}, \frac{-1}{4}, \frac{-1}{4}, \frac{-3}{4}, \frac{-321}{128}, \frac{3013}{128} \right]$$
$$\left[0,0,0,0,0,0, \frac{1}{2}, 0, 0, 0, \frac{1}{2}, 0, \frac{1}{2}, 0, 0, 0, 0, \frac{-1}{2}, \frac{-417}{64}, 0, \right.$$
$$\left. \frac{1}{4}, \frac{-1}{4}, \frac{-1}{4}, \frac{-1}{4}, \frac{-1}{4}, \frac{-1}{4}, \frac{1}{4}, \frac{465}{64} \right]$$
$$\left[0,0,0,0,0,0,0,0,0,0,0, \frac{-1}{4}, 0, 0, 0, 0, 0, 0, 0, 0, \frac{385}{128}, 0, \right.$$
$$\left. 0, 0, 0, 0, 0, 0, 0, 0, \frac{-417}{128} \right]$$
$$[0,0,0,0,0,0,0,0,0,0,0,0,0,0,0,0,0,0,0,-1,0,0,$$
$$0,0,0,0,0,0,1]$$

$$S = \begin{bmatrix} C & C'_\infty (J'_\infty)^5 \\ \vdots & \vdots \\ C J^5 & C'_\infty \end{bmatrix}$$
$$= [0,0,0,0,0,0,0,0,0,0,0,0,0,0,0,0,0,2,0,0,0,$$
$$0,0,0,0,1,-1,-1]$$
$$\left[1,0,0,0,0,0,0,0,0,0,0, \frac{9}{4}, 0, 0, 0, 0, 0, 2, -1, \frac{-385}{128}, 0, \right.$$
$$\left. 0, 0, 0, 0, 1, -1, 0, \frac{417}{128} \right]$$
$$[1,0,0,1,0,0,0,0,0,0,2,0,0,0,0,0,2,-1,0,0,0,$$
$$0,0,0,1,-1,0,0]$$
$$[1,1,0,1,1,0,0,0,0,0,1,0,0,0,0,0,1,0,0,0,0,$$
$$0,0,0,1,0,0,0]$$
$$[0,0,0,0,0,0,0,0,0,0,2,0,0,0,0,0,2,0,0,0,0,$$
$$0,0,0,1,-1,-1,-1]$$
$$\left[1,0,0,0,0,0,0,0,0,0, \frac{9}{4}, \frac{-3}{2}, 0, 0, 0, 0, 2, -1, \frac{-385}{128}, \frac{689}{32}, \right.$$
$$\left. 0, 0, 0, 0, 1, -1, 0, \frac{417}{128}, \frac{-3173}{128} \right]$$
$$[1,0,0,-1,0,0,0,0,0,2,-10,0,0,0,2,-1,0,0,0,$$
$$0,0,0,1,-1,0,0,0]$$

$[1, 1, 1, -1, -1, 0, 0, 0, 0, 1, 0, 0, 0, 0, 0, 1, 0, 0, 0, 0, 0,$
$0, 0, 1, 0, 0, 0, 0]$

$[0, 0, 0, 0, 0, 0, 0, 0, 0, 2, 0, 0, 0, 0, 0, 2, 0, 0, 0, 0, 0,$
$0, 0, 1, -1, -1, -1, -1]$

$$\left[1, 0, 0, 0, 0, 0, 0, 0, \frac{9}{4}, \frac{-3}{2}, \frac{-385}{128}, 0, 0, 0, 2, -1, \frac{-385}{128}, \frac{689}{32}, \right.$$
$$\left. \frac{-3621}{64}, 0, 0, 0, 1, -1, 0, \frac{417}{128}, \frac{-3173}{128}, \frac{8683}{128} \right]$$

$[1, 0, 0, 1, 0, 0, 0, 0, 2, -1, 0, 0, 0, 0, 2, -1, 0, 0, 0, 0, 0,$
$0, 1, -1, 0, 0, 0, 0]$

$[1, 1, 2, 1, 1, 0, 0, 0, 1, 0, 0, 0, 0, 0, 1, 0, 0, 0, 0, 0, 0,$
$0, 1, 0, 0, 0, 0, 0]$

$[0, 0, 0, 0, 0, 0, 0, 0, 2, 0, 0, 0, 0, 0, 2, 0, 0, 0, 0, 0, 0,$
$0, 1, -1, -1, -1, -1, -1]$

$$\left[1, 0, 0, 0, 0, 0, 0, 0, \frac{9}{4}, \frac{-3}{2}, \frac{-385}{128}, \frac{705}{32}, 0, 0, 2, -1, \frac{-385}{128}, \frac{689}{32}, \right.$$
$$\left. \frac{-3621}{64}, \frac{1494}{16}, 0, 0, 1, -1, 0, \frac{417}{128}, \frac{-3173}{128}, \frac{8683}{128}, \frac{-9731}{128} \right]$$

$[1, 0, 0, -1, 0, 0, 0, 2, -1, 0, 0, 0, 0, 2, -1, 0, 0, 0, 0, 0,$
$0, 1, -1, 0, 0, 0, 0, 0]$

$[1, 1, 3, -1, -1, 0, 0, 1, 0, 0, 0, 0, 0, 1, 0, 0, 0, 0, 0, 0, 0,$
$1, 0, 0, 0, 0, 0, 0]$

$[0, 0, 0, 0, 0, 0, 0, 2, 0, 0, 0, 0, 0, 2, 0, 0, 0, 0, -2, 0, 0,$
$1, -1, -1, -1, -1, -1, 1]$

$$\left[1, 0, 0, 0, 0, 0, \frac{9}{4}, \frac{-3}{2}, \frac{-385}{128}, \frac{705}{32}, \frac{-3605}{64}, 0, 2, -1, \frac{-385}{128}, \right.$$
$$\frac{689}{32}, \frac{-3621}{64}, \frac{1493}{16}, \frac{-8623}{128}, 0, 1, -1, 0, \frac{417}{128}, \frac{-3173}{128}, \frac{8683}{128},$$
$$\left. \frac{-9731}{128}, \frac{-1089}{64} \right]$$

$[1, 0, 0, 1, 0, 0, 2, -1, 0, 0, 0, 0, 2, -1, 0, 0, 0, 0, 0, 0, 1,$
$-1, 0, 0, 0, 0, 0, 0]$

$[1, 1, 4, 1, 1, 0, 1, 0, 0, 0, 0, 0, 1, 0, 0, 0, 0, 0, 0, 0, 1,$
$0, 0, 0, 0, 0, 0, 0]$

$$\left[0, 0, 0, 0, 0, 0, 2, 0, 0, 0, 1, 0, 2, 0, 0, 0, 0, -2, \frac{-449}{32}, 0, \right.$$

$$1, -1, -1, -1, -1, -1, 1, \frac{385}{32}\Bigg]$$

$$\Bigg[1, 0, 0, 0, 0, \frac{9}{4}, \frac{-3}{2}, \frac{-385}{128}, \frac{705}{32}, \frac{-3605}{64}, \frac{1485}{16}, 2, -1, \frac{-385}{128},$$

$$\frac{689}{32}, \frac{-3621}{64}, \frac{1493}{16}, \frac{-8623}{128}, \frac{-1215}{32}, 1, -1, 0, \frac{417}{128}, \frac{-3173}{128}, \frac{8683}{128},$$

$$\frac{-9731}{128}, \frac{-1089}{64}, \frac{5075}{64}\Bigg]$$

$$[1, 0, 0, -1, 0, 2, -1, 0, 0, 0, 0, 2, -1, 0, 0, 0, 0, 0, 0, 1,$$
$$-1, 0, 0, 0, 0, 0, 0, 0]$$

$$[1, 1, 5, -1, -1, 1, 0, 0, 0, 0, 0, 1, 0, 0, 0, 0, 0, 0, 0, 1, 0,$$
$$0, 0, 0, 0, 0, 0, 0].$$

so from

$$Z' = \begin{bmatrix} Z_1 \\ Z_2' \end{bmatrix} = \begin{bmatrix} I_n & 0 \\ 0 & (J'_\infty)^{q_1-1} \end{bmatrix} \begin{bmatrix} S \\ V \end{bmatrix}^{-1} [0, \ldots, 0, I]^T$$

$$= \begin{bmatrix}
0 & -2 & -3 & -1 \\
0 & 0 & 0 & \frac{1}{2} \\
0 & 0 & 0 & -1 \\
0 & 0 & -1 & 0 \\
0 & 0 & 2 & \frac{-1}{2} \\
0 & 0 & 0 & 0 \\
0 & 0 & 0 & 0 \\
0 & 0 & 0 & 0 \\
0 & 0 & 0 & 0 \\
0 & 0 & 0 & 0 \\
0 & 0 & 0 & 0 \\
1 & \frac{-5}{2} & \frac{-9}{2} & \frac{3}{4} \\
0 & 1 & 1 & -1 \\
0 & 0 & 0 & 0 \\
0 & 0 & 0 & 0 \\
0 & 0 & 0 & 0 \\
0 & 0 & 0 & 0 \\
0 & 0 & 0 & 0 \\
0 & 0 & 0 & 0 \\
-1 & \frac{9}{2} & \frac{17}{2} & \frac{1}{4} \\
1 & \frac{-3}{2} & \frac{-3}{2} & \frac{7}{4} \\
0 & 1 & 1 & 0 \\
0 & 0 & 0 & 0 \\
0 & 0 & 0 & 0 \\
0 & 0 & 0 & 0 \\
0 & 0 & 0 & 0 \\
0 & 0 & 0 & 0 \\
0 & 0 & 0 & 0
\end{bmatrix}.$$

(9.19)

and from

$$
P^{-1}Z_2' = \begin{bmatrix}
1 & \frac{-5}{2} & \frac{-9}{2} & \frac{3}{4} \\
0 & 1 & 1 & -1 \\
-1 & \frac{9}{2} & \frac{17}{2} & \frac{1}{4} \\
1 & \frac{-3}{2} & \frac{-3}{2} & \frac{7}{4} \\
0 & 1 & 1 & 0 \\
0 & 0 & 0 & 0 \\
0 & 0 & 0 & 0 \\
0 & 0 & 0 & 0 \\
0 & 0 & 0 & 0 \\
0 & 0 & 0 & 0 \\
0 & 0 & 0 & 0 \\
0 & 0 & 0 & 0 \\
0 & 0 & 0 & 0 \\
0 & 0 & 0 & 0 \\
0 & 0 & 0 & 0 \\
0 & 0 & 0 & 0 \\
0 & 0 & 0 & 0 \\
0 & 0 & 0 & 0 \\
0 & 0 & 0 & 0 \\
0 & 0 & 0 & 0 \\
0 & 0 & 0 & 0 \\
0 & 0 & 0 & 0 \\
0 & 0 & 0 & 0 \\
0 & 0 & 0 & 0 \\
0 & 0 & 0 & 0
\end{bmatrix}
$$

one obtains

$$
Z_2 = \begin{bmatrix}
1 & \frac{-5}{2} & \frac{-9}{2} & \frac{3}{4} \\
0 & 1 & 1 & -1 \\
-1 & \frac{9}{2} & \frac{17}{2} & \frac{1}{4} \\
1 & \frac{-3}{2} & \frac{-3}{2} & \frac{7}{4} \\
0 & 1 & 1 & 0
\end{bmatrix}.
$$

Hence, by Theorem 9.3.1 we have constructed the refined resolvent decomposition for the regular polynomial matrix $A(s)$ as follows

$$
A^{-1}(s) = [C, C_\infty] \left[\begin{array}{c|c} I_n s - J & 0 \\ \hline 0 & J_\infty s - I_\mu \end{array} \right]^{-1} \begin{bmatrix} Z_1 \\ Z_2 \end{bmatrix}
$$

$$
= \left[\begin{array}{ccccc|ccccc}
0 & 0 & 0 & 0 & 0 & 0 & 2 & 0 & 1 & -1 \\
1 & 0 & 0 & 0 & 0 & 2 & -1 & 1 & -1 & 0 \\
1 & 0 & 0 & 1 & 0 & 2 & -1 & 1 & -1 & 0 \\
1 & 1 & 0 & 1 & 1 & 1 & 0 & 1 & 0 & 0
\end{array} \right]
$$

$$
\begin{bmatrix}
s-1 & 0 & 0 & 0 & 0 & 0 & 0 & 0 & 0 & 0 \\
0 & s-1 & -1 & 0 & 0 & 0 & 0 & 0 & 0 & 0 \\
0 & 0 & s-1 & 0 & 0 & 0 & 0 & 0 & 0 & 0 \\
0 & 0 & 0 & s+1 & 0 & 0 & 0 & 0 & 0 & 0 \\
0 & 0 & 0 & 0 & s+1 & 0 & 0 & 0 & 0 & 0 \\
\hline
0 & 0 & 0 & 0 & 0 & -1 & s & 0 & 0 & 0 \\
0 & 0 & 0 & 0 & 0 & 0 & -1 & 0 & 0 & 0 \\
0 & 0 & 0 & 0 & 0 & 0 & 0 & -1 & s & 0 \\
0 & 0 & 0 & 0 & 0 & 0 & 0 & 0 & -1 & s \\
0 & 0 & 0 & 0 & 0 & 0 & 0 & 0 & 0 & -1
\end{bmatrix}^{-1}
$$

$$
\times
\begin{bmatrix}
0 & -2 & -3 & -1 \\
0 & 0 & 0 & \frac{1}{2} \\
0 & 0 & 0 & -1 \\
0 & 0 & -1 & 0 \\
0 & 0 & 2 & \frac{-1}{2} \\
\hline
1 & \frac{-5}{2} & \frac{-9}{2} & \frac{3}{4} \\
0 & 1 & 1 & -1 \\
-1 & \frac{9}{2} & \frac{17}{2} & \frac{1}{4} \\
1 & \frac{-3}{2} & \frac{-3}{2} & \frac{7}{4} \\
0 & 1 & 1 & 0
\end{bmatrix}.
$$

Based on Algorithm 9.4.1, a procedure in the symbolic computational package Maple has been developed and has been implemented to carry out the above calculations. Note that in the above refined resolvent decomposition the matrices C_∞, J_∞, and Z_2 have the minimal dimensions of 4×5, 5×5, and 5×4. Compared to the resolvent decomposition of Gohberg et al. [17] given by

$$
A^{-1}(s) = [C, C'_\infty]
\begin{bmatrix}
I_n s - J & 0 \\
0 & J'_\infty s - I_v
\end{bmatrix}^{-1}
\begin{bmatrix}
Z_1 \\
Z'_2
\end{bmatrix},
$$

where $C'_\infty \in \Re^{4 \times 23}$, $J'_\infty \in \Re^{23 \times 23}$, and $Z'_2 \in \Re^{23 \times 4}$ are given by Eqs. (9.17), (9.18), (9.19), respectively, due to the fact that the redundant information has been deleted, the above refined resolvent decomposition is obviously in a much more concise form, which will definitely bring some convenience when applied to the solution of the regular PMDs.

9.5 Conclusions

So far there have been two special resolvent decompositions proposed in the literature through which the solution of a PMD may be expressed. These are based on two different interpretations of the notion of infinite Jordan pair, the first being due to Gohberg et al. [17], and the second due to Vardulakis [13]. The resolvent decomposition proposed by Gohberg et al. [17] uses a certain redundant system structure that results in overly large dimensions of the infinite Jordan pair, though it is relatively simple to calculate the infinite Jordan pair. On the other hand, the approach proposed by Vardulakis [13] uses only the relevant system structure, without using redundant information, and the resulting infinite Jordan pair is of minimal dimensions. It is, however, relatively more difficult to compute the required special realizations.

In this chapter, it is established that the redundant information contained in the infinite Jordan pair defined by Gohberg et al. [17] can be deleted through a certain transformation. Based on this, a natural connection between the infinite Jordan pairs defined by Gohberg et al. [17] and that of Vardulakis [13] has been exploited. This facilitates a refinement of the resolvent decomposition. This resulting resolvent decomposition more precisely reflects the relevant system structure and thereby inherits the advantages of both the decompositions of Gohberg et al. [17] and Vardulakis [13].

In the proposed approach the matrices Z, Z_∞ in Eq. (9.3) are formulated explicitly, which means this method is constructive. The main idea in this proposed approach is to calculate an elementary matrix P, which is very easy to obtain, to delete the redundant information, then to propose the refined resolvent decomposition. This elementary matrix has the effect of deleting the redundant information in two ways. First, it deletes the redundant information in those blocks in the infinite Jordan pair of Gohberg et al. [17] that correspond to the infinite zeros and bring them into the correct sizes. Second, it deletes the whole blocks in the infinite Jordan pair of Gohberg et al. [17] that correspond to the infinite poles and the whole blocks that are not dynamically important. This elementary matrix serves to transform the partitioned block matrix in Z' that corresponds to the redundant information into zero, the resulting refined resolvent decomposition is thus of minimal dimensions. Further, by using this elementary matrix the mechanism of decoupling in the solution of Gohberg et al. [17] is explained clearly. This refined resolvent decomposition facilitates computation of the inverse matrix of $A(s)$ due to the fact that the dimensions of the matrices used are minimal. Once the refined resolvent decomposition is obtained, the generalized infinite Jordan pair and the elementary matrix P are no longer needed in the calculation of the solution of the regular PMD. This therefore presents another merit to this method, which is algorithmically attractive when applied in actual computation.

Based on this refined resolvent decomposition, the complete solution of regular PMDs has then been investigated. This solution presents the zero input response and the zero state response precisely and takes into account the impulsive properties of the system. An algorithm, which has already been implemented in the symbolic computational package Maple, for the investigation of this refined resolvent decomposition is provided.

Compared with the complete solution of regular PMDs given in Chapter 8, where the solution is proposed based on any resolvent decomposition, the resolvent decomposition obtained in this chapter is minimal, the built solution is thus specific to this refined resolvent decomposition.

Frequency structures of generalized companion form and application to the solution of regular PMDs

10

10.1 Introduction

Consider a regular *polynomial matrix description* (PMD) described by

$$A(\rho)\beta(t) = B(\rho)u(t), \quad t \geq 0 \tag{10.1}$$

where $\rho := d/dt$ is the differential operator, $A(\rho) = A_q\rho^q + A_{q-1}\rho^{q-1} + \cdots + A_1\rho + A_0 \in R[\rho]^{r\times r}$, $\mathrm{rank}_{R[\rho]}A(\rho) = r$, $A_i \in R^{r\times r}$, $i = 0, 1, 2, \ldots, q$, $q \geq 1$, $B(\rho) = B_l\rho^l + B_{l-1}\rho^{l-1} + \cdots + B_1\rho + B_0 \in R[\rho]^{r\times m}$, $B_j \in R^{r\times m}$, $j = 0, 1, 2, \ldots, l$, $l \geq 0$, $\beta(t) : [0, +\infty) \to R^r$ is the *pseudo-state* of the PMDs, $u(t) : [0, +\infty) \to R^m$ is a p times piecewise continuously differentiable function called the *input* of the PMD.

A special case of Eq. (10.1) is the so-called *generalized state space system* (GSSS) in the following form

$$(\rho E - A)x(t) = Bu(t), \quad t \geq 0. \tag{10.2}$$

The solution of the above GSSS is well-known from the work of Gantmacher [4], Verghese [109], Campbell [7], Yip and Sincovec [14], and Cobb [10]. Most of these results are based on Gantmacher's [4] analysis of the canonical form of the *matrix pencil sE − A* called the *Weierstrass canonical form*. The convenience in actual computation can easily be seen, for only certain constant matrix transformations are necessary to bring the matrix pencil $sE - A$ to its Weierstrass canonical form, and from this the solution of the GSSS can be formulated.

More recently, many authors such as Gohberg et al. [17], Vardulakis and Fragulis [110], Vardulakis [13], and Fragulis [12] have discussed the solution of the regular PMD (10.1). These solutions are all based on the *resolvent decomposition* [13, 17] of the regular polynomial matrix $A(s)$. On the one hand, such treatments have the immediate advantage that they separate the system behavior into the slow (smooth) response and the fast (impulsive) response, which may provide a deep insight into the system structure. On the other hand, such treatments bring some inconvenience for the actual computation since the classical methods of determining the finite Jordan pairs or the infinite Jordan pairs have to transform the polynomial matrix $A(s)$ into its Smith-McMillan form or its Smith-McMillan form at infinity. However, such transformations

A Generalized Framework of Linear Multivariable Control. http://dx.doi.org/10.1016/B978-0-08-101946-7.00010-X

are not stable in numerical computation terms, for they result in an extraordinarily large number of polynomial manipulations.

For simplicity, it is usual to assume that the initial conditions of the pseudo-state $\beta(t)$ and the input $u(t)$ are all zero. However, in certain real cases, the initial conditions of the state might result from a random disturbance entering the system, and a feedback controller is thus called for. Since the precise value of the initial condition of the state is unpredictable and the control is likely to depend on those initial conditions of the state, it is somewhat stronger than necessary to assume that all the initial conditions of the state and the control are zero.

Since in Eq. (10.1) $B(\rho)$ contains the differential operator, not only the nonzero initial conditions of $\beta(t)$ but also the nonzero initial conditions of $u(t)$ should be considered, for both of them can create the impulsive modes as well as the slow response to the system.

In this chapter, a novel approach is presented to determine the complete solution of regular PMDs, which takes into account not only the initial conditions on $\beta(t)$ but also the initial conditions on $u(t)$. One kind of linearization [17] of the regular polynomial matrix $A(s)$ is the so-called *generalized companion matrix*, which is in fact a regular matrix pencil. Recently in [30], a realization approach was suggested that reduced high-order linear differential equations to the first-order system representations by using such a generalized companion matrix. However, the complete structures of this companion matrix were not discussed in [30]. It is known that the Weierstrass canonical form of this matrix pencil can easily be obtained by certain constant matrix transformations. In this chapter certain properties of this companion form are established, and a special *resolvent decomposition* of $A(s)$ is proposed that is based on the Weierstrass canonical form. The solution of a regular PMD is then formulated from this resolvent decomposition. An obvious advantage of the approach adopted here is that it immediately avoids the polynomial matrix transformation necessary to obtain the finite and infinite Jordan pairs of $A(s)$, and only requires the constant matrix transformation to obtain the Weierstrass canonical form of the generalized companion form, which is less sensitive than the former in computational terms. Since numerically efficient algorithms to generate the canonical form of a matrix pencil are well developed (see, e.g., [19, 20]), the formula proposed here is more attractive in computational terms than the previously known results.

10.2   The frequency structures of generalized companion form and a new resolvent decomposition

Let $A(s) = \sum_{i=0}^{q} A_i s^i$ be a regular $r \times r$ polynomial matrix. The generalized companion polynomial matrix $C_A(s)$ is defined as follows

$$
C_A(s) := \begin{bmatrix} I_r & 0 & \cdots & 0 & 0 \\ 0 & I_r & \cdots & 0 & 0 \\ \vdots & \vdots & \ddots & \vdots & \vdots \\ 0 & 0 & \cdots & I_r & 0 \\ 0 & 0 & \cdots & 0 & A_q \end{bmatrix} s + \begin{bmatrix} 0 & -I_r & 0 & \cdots & 0 \\ 0 & 0 & -I_r & \cdots & 0 \\ \vdots & \vdots & \vdots & \ddots & \vdots \\ 0 & 0 & 0 & \cdots & -I_r \\ A_0 & A_1 & A_2 & \cdots & A_{q-1} \end{bmatrix} \in R^{qr \times qr},
$$

where $I_r \in R^{r \times r}$ denotes the identity matrix. $C_A(s)$ is in fact a linearization of $A(s)$ for it satisfies

$$E(s)C_A(s) = \begin{bmatrix} A(s) & 0_{r \times (q-1)r} \\ 0_{(q-1)r \times r} & I_{(q-1)r \times (q-1)r} \end{bmatrix} F(s), \tag{10.3}$$

where

$$E(s) := \begin{bmatrix} A_1 + sA_2 + \cdots + s^{q-1}A_q & A_2 + sA_3 + \cdots + s^{q-2}A_q & \cdots & A_{q-1} + sA_q & I_r \\ -I_r & 0 & \cdots & 0 & 0 \\ 0 & -I_r & \cdots & 0 & 0 \\ \vdots & \vdots & \ddots & \vdots & \vdots \\ 0 & 0 & \cdots & -I_r & 0 \end{bmatrix} \in R^{qr \times qr},$$

$$F(s) := \begin{bmatrix} I_r & 0 & 0 & \cdots & 0 & 0 \\ -sI_r & I_r & 0 & \cdots & 0 & 0 \\ 0 & -sI_r & I_r & \cdots & 0 & 0 \\ \vdots & \vdots & \vdots & \ddots & \vdots & \vdots \\ 0 & 0 & 0 & \cdots & -sI_r & I_r \end{bmatrix} \in R^{qr \times qr}.$$

$C_A(s)$ is indeed a matrix pencil with a specially simple form. Such linearization originally appeared in [17], and was later used by [30]. Viewing the fact that $\det E(s) = 1$ and $\det F(s) = 1$, from Eq. (10.3) we deduce that $A(s)$ is regular as a polynomial matrix if and only if $C_A(s)$ is regular as a matrix pencil.

Proposition 10.2.1 (The Weierstrass canonical form of $C_A(s)$, [4]). *For the regular matrix pencil $C_A(s)$, there exist nonsingular constant matrices M and N such that*

$$MC_A(s)N = \begin{bmatrix} sI_n - J & 0 \\ 0 & sJ_\infty - I_v \end{bmatrix} := P(s), \tag{10.4}$$

where J is in Jordan canonical form with the finite zeros [5] of $C_A(s), J_\infty :=$ block $diag[J_{\infty 1}, J_{\infty 2}, \ldots, J_{\infty k}], J_{\infty i}$ ($i = 1, \ldots, k$) are Jordan matrices with the size of $q_i \times q_i, q_1 \leq q_2 \leq \cdots \leq q_k, v = \sum_{i=1}^{k} q_i$, and J_∞ is nilpotent with nil-potency q_k.

It is noted that the matrix $sI_n - J$ contains the *finite elementary divisors* of $C_A(s)$, and the matrix $sJ_\infty - I_v$ contains the *infinite elementary divisors* [18] of $C_A(s)$. It is also noted in [111] that the matrix pencil $C_A(s)$ has the same finite elementary divisors as $A(s)$. For the relationship between the finite frequency structure of $A(s)$ and that of its companion matrix $C_A(s)$, the following result is obvious.

Proposition 10.2.2. *The Smith-McMillan form of a matrix $H(s)$ is denoted by $S_H^C(s)$. We have*

$$S_{C_A}^C(s) = \begin{bmatrix} I_{(q-1)r} & \\ & S_A^C(s) \end{bmatrix} = \begin{bmatrix} I_v & \\ & S_{(sI_n-J)}^C(s) \end{bmatrix}.$$

The above proposition details the relationship between the finite frequency structure of $A(s)$ and that of its companion form $C_A(s)$, which is self-evident. The relationship between the infinite frequency structure of $A(s)$ and that of $C_A(s)$ is rather more obscure, as the transformation used in Proposition 10.2.1 together with the transformation of linearization do not preserve the infinite frequency structures of $C_A(s)$.

Consequently there exists a significant difference between $S_A^\infty(s)$ and $S_{C_A}^\infty(s)$ as noted by [20]. The following important original result, on which the contribution of this chapter is built, explains this difference.

Theorem 10.2.1. *The Smith-McMillan form at infinity of a regular polynomial matrix $A(s)$ is $S_A^\infty(s) = \text{diag}\{s^q, s^{q-q_1}, \ldots, s^{q-q_k}\}$ if and only if the Jordan block matrices $J_{\infty i}(i = 1, \ldots, k)$ in the Weierstrass canonical form of its generalized companion matrix $C_A(s)$ are of sizes $q_i \times q_i$.*

Proof. If $A(s) = \sum_{i=0}^q A_i s^i$, its *dual* is defined [13, 18] as $D_A(w) = w^q \sum_{i=0}^q A_i \frac{1}{w^i}$. The infinite frequency structure of $A(s)$ is displayed by the finite frequency structure of its dual $D_A(\frac{1}{w})$ at $w = 0$ [13]. That is

$$S_A^\infty(s) = s^q S_{D_A}^0\left(\frac{1}{s}\right),$$

where $S_{D_A}^0(w)$ denotes the *local* Smith form at $w = 0$ of the dual of $A(s)$. Thus

$$S_A^\infty(s) = \text{diag}\{s^q, s^{q-q_1}, \ldots, s^{q-q_k}\}$$

if and only if

$$S_{D_A}^0(w) = \text{diag}\{1, w^{q_1}, \ldots, w^{q_k}\}.$$

From Eqs. (10.3), (10.4), we have

$$ME^{-1}(s)\begin{bmatrix} A(s) & 0_{r\times(q-1)r} \\ 0_{(q-1)r\times r} & I_{(q-1)r\times(q-1)r} \end{bmatrix}F(s)N = \begin{bmatrix} sI_n - J & 0 \\ 0 & sJ_\infty - I_v \end{bmatrix}. \quad (10.5)$$

Because

$$E(s) = s^{q-1}D_E\left(\frac{1}{s}\right),$$

so

$$E^{-1}(s) = \frac{1}{s^{q-1}}D_E^{-1}\left(\frac{1}{s}\right),$$

and

$$A(s) = s^q D_A\left(\frac{1}{s}\right),$$

substituting these into Eq. (10.5), we obtain

$$M\frac{1}{s^{q-1}}D_E^{-1}\left(\frac{1}{s}\right)\begin{bmatrix} s^q D_A(\frac{1}{s}) & 0_{r\times(q-1)r} \\ 0_{(q-1)r\times r} & I_{(q-1)r\times(q-1)r} \end{bmatrix}F(s)N = s\begin{bmatrix} I_n - \frac{1}{s}J & 0 \\ 0 & J_\infty - \frac{1}{s}I_v \end{bmatrix},$$

or alternatively since $w = \frac{1}{s}$

$$MD_E^{-1}(w)\begin{bmatrix} D_A(w) & 0_{r\times(q-1)r} \\ 0_{(q-1)r\times r} & w^q I_{(q-1)r\times(q-1)r} \end{bmatrix}F\left(\frac{1}{w}\right)N = \begin{bmatrix} I_n - wJ & 0 \\ 0 & J_\infty - wI_v \end{bmatrix}.$$

Now we denote

$$Q(w) := D_E^{-1}(w) \begin{bmatrix} D_A(w) & 0_{r \times (q-1)r} \\ 0_{(q-1)r \times r} & w^q I_{(q-1)r \times (q-1)r} \end{bmatrix} F\left(\frac{1}{w}\right),$$

and

$$Q'(w) := \begin{bmatrix} I_n - wJ & 0 \\ 0 & J_\infty - wI_v \end{bmatrix}.$$

Since M and N are both constant nonsingular matrices, we have

$$S_{Q(w)}^0(w) = S_{Q'(w)}^0(w).$$

By direct calculations, we find that

$D_E(w)$

$$= \begin{bmatrix} A_1 w^{q-1} + A_2 w^{q-2} + \cdots + A_q & A_2 w^{q-1} + A_3 w^{q-2} + \cdots + A_q w & \cdots & A_{q-1} w^{q-1} + A_q w^{q-2} & w^{q-1} I_r \\ -w^{q-1} I_r & 0 & \cdots & 0 & 0 \\ 0 & -w^{q-1} I_r & \cdots & 0 & 0 \\ \vdots & \vdots & \ddots & \vdots & \vdots \\ 0 & 0 & \cdots & -w^{q-1} I_r & 0 \end{bmatrix}.$$

$D_E^{-1}(w)$

$$= \begin{bmatrix} 0 & -\frac{1}{w^{q-1}} I_r & 0 & \cdots & 0 \\ 0 & 0 & -\frac{1}{w^{q-1}} I_r & \cdots & 0 \\ 0 & 0 & 0 & \cdots & 0 \\ \vdots & \vdots & \vdots & \ddots & \vdots \\ 0 & 0 & 0 & \cdots & -\frac{1}{w^{q-1}} I_r \\ \frac{1}{w^{q-1}} I_r & \frac{A_1 w^{q-1} + A_2 w^{q-2} + \cdots + A_q}{w^{2(q-1)}} & \frac{A_2 w^{q-1} + A_3 w^{q-2} + \cdots + A_q w}{w^{2(q-1)}} & \cdots & \frac{A_{q-1} w^{q-1} + A_q w^{q-2}}{w^{2(q-1)}} \end{bmatrix}.$$

Now

$$F\left(\frac{1}{w}\right) = \begin{bmatrix} I_r & 0 & 0 & \cdots & 0 & 0 \\ -\frac{1}{w} I_r & I_r & 0 & \cdots & 0 & 0 \\ 0 & -\frac{1}{w} I_r & I_r & \cdots & 0 & 0 \\ \vdots & \vdots & \vdots & \ddots & \vdots & \vdots \\ 0 & 0 & 0 & \cdots & -\frac{1}{w} I_r & I_r \end{bmatrix},$$

and so

$$Q(w) = D_E^{-1}(w) \begin{bmatrix} D_A(w) & 0 \\ 0 & w^q I_{(q-1)r} \end{bmatrix} F\left(\frac{1}{w}\right)$$

$$= \begin{bmatrix} 0 & -\frac{1}{w^{q-1}} I_r & 0 & \cdots & 0 \\ 0 & 0 & -\frac{1}{w^{q-1}} I_r & \cdots & 0 \\ 0 & 0 & 0 & \cdots & 0 \\ \vdots & \vdots & \vdots & \ddots & \vdots \\ 0 & 0 & 0 & \cdots & -\frac{1}{w^{q-1}} I_r \\ \frac{1}{w^{q-1}} I_r & \frac{A_1 w^{q-1} + A_2 w^{q-2} + \cdots + A_q}{w^{2(q-1)}} & \frac{A_2 w^{q-1} + A_3 w^{q-2} + \cdots + A_q w}{w^{2(q-1)}} & \cdots & \frac{A_{q-1} w^{q-1} + A_q w^{q-2}}{w^{2(q-1)}} \end{bmatrix}$$

$$\text{diag}\{D_A(w), w^q I_r, \ldots, w^q I_r, w^q I_r\} \begin{bmatrix} I_r & 0 & 0 & \cdots & 0 & 0 \\ -\frac{1}{w}I_r & I_r & 0 & \cdots & 0 & 0 \\ 0 & -\frac{1}{w}I_r & I_r & \cdots & 0 & 0 \\ \vdots & \vdots & \vdots & \ddots & \vdots & \vdots \\ 0 & 0 & 0 & \cdots & -\frac{1}{w}I_r & I_r \end{bmatrix}$$

$$= \begin{bmatrix} I_r & -wI_r & 0 & \cdots & 0 & 0 \\ 0 & I_r & -wI_r & \cdots & 0 & 0 \\ 0 & 0 & I_r & \cdots & 0 & 0 \\ \vdots & \vdots & \vdots & \ddots & \vdots & \vdots \\ 0 & 0 & 0 & \cdots & I_r & -wI_r \\ A_0 w & A_1 w & A_2 w & \cdots & A_{q-2}w & A_{q-1}w + A_q \end{bmatrix}$$

It is seen that by some unimodular transformations, $Q(w)$ can be brought to a diagonal block matrix

$$\text{diag}\{\underbrace{I_r, \ldots, I_r}_{q-1}, D_A(w)\},$$

subsequently

$$S^0_{Q(w)}(w) = \text{diag}\{1, \ldots, 1, S^0_{D_A}(w)\},$$

that is to say

$$\begin{bmatrix} I_{(q-1)r} & \\ & S^0_{D_A}(w) \end{bmatrix} = \begin{bmatrix} I_n & \\ & S^0_{(-wI_v + J_\infty)}(w) \end{bmatrix}.$$

Now by noticing that

$$S^0_{(-wI_v + J_\infty)}(w) = \text{diag}\{\underbrace{1, \ldots, 1}_{v-k}, w^{q_1}, \ldots, w^{q_k}\},$$

if and only if the associated Jordan block matrices $J_{\infty i}(i = 1, \ldots, k)$ are of sizes $q_i \times q_i$, we finally obtain the required result. $\qquad\qquad\square$

Remark 10.2.1. If in Theorem 10.2.1 it happens that $q_1 = q_2 = \cdots = q_h$ for some $h \le k$, then in the reconstruction of $S^\infty_A(s)$ from the Weierstrass canonical form of $C_A(s)$ it will be necessary to include h polynomials of the form s^{q_1}. If the dimension of $A(s)$ is known (which will usually be the case) then h is simply determined. In case the dimension of $A(s)$ is not known, this number h can be determined from the formula

$$\sharp\{\text{finite zeros}\} + \sharp\{\text{infinite zeros}\} + \sharp\{\text{infinite poles}\} = \dim C_A(s),$$

where $\sharp$ denotes the number of such poles or zeros counted according to the algebraic multiplicities alone (i.e., taking no account of the geometric multiplicities).

Theorem 10.2.1 relates the infinite frequency structure of a polynomial matrix with the associated Weierstrass canonical matrix structure in a very natural way. Thus together with Propositions 10.2.1 and 10.2.2 one can immediately, from the Weierstrass canonical form of the generalized companion matrix, give characterizations of the finite and infinite frequency structure of the polynomial matrix $A(s)$. These results can be considered as the generalizations of the associated properties from the matrix pencil case to the polynomial matrix case. The above results are also interesting from the numerical computation point of view. They suggest an alternative way to find the finite and infinite frequency structure of a regular polynomial matrix.

Since the finite and infinite frequency structure of a regular polynomial matrix are completely characterized by the Weierstrass canonical form of its companion matrix, the only thing we need to do is to transform the easily formed companion matrix (a regular matrix pencil) into its Weierstrass canonical form by some constant matrix transformation. Compared to the classical methods, which are by means of the polynomial matrix transformations, the method proposed here will be less sensitive to data perturbations and rounding errors.

From Gohberg et al. [17] and Vardulakis [13], we can see that the *resolvent decomposition* of the regular polynomial matrix $A(s)$ plays a key role in formulating the solution of the regular PMD. The above linearization of $A(s)$ enables us to establish the following resolvent decomposition, which is based on the Weierstrass canonical form of $C_A(s)$ and is different from that of Gohberg et al. [17] and Vardulakis [13].

Theorem 10.2.2 (A *resolvent decomposition of* $A(s)$). *Let* $A(s) = \sum_{i=0}^{q} A_i s^i \in R[s]^{r \times r}$ *be regular and its companion matrix be* $C_A(s)$. *If the constant matrices* $M \in R^{qr \times qr}$ *and* $N \in R^{qr \times qr}$ *are those that reduce* $C_A(s)$ *into its Weierstrass canonical form (6.4) and M and N are partitioned as follows:*

$$
M := \begin{bmatrix} M_{11} & M_{12} & \cdots & M_{1q} \\ M_{21} & M_{22} & \cdots & M_{2q} \\ \vdots & \vdots & \vdots & \vdots \\ M_{q1} & M_{q2} & \cdots & M_{qq} \end{bmatrix}, \quad \text{where } M_{ij} \in R^{r \times r}, i = 1, \ldots, q; \; j = 1, \ldots, q;
$$

$$
N := \begin{bmatrix} N_{11} & N_{12} & \cdots & N_{1q} \\ N_{21} & N_{22} & \cdots & N_{2q} \\ \vdots & \vdots & \vdots & \vdots \\ N_{q1} & N_{q2} & \cdots & N_{qq} \end{bmatrix}, \quad \text{where } N_{ij} \in R^{r \times r}, i = 1, \ldots, q; \; j = 1, \ldots, q,
$$

then

$$
A^{-1}(s) = [N_{11}, N_{12}, \ldots, N_{1q}] \begin{bmatrix} sI_n - J & 0 \\ 0 & sJ_\infty - I_v \end{bmatrix}^{-1} \begin{bmatrix} M_{1q} \\ M_{2q} \\ \vdots \\ M_{qq} \end{bmatrix}. \tag{10.6}
$$

Proof. Let $E_q(s) = A_q$, $E_{i-1}(s) = A_{i-1} + sE_i(s)$, $i = q, \ldots, 2$. We observe that in Eq. (10.3)

$$
E(s) = \begin{bmatrix} E_1(s) & E_2(s) & \cdots & E_{q-1}(s) & I_r \\ -I_r & 0 & \cdots & 0 & 0 \\ 0 & -I_r & \cdots & 0 & 0 \\ \vdots & \vdots & \ddots & \vdots & \vdots \\ 0 & 0 & \cdots & -I_r & 0 \end{bmatrix},
$$

so

$$
E^{-1}(s) = \begin{bmatrix} 0 & -I_r & 0 & \cdots & 0 \\ 0 & 0 & -I_r & \cdots & 0 \\ \vdots & \vdots & \vdots & \ddots & \vdots \\ 0 & 0 & 0 & \cdots & -I_r \\ I_r & E_1(s) & E_2(s) & \cdots & E_{q-1}(s) \end{bmatrix}.
$$

From Eqs. (10.3), (10.4), we have

$$
ME^{-1}(s) \begin{bmatrix} A(s) & 0_{r \times (q-1)r} \\ 0_{(q-1)r \times r} & 0_{(q-1)r \times (q-1)r} \end{bmatrix} F(s)N = \begin{bmatrix} sI_n - J & 0 \\ 0 & sJ_\infty - I_v \end{bmatrix},
$$

so

$$
\begin{bmatrix} A^{-1}(s) & 0 \\ 0 & I_{(q-1)r} \end{bmatrix} = F(s)N \begin{bmatrix} sI_n - J & 0 \\ 0 & sJ_\infty - I_v \end{bmatrix}^{-1} ME^{-1}(s). \tag{10.7}
$$

In fact

$$
F(s)N = \begin{bmatrix} N_{11} & N_{12} & N_{13} & \cdots & N_{1q} \\ \hline N_{21} - sN_{11} & N_{22} - sN_{12} & N_{23} - sN_{13} & \cdots & N_{2q} - sN_{1q} \\ \vdots & \vdots & \vdots & \vdots & \vdots \\ N_{q1} - sN_{(q-1)1} & N_{q2} - sN_{(q-1)2} & N_{q3} - sN_{(q-1)3} & \cdots & N_{qq} - sN_{(q-1)q} \end{bmatrix}
$$

$$
ME^{-1}(s) = \begin{bmatrix} M_{1q} & M_{1q}E_1(s) - M_{11} & M_{1q}E_2(s) - M_{12} & \cdots & M_{1q}E_{q-1}(s) - M_{1(q-1)} \\ M_{2q} & M_{2q}E_1(s) - M_{21} & M_{2q}E_2(s) - M_{22} & \cdots & M_{2q}E_{q-1}(s) - M_{2(q-1)} \\ \vdots & \vdots & \vdots & \vdots & \vdots \\ M_{qq} & M_{qq}E_1(s) - M_{q1} & M_{qq}E_2(s) - M_{q2} & \cdots & M_{qq}E_{q-1}(s) - M_{q(q-1)} \end{bmatrix}.
$$

According to the partition of the matrices in Eq. (10.7), we obtain the desired result. □

The difference between the above resolvent decomposition of $A(s)$ and that of Gohberg et al. [17] and Vardulakis [13] lies in the fact it is produced on the basis of the Weierstrass canonical form of the companion matrix $C_A(s)$ rather than on that of the finite and the infinite Jordan pairs of the original polynomial matrix. As stated previously, such treatment has an obvious advantage in computation. Although the dimension of $C_A(s)$ is larger than that of $A(s)$, $C_A(s)$ is in fact sparse and has the special structure that facilities the computation. Computation of the finite and infinite

Jordan pairs, however, is difficult. In the classical method, to find the finite and infinite Jordan matrices, one has to use unimodular matrix manipulations, numerical stability is thus lost because pivoting is based on the powers of the variable rather than on the coefficients as the method proposed here. This advantage in computation is well documented in, e.g., [1]. It should also be noted that the above proposed resolvent decomposition is not minimal due to the fact that

$$\sharp\{\text{finite zeros}\} + \sharp\{\text{infinite zeros}\} + \sharp\{\text{infinite poles}\} = \dim C_A(s),$$

where $\sharp$ denotes the number of such poles or zeros counted according to the algebraic multiplicities, some redundant information brought from the infinite poles and infinite zeros is contained in it. Further to get rid of this redundant information, one can use the method proposed in Chapter 9.

Example 10.2.1. Consider the regular polynomial matrix

$$A(s) = \sum_{i=0}^{3} A_i s^i = \begin{bmatrix} 1 & s^3 & 0 \\ 0 & 1 & s \\ 0 & 0 & s^2 \end{bmatrix},$$

with

$$A_0 = \begin{bmatrix} 1 & 0 & 0 \\ 0 & 1 & 0 \\ 0 & 0 & 0 \end{bmatrix}, \quad A_1 = \begin{bmatrix} 0 & 0 & 0 \\ 0 & 0 & 1 \\ 0 & 0 & 0 \end{bmatrix}, \quad A_2 = \begin{bmatrix} 0 & 0 & 0 \\ 0 & 0 & 0 \\ 0 & 0 & 1 \end{bmatrix}, \quad A_3 = \begin{bmatrix} 0 & 1 & 0 \\ 0 & 0 & 0 \\ 0 & 0 & 0 \end{bmatrix},$$

its companion matrix $C_A(s)$ is

$$C_A(s) = \begin{bmatrix} I_3 & 0 & 0 \\ 0 & I_3 & 0 \\ 0 & 0 & A_3 \end{bmatrix} s + \begin{bmatrix} 0 & -I_3 & 0 \\ 0 & 0 & -I_3 \\ A_0 & A_1 & A_2 \end{bmatrix}$$

$$= \begin{bmatrix} s & 0 & 0 & -1 & 0 & 0 & 0 & 0 & 0 \\ 0 & s & 0 & 0 & -1 & 0 & 0 & 0 & 0 \\ 0 & 0 & s & 0 & 0 & -1 & 0 & 0 & 0 \\ 0 & 0 & 0 & s & 0 & 0 & -1 & 0 & 0 \\ 0 & 0 & 0 & 0 & s & 0 & 0 & -1 & 0 \\ 0 & 0 & 0 & 0 & 0 & s & 0 & 0 & -1 \\ 1 & 0 & 0 & 0 & 0 & 0 & 0 & s & 0 \\ 0 & 1 & 0 & 0 & 0 & 1 & 0 & 0 & 0 \\ 0 & 0 & 0 & 0 & 0 & 0 & 0 & 0 & -1 \end{bmatrix}.$$

It is easy to obtain

$$S_A^C(s) = \text{diag}\{1, 1, s^2\}, \quad S_{C_A}^C(s) = \text{diag}\{1, 1, 1, 1, 1, 1, 1, 1, s^2\}.$$

$$S_A^\infty(s) = \text{diag}\left\{s^3, s^2, \frac{1}{s^3}\right\},$$

but

$$S_{C_A}^\infty(s) = \text{diag}\left\{s, s, s, s, s, s, s, 1, \frac{1}{s^5}\right\}.$$

By some simple elementary operations of the rows and the columns, $C_A(s)$ is brought to its Weierstrass canonical form

$$
P(s) = \left[\begin{array}{cc|ccccccc}
s & -1 & 0 & 0 & 0 & 0 & 0 & 0 & 0 \\
0 & s & 0 & 0 & 0 & 0 & 0 & 0 & 0 \\
\hline
0 & 0 & -1 & 0 & 0 & 0 & 0 & 0 & 0 \\
0 & 0 & 0 & -1 & s & 0 & 0 & 0 & 0 \\
0 & 0 & 0 & 0 & -1 & s & 0 & 0 & 0 \\
0 & 0 & 0 & 0 & 0 & -1 & s & 0 & 0 \\
0 & 0 & 0 & 0 & 0 & 0 & -1 & s & 0 \\
0 & 0 & 0 & 0 & 0 & 0 & 0 & -1 & s \\
0 & 0 & 0 & 0 & 0 & 0 & 0 & 0 & -1
\end{array}\right]
$$

The elementary operating matrices M and N are as follows

$$
M = \left[\begin{array}{cccccc|ccc}
0 & 0 & 1 & 0 & 0 & -1 & 0 & 0 & -1 \\
0 & 0 & 0 & 0 & 0 & 1 & 0 & 0 & 1 \\
0 & 0 & 0 & 0 & 0 & 0 & 0 & 0 & -1 \\
0 & 0 & 0 & -1 & 0 & 0 & 0 & 0 & 0 \\
-1 & 0 & 0 & 0 & 0 & 0 & 0 & 0 & 0 \\
0 & 0 & 0 & 0 & 0 & 0 & 1 & 0 & 0 \\
0 & 0 & 0 & 0 & 1 & 0 & 0 & 0 & 0 \\
0 & 1 & 0 & 0 & 0 & 1 & 0 & 0 & 1 \\
0 & 0 & 0 & 0 & 0 & 0 & 0 & -1 & 0
\end{array}\right],
$$

$$
N = \left[\begin{array}{ccc|cccccc}
0 & 0 & 0 & 0 & 0 & -1 & 0 & 0 & 0 \\
0 & -1 & 0 & 0 & 0 & 0 & 0 & 0 & 1 \\
1 & 1 & 0 & 0 & 0 & 0 & 0 & 0 & 0 \\
\hline
0 & 0 & 0 & 0 & -1 & 0 & 0 & 0 & 0 \\
0 & 0 & 0 & 0 & 0 & 0 & 0 & 1 & 0 \\
0 & 1 & 0 & 0 & 0 & 0 & 0 & 0 & 0 \\
0 & 0 & 0 & -1 & 0 & 0 & 0 & 0 & 0 \\
0 & 0 & 0 & 0 & 0 & 0 & 1 & 0 & 0 \\
0 & 0 & 1 & 0 & 0 & 0 & 0 & 0 & 0
\end{array}\right].
$$

Thus the resolvent decomposition is then

$$
A^{-1}(s) = \left[\begin{array}{ccccccccc}
0 & 0 & 0 & 0 & 0 & -1 & 0 & 0 & 0 \\
0 & -1 & 0 & 0 & 0 & 0 & 0 & 0 & 1 \\
1 & 1 & 0 & 0 & 0 & 0 & 0 & 0 & 0
\end{array}\right]
$$

$$\times \left[\begin{array}{cc|ccccccc} s & -1 & 0 & 0 & 0 & 0 & 0 & 0 & 0 \\ 0 & s & 0 & 0 & 0 & 0 & 0 & 0 & 0 \\ \hline 0 & 0 & -1 & 0 & 0 & 0 & 0 & 0 & 0 \\ 0 & 0 & 0 & -1 & s & 0 & 0 & 0 & 0 \\ 0 & 0 & 0 & 0 & -1 & s & 0 & 0 & 0 \\ 0 & 0 & 0 & 0 & 0 & -1 & s & 0 & 0 \\ 0 & 0 & 0 & 0 & 0 & 0 & -1 & s & 0 \\ 0 & 0 & 0 & 0 & 0 & 0 & 0 & -1 & s \\ 0 & 0 & 0 & 0 & 0 & 0 & 0 & 0 & -1 \end{array}\right]^{-1} \left[\begin{array}{ccc} 0 & 0 & -1 \\ 0 & 0 & 1 \\ 0 & 0 & -1 \\ 0 & 0 & 0 \\ 0 & 0 & 0 \\ 1 & 0 & 0 \\ 0 & 0 & 0 \\ 0 & 0 & 1 \\ 0 & -1 & 0 \end{array}\right].$$

which may be verified by direct calculation.

10.3 Application to the solution of regular PMDs

Consider the regular PMDs given by Eq. (10.1). Let the initial values of $u(t)$ and its $(l-1)$-derivatives at $t=0$ be $u(0)$, $u^{(1)}(0)\cdots u^{(l-1)}(0)$ and the initial values of $\beta(t)$ and its $(q-1)$-derivatives at $t=0$ be $\beta(0)$, $\beta^{(1)}(0)\cdots\beta^{(q-1)}(0)$. Taking the Laplace transformation of Eq. (10.1), we obtain

$$A(s)\hat{\beta}(s) - \hat{\alpha}_\beta(s) = B(s)\hat{u}(s) - \hat{\alpha}_u(s), \tag{10.8}$$

where $\hat{\beta}(s) := \int_0^{+\infty} \beta(t)e^{-st}\,dt$, $\hat{u}(s) := \int_0^{+\infty} u(t)e^{-st}\,dt$. Because of the Laplace-transformation rule

$$L\left\{\frac{d^i}{dt^i}\beta(t)\right\} = s^i\hat{\beta}(s) - s^{i-1}\beta(0) - \cdots - s\beta^{(i-2)}(0) - \beta^{(i-1)}(0); \quad i = 0, 1, \ldots$$

$$L\left\{\frac{d^j}{dt^j}u(t)\right\} = s^j\hat{u}(s) - s^{j-1}u(0) - \cdots - su^{(j-2)}(0) - u^{(j-1)}(0); \quad j = 0, 1, \ldots$$

$\hat{\alpha}_\beta(s), \hat{\alpha}_u(s)$ can be written [107] as follows

$$\hat{\alpha}_\beta(s) = [s^{q-1}I_r, s^{q-2}I_r, \ldots, sI_r, I_r] \begin{bmatrix} A_q & 0 & \cdots & 0 \\ A_{q-1} & A_q & \cdots & 0 \\ \vdots & \vdots & \ddots & \vdots \\ A_1 & A_2 & \cdots & A_q \end{bmatrix} \begin{bmatrix} \beta(0) \\ \beta^{(1)}(0) \\ \vdots \\ \beta^{(q-1)}(0) \end{bmatrix}$$

$$\hat{\alpha}_u(s) = [s^{l-1}I_r, s^{l-2}I_r, \ldots, sI_r, I_r] \begin{bmatrix} B_l & 0 & \cdots & 0 \\ B_{l-1} & B_l & \cdots & 0 \\ \vdots & \vdots & \ddots & \vdots \\ B_1 & B_2 & \cdots & B_l \end{bmatrix} \begin{bmatrix} u(0) \\ u^{(1)}(0) \\ \vdots \\ u^{(l-1)}(0) \end{bmatrix}.$$

Now we repartition

$$\hat{M} := \begin{bmatrix} M_{1q} \\ M_{2q} \\ \vdots \\ M_{qq} \end{bmatrix} := \begin{bmatrix} M_n \\ M_\infty \end{bmatrix}, \quad \text{where } M_n \in R^{n \times r}, \ M_\infty \in R^{v \times r},$$

and

$$\hat{N} := [N_{11}, N_{12}, \ldots, N_{1q}] := [N_n, N_\infty], \quad \text{where } N_n \in R^{r \times n}, \ N_\infty \in R^{r \times v}.$$

Note that

$$n + v = qr.$$

According to Theorem 10.2.2, we have

$$A^{-1}(s) = N_\infty(sJ_\infty - I_v)^{-1}M_\infty + N_n(sI_n - J)^{-1}M_n.$$

Similar to [13], we obtain

$$A^{-1}(s)\hat{\alpha}_\beta(s) = [N_n, N_\infty] \left[\begin{array}{c|c} sI_n - J & 0_{n,v} \\ \hline 0_{v,n} & sJ_\infty - I_v \end{array} \right]^{-1} \left[\begin{array}{c} x_{s\beta}(0) \\ J_\infty x_{f\beta}(0) \end{array} \right],$$

where

$$\left[\begin{array}{c} x_{s\beta}(0) \\ J_\infty x_{f\beta}(0) \end{array} \right] = \left[\begin{array}{c|c} J^{q-1}M_n, \ldots, M_n & 0_{n,qr} \\ \hline 0_{v,qr} & J_\infty M_\infty, \ldots, J_\infty^q M_\infty \end{array} \right]$$

$$\times \left[\begin{array}{ccc|ccc} A_q & 0 & \cdots & 0 \\ A_{q-1} & A_q & \cdots & 0 \\ \vdots & \vdots & \ddots & \vdots \\ A_1 & A_2 & \cdots & A_q \\ \hline A_0 & A_1 & \cdots & A_{q-1} \\ 0 & A_0 & \cdots & A_{q-2} \\ \vdots & \vdots & \ddots & \vdots \\ 0 & 0 & \cdots & A_0 \end{array} \right] \left[\begin{array}{c} \beta(0) \\ \beta^{(1)}(0) \\ \vdots \\ \beta^{(q-1)}(0) \end{array} \right].$$

Similarly,

$$A^{-1}(s)\hat{\alpha}_u(s) = [N_n, N_\infty] \left[\begin{array}{c|c} sI_n - J & 0_{n,v} \\ \hline 0_{v,n} & sJ_\infty - I_v \end{array} \right]^{-1} \left[\begin{array}{c} x_{su}(0) \\ J_\infty x_{fu}(0) \end{array} \right],$$

where

$$\left[\begin{array}{c} x_{su}(0) \\ J_\infty x_{fu}(0) \end{array} \right] = \left[\begin{array}{c|c} J^{l-1}M_n, \ldots, M_n & 0_{n,lr} \\ \hline 0_{v,lr} & J_\infty M_\infty, \ldots, J_\infty^l M_\infty \end{array} \right]$$

$$\times \begin{bmatrix} \begin{array}{ccccc} B_l & 0 & \cdots & 0 \\ B_{l-1} & B_l & \cdots & 0 \\ \vdots & \vdots & \ddots & \vdots \\ B_1 & B_2 & \cdots & B_l \\ \hline B_0 & B_1 & \cdots & B_{l-1} \\ 0 & B_0 & \cdots & B_{l-2} \\ \vdots & \vdots & \ddots & \vdots \\ 0 & 0 & \cdots & B_0 \end{array} \end{bmatrix} \begin{bmatrix} u(0) \\ u^{(1)}(0) \\ \vdots \\ u^{(l-1)}(0) \end{bmatrix}.$$

We also obtain

$$A^{-1}(s)B(s)\hat{u}(s) = [N_n, N_\infty]\Psi\Phi \begin{bmatrix} I_m \\ sI_m \\ \vdots \\ s^{q_k-q+l}I_m \end{bmatrix} \hat{u}(s) + N_n[sI_n - J]^{-1}\Omega\hat{u}(s),$$

where

$$\Psi := \left[\begin{array}{c|c} J^{l-1}M_n, J^{l-2}M_n, \ldots, M_n & 0_{n,(q_k-q+1)r} \\ \hline 0_{v,lr} & M_\infty, J_\infty M_\infty, \ldots, J_\infty^{q_k-q}M_\infty \end{array} \right]$$

$$\Phi := \begin{bmatrix} \begin{array}{cccccccc} B_l & 0 & \cdots & 0 & 0 & 0 & \cdots & 0 \\ B_{l-1} & B_l & \cdots & 0 & 0 & 0 & \cdots & 0 \\ \vdots & \vdots & \ddots & \vdots & \vdots & \vdots & \ddots & \vdots \\ B_1 & B_2 & \cdots & B_l & 0 & 0 & \cdots & 0 \\ \hline B_0 & B_1 & \cdots & B_{l-1} & B_l & 0 & \cdots & 0 \\ 0 & B_0 & \cdots & B_{l-2} & B_{l-1} & B_l & \cdots & 0 \\ \vdots & \vdots & \ddots & \vdots & \vdots & \ddots & \ddots & \vdots \\ 0 & 0 & \cdots & B_0 & B_1 & \cdots & B_{l-1} & B_l \end{array} \end{bmatrix} \qquad (10.9)$$

$$\Omega := J^l M_n B_l + J^{l-1}M_n B_{l-1} + \cdots + J M_n B_1 + M_n B_0.$$

Now Eq. (10.8) can be written as

$$\hat{\beta}(s) = A^{-1}(s)\hat{\alpha}_\beta(s) + A^{-1}(s)B(s)u(s) - A^{-1}(s)\hat{\alpha}_u(s). \qquad (10.10)$$

By taking the inverse Laplace transformation of the above, we finally obtain the following result.

Theorem 10.3.1. *The complete solution of the regular PMDs (Eq. 10.1) is*

$$\beta(t) = N_n e^{Jt}x_{s\beta}(0) - N_n e^{Jt}x_{su}(0)$$

$$- N_\infty \sum_{i=1}^{q_k-q} \delta(t)^{(i-1)} J_\infty^{i-1}(J_\infty x_{f\beta}(0)) + N_\infty \sum_{i=1}^{q_k-q} \delta(t)^{(i-1)} J_\infty^{i-1}(J_\infty x_{fu}(0))$$

$$+ \int_0^t N_n e^{J(t-\tau)} \Omega u(\tau) d\tau + [N_n, N_\infty] \Psi \Phi \begin{bmatrix} u(t) \\ u(t)^{(1)} \\ \vdots \\ u(t)^{(q_k-q+l)} \end{bmatrix}, \quad t \geq 0.$$

$$(10.11)$$

From the above, it is clearly seen that the nonzero initial conditions of $\beta(t)$ and $u(t)$ both contribute to each of the corresponding *slow (smooth) zero input response*

$$N_n e^{Jt} x_{s\beta}(0), \quad N_n e^{Jt} x_{su}(0)$$

and the *fast (impulse) response*

$$N_\infty \sum_{i=1}^{q_k-q} \delta(t)^{(i-1)} J_\infty^{i-1}(J_\infty x_{f\beta}(0)), \quad N_\infty \sum_{i=1}^{q_k-q} \delta(t)^{(i-1)} J_\infty^{i-1}(J_\infty x_{fu}(0)).$$

The above solution can be seen to be a natural generalization of that of GSSSs. However, in the case of the GSSSs, the initial conditions of $u(t)$ do not affect the solution structure, for wherein B is independent of the differential operator.

Example 10.3.1 (continued). Consider the following regular PMD

$$A(\rho)\beta(t) = B(\rho)u(t), \tag{10.12}$$

where $A(\rho)$ is given by Example 10.2.1, i.e.,

$$A(\rho) = \sum_{i=0}^{3} A_i \rho^i = \begin{bmatrix} 1 & \rho^3 & 0 \\ 0 & 1 & \rho \\ 0 & 0 & \rho^2 \end{bmatrix},$$

with

$$A_0 = \begin{bmatrix} 1 & 0 & 0 \\ 0 & 1 & 0 \\ 0 & 0 & 0 \end{bmatrix}, \quad A_1 = \begin{bmatrix} 0 & 0 & 0 \\ 0 & 0 & 1 \\ 0 & 0 & 0 \end{bmatrix}, \quad A_2 = \begin{bmatrix} 0 & 0 & 0 \\ 0 & 0 & 0 \\ 0 & 0 & 1 \end{bmatrix}, \quad A_3 = \begin{bmatrix} 0 & 1 & 0 \\ 0 & 0 & 0 \\ 0 & 0 & 0 \end{bmatrix},$$

and

$$B(\rho) = \sum_{i=0}^{2} B_i \rho^i = \begin{bmatrix} \rho & \rho^2 \\ 0 & \rho \\ 1 & \rho \end{bmatrix},$$

with

$$B_0 = \begin{bmatrix} 0 & 0 \\ 0 & 0 \\ 1 & 0 \end{bmatrix}, \quad B_1 = \begin{bmatrix} 1 & 0 \\ 0 & 1 \\ 1 & 1 \end{bmatrix}, \quad B_2 = \begin{bmatrix} 0 & 1 \\ 0 & 0 \\ 0 & 0 \end{bmatrix}.$$

$\beta(t) = [\beta_1(t), \beta_2(t), \beta_3(t)]^T \in R^3$ is the pseudo-state, $u(t) = [u_1(t), u_2(t)]^T \in R^2$ is the input. The initial values of $\beta(t)$ and its derivatives at $t = 0$ are assumed to be

$$\beta(0), \beta^{(1)}(0), \beta^{(2)}(0), \ldots,$$

The initial values of $u(t)$ and its derivatives at $t = 0$ are assumed as to be

$$u(0), u^{(1)}(0), u^{(2)}(0), \ldots.$$

It is seen that

$$r = 3, \ l = 2, \ m = 2, \ q = 3.$$

Recall Example 10.2.1, where the resolvent decomposition $A^{-1}(s)$ is established on the basis of the Weierstrass canonical form of the companion matrix $C_A(s)$. It is found in Example 10.2.1 that

$$A^{-1}(s) = [N_n, N_\infty] \left[\begin{array}{c|c} sI_n - J & 0_{n,v} \\ \hline 0_{v,n} & sJ_\infty - I_v \end{array} \right]^{-1} \left[\begin{array}{c} M_n \\ M_\infty \end{array} \right]$$

$$= \left[\begin{array}{cc|cccccc} 0 & 0 & 0 & 0 & 0 & -1 & 0 & 0 & 0 \\ 0 & -1 & 0 & 0 & 0 & 0 & 0 & 0 & 1 \\ 1 & 1 & 0 & 0 & 0 & 0 & 0 & 0 & 0 \end{array} \right]$$

$$\times \left[\begin{array}{cc|ccccccc} s & -1 & 0 & 0 & 0 & 0 & 0 & 0 & 0 \\ 0 & s & 0 & 0 & 0 & 0 & 0 & 0 & 0 \\ \hline 0 & 0 & -1 & 0 & 0 & 0 & 0 & 0 & 0 \\ 0 & 0 & 0 & -1 & s & 0 & 0 & 0 & 0 \\ 0 & 0 & 0 & 0 & -1 & s & 0 & 0 & 0 \\ 0 & 0 & 0 & 0 & 0 & -1 & s & 0 & 0 \\ 0 & 0 & 0 & 0 & 0 & 0 & -1 & s & 0 \\ 0 & 0 & 0 & 0 & 0 & 0 & 0 & -1 & s \\ 0 & 0 & 0 & 0 & 0 & 0 & 0 & 0 & -1 \end{array} \right]^{-1} \left[\begin{array}{ccc} 0 & 0 & -1 \\ 0 & 0 & 1 \\ \hline 0 & 0 & -1 \\ 0 & 0 & 0 \\ 0 & 0 & 0 \\ 1 & 0 & 0 \\ 0 & 0 & 0 \\ 0 & 0 & 1 \\ 0 & -1 & 0 \end{array} \right].$$

This gives

$$N_n = \left[\begin{array}{cc} 0 & 0 \\ 0 & -1 \\ 1 & 1 \end{array} \right], \quad N_\infty = \left[\begin{array}{ccccccc} 0 & 0 & 0 & -1 & 0 & 0 & 0 \\ 0 & 0 & 0 & 0 & 0 & 0 & 1 \\ 0 & 0 & 0 & 0 & 0 & 0 & 0 \end{array} \right],$$

$$J = \left[\begin{array}{cc} 0 & 1 \\ 0 & 0 \end{array} \right], \quad n = 2;$$

$$M_n = \left[\begin{array}{ccc} 0 & 0 & -1 \\ 0 & 0 & 1 \end{array} \right], \quad M_\infty = \left[\begin{array}{ccc} 0 & 0 & -1 \\ 0 & 0 & 0 \\ 0 & 0 & 0 \\ 1 & 0 & 0 \\ 0 & 0 & 0 \\ 0 & 0 & 1 \\ 0 & -1 & 0 \end{array} \right],$$

$$\mu = 7, \quad J_\infty = \text{block diag}\{J_{\infty 1}, J_{\infty 2}\},$$

where $J_{\infty i}(i = 1, 2)$ are Jordan matrices with sizes of $q_1 = 1$, $q_2 = 6$ respectively. It is noted that

$$
\begin{bmatrix} x_{s\beta}(0) \\ J_\infty x_{f\beta}(0) \end{bmatrix} = \begin{bmatrix} J^2 M_n, J M_n, M_n & 0_{2,9} \\ \hline 0_{7,9} & J_\infty M_\infty, J_\infty^2 M_\infty, J_\infty^3 M_\infty \end{bmatrix} \begin{bmatrix} A_3 & 0 & 0 \\ A_2 & A_3 & 0 \\ A_1 & A_2 & A_3 \\ A_0 & A_1 & A_2 \\ 0 & A_0 & A_1 \\ 0 & 0 & A_0 \end{bmatrix} \begin{bmatrix} \beta(0) \\ \beta^{(1)}(0) \\ \beta^{(2)}(0) \end{bmatrix},
$$

$$
\begin{bmatrix} x_{su}(0) \\ J_\infty x_{fu}(0) \end{bmatrix} = \begin{bmatrix} J M_n, M_n & 0_{2,6} \\ \hline 0_{7,6} & J_\infty M_\infty, J_\infty^2 M_\infty \end{bmatrix} \begin{bmatrix} B_2 & 0 \\ B_1 & B_2 \\ B_0 & B_1 \\ 0 & B_0 \end{bmatrix} \begin{bmatrix} u(0) \\ u^{(1)}(0) \end{bmatrix}.
$$

and that

$$
\Psi = \begin{bmatrix} J M_n, M_n & 0_{2,12} \\ \hline 0_{7,6} & M_\infty, J_\infty M_\infty, J_\infty^2 M_\infty, J_\infty^3 M_\infty \end{bmatrix},
$$

$$
\Phi = \begin{bmatrix} B_2 & 0 & 0 & 0 & 0 & 0 \\ B_1 & B_2 & 0 & 0 & 0 & 0 \\ B_0 & B_1 & B_2 & 0 & 0 & 0 \\ 0 & B_0 & B_1 & B_2 & 0 & 0 \\ 0 & 0 & B_0 & B_1 & B_2 & 0 \\ 0 & 0 & 0 & B_0 & B_1 & B_2 \end{bmatrix},
$$

$$
\Omega = J^2 M_n B_2 + J M_n B_1 + M_n B_0.
$$

By Theorem 10.3.1, one finally obtains the complete solution of the regular PMD (10.12)

$$
\beta(t) = \begin{bmatrix} \beta_1(t) \\ \beta_2(t) \\ \beta_3(t) \end{bmatrix}
$$

given by

$$
\beta_1(t) = (-\beta_2^{(2)}(0) - u_1^{(1)}(0))\delta(t) - (\beta_2^{(1)}(0) + u_1(0)
$$
$$
+ u_2^{(1)}(0))\delta^{(1)}(t) - (\beta_2(0) + \beta_3^{(1)}(0) - u_2^{(1)}(0))\delta^{(2)}(t)
$$
$$
+ (u_1^{(1)}(t) + u_1^{(2)}(t) + u_2^{(2)}(t) + u_2^{(3)}(t) - u_2^{(4)}(t))
$$

$$
\beta_2(t) = -\beta_3^{(1)}(0) + u_2(0) - \int_0^t u_1(\tau)d\tau - u_2(t) + u_2^{(1)}(t)
$$

$$
\beta_3(t) = \beta_3(0) - \beta_3^{(1)}(0) + (t + 1)\beta_3^{(1)}(0) + u_2(0) - (t + 1)u_2(0)
$$
$$
+ \int_0^t [(t - \tau)u_1(\tau) + u_2(\tau)]d\tau.
$$

A computational language package in Maple has been developed and is used to obtain the above solution.

10.4 Conclusions

In this chapter, we have investigated the relationship between the finite and infinite frequency structure of a regular polynomial matrix and that of a simply determined companion matrix. It has been shown that in the Weierstrass canonical form of this generalized companion matrix, the finite Jordan block matrices determine the finite zeros of the original polynomial matrix and the sizes of the infinite Jordan block matrices determine the infinite frequency structure of the original polynomial matrix and vice versa. A resolvent decomposition of $A(s)$ has been proposed based on the Weierstrass canonical form of the companion matrix, which is easier to obtain than the finite Jordan pair and the infinite Jordan pair [13, 17]. Subsequently a solution procedure has been developed by using this resolvent decomposition.

The results of this chapter enable one to give characterizations of the finite and infinite frequency structure of the polynomial matrix $A(s)$ simply from the Weierstrass canonical form of the generalized companion matrix. These results can be considered as the generalizations of the associated properties from the matrix pencil case to the polynomial matrix case. From the numerical computation point of view, these results suggest an alternative way of finding the finite and infinite frequency structure of a regular polynomial matrix. Since the finite and infinite frequency structure of a regular polynomial matrix are completely characterized by the Weierstrass canonical form of its companion matrix, the only thing we need to do is to transform the easily formed companion matrix (a regular matrix pencil) into its Weierstrass canonical form by some constant matrix transformation. Compared to the classical methods [13, 17], which are by means of the polynomial matrix transformations, the method proposed here will be less sensitive to data perturbations and rounding errors. Based on this observation, the proposed solution procedure is thus more attractive in actual computation than the classical solution methods [13, 17], for only certain constant matrix transformations are necessary to bring the matrix pencil $C_A(s)$ to its Weierstrass canonical form, and from this the solution of the regular PMD can be formulated.

A generalized chain-scattering representation and its algebraic system properties

11

11.1 Introduction

In classical network theory, a circuit representation called the *chain matrix* [21] has been widely used to deal with the cascade connection of circuits arising in analysis and synthesis problems. Recently, Kimura [22] developed the *chain-scattering representations* (CSRs), which was subsequently used to provide a unified framework for H^∞ control theory. The CSR is in fact an alternative way of representing a plant. Compared to the usual transfer function formulation, it has some remarkable properties. One is its *cascade structure*, which enables *feedback* to be represented as a matrix multiplication. The other is the *symmetry (duality)* between the CSR and its inverse called the *dual chain-scattering representation* (DCSR). Due to these characteristic features, it has successfully been used in several areas of control system design [22–24].

Consider a plant P (Fig. 11.1) with two kinds of inputs (w, u) and two kinds of outputs (z, y) represented by

$$
\begin{bmatrix} z \\ y \end{bmatrix} = P \begin{bmatrix} w \\ u \end{bmatrix} = \begin{bmatrix} P_{11} & P_{12} \\ P_{21} & P_{22} \end{bmatrix} \begin{bmatrix} w \\ u \end{bmatrix},
\tag{11.1}
$$

where $P_{ij}(i = 1, 2; \ j = 1, 2)$ are all rational matrices with dimensions $m_i \times k_j(i = 1, 2; \ j = 1, 2)$.

In order to compute the CSR and DCSR of Eq. (11.1), P_{21} and P_{12} are generally assumed to be invertible [22]. However, for general plants, if neither of P_{12} and P_{21} is invertible, one cannot use CSR and DCSR directly. Although the approach was extended by Kimura [22] to the case in which P_{12} is column full rank and P_{21} is row full rank by Z the plant, a systematic treatment is still needed in the general setting when neither of these conditions is satisfied.

In this chapter, we consider general plants, and therefore make no assumption about the rank of P_{12} and P_{21}. From an input-output consistency point of view, the conditions under which the CSRs exist are developed. These conditions essentially relax those assumptions that were generally put on the rank of P_{21} and P_{12} and make it possible to extend the applications of this approach from the regular cases of the 1-block case, the 2-block case and the 4-block case [25] to the general case. Based on this, the chain-scattering matrices are formulated into general parameterized forms by using the

A Generalized Framework of Linear Multivariable Control. http://dx.doi.org/10.1016/B978-0-08-101946-7.00011-1

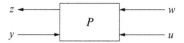

Fig. 11.1 Input-output representation.

matrix generalized inverse. Subsequently the cascade structure property, the symmetry and the realizations of them are investigated.

Some preliminary results related to the generalized inverse of a matrix are necessary for what we shall present in the sequel.

Definition 11.1.1 ([112]). For every rational matrix $A \in \Re(s)^{n \times m}$, a unique matrix $A^{\dagger} \in \Re(s)^{n \times m}$, which is called the *Moore-Penrose inverse*, exists satisfying

(i) $AA^{\dagger}A = A$,
(ii) $A^{\dagger}AA^{\dagger} = A^{\dagger}$,
(iii) $(AA^{\dagger})^T = AA^{\dagger}$,
(iv) $(A^{\dagger}A)^T = A^{\dagger}A$,

where A^T denotes the conjugate transpose of A. In the special case that A is a square nonsingular matrix, the Moore-Penrose inverse of A is simply its inverse, i.e., $A^{\dagger} = A^{-1}$. In case a matrix A^- satisfies only the first condition, it is called a {1}-*inverse*. {1}-inverse are not unique but play an important role in the following development.

Lemma 11.1.1 ([112]). *A matrix equation $M_z = f$, with $M \in \Re(s)^{m \times n}$, is consistent (can be solved) for z if and only if $(I - MM^-)f = 0$ for any {1}-inverse M^-. If the equation is consistent, its general solution can be written in the form $z = M^-f + (I_n - M^-M)h$, where I_n is the $n \times n$ identity matrix and h is an arbitrary $n \times 1$ rational vector.*

Lemma 11.1.2 ([112]). *The general form of any {1}-inverse of a rational matrix M is*

$$M^G = M^- + K_M - M^-MK_MMM^-,$$

where M^- is a particular {1}-inverse of M such as its Moore-Penrose inverse and K_M is an arbitrary rational matrix with appropriate dimension.

11.2 Input-output consistency and GCSR

In relation to the conditions under which the *generalized chain-scattering representation* (GCSR) and the *dual generalized chain-scattering representation* (DGCSR) exits, we give the following definition.

Definition 11.2.1. An input-output pair (u, y) $((w, z))$ is said to be *consistent* about $w(u)$ for the plant P if there exists at least one input $w(u)$ satisfying

$$P_{21}w = y - P_{22}u, \quad P_{12}u = z - P_{11}w.$$

In this case, P is said to be *input-output consistent* about $w(u)$ with respect to the input-output pair (u, y) $((w, z))$.

The following results state the characterization of the consistency.

Theorem 11.2.1. *The plant P is input-output consistent about w with respect to the input-output pair (u, y) if and only if one of the following equivalent conditions hold*

(a) $(I - P_{21}P_{21}^-)[-P_{22}, I]\begin{bmatrix} u \\ y \end{bmatrix} = 0,$

(b) $\mathrm{rank}[P_{21}, y - P_{22}u] = \mathrm{rank}P_{21},$

(c) $\mathrm{Im}P_{21} = \mathrm{Im}[-P_{22}, I]\begin{bmatrix} u \\ y \end{bmatrix},$

(d) $\begin{bmatrix} u \\ y \end{bmatrix} \in \mathrm{Ker}(I - P_{21}P_{21}^-)[-P_{22}, I],$

where P_{21}^- is any $\{1\}$-inverse of the rational matrix P_{21}.

Proof. (a) It is obvious that P is input-output consistent about w with an input-output pair (u, y) if and only if

$$P_{21}w = y - P_{22}u$$

is consistent. By Lemma 11.1.1 this is true if and only if

$$(I - P_{21}P_{21}^-)[-P_{22}, I]\begin{bmatrix} u \\ y \end{bmatrix} = 0.$$

Now we need to show that this condition is independent of the particular $\{1\}$-inverse used. Let P_{21}^g be any other $\{1\}$-inverse of P_{21}. By Lemma 11.1.2, it follows that

$$P_{21}^g = P_{21}^- + K - P_{21}^- P_{21}KP_{21}P_{21}^-$$

for a suitable choice of the arbitrary matrix K. Hence

$$(I - P_{21}P_{21}^g)[-P_{22}, I]\begin{bmatrix} u \\ y \end{bmatrix} = (I - P_{21}P_{21}^- - P_{21}K + P_{21}KP_{21}P_{21}^-)[-P_{22}, I]\begin{bmatrix} u \\ y \end{bmatrix}$$

$$= (I - P_{21}K)(I - P_{21}P_{21}^-)[-P_{22}, I]\begin{bmatrix} u \\ y \end{bmatrix}$$

$$= 0,$$

whenever

$$(I - P_{21}P_{21}^-)[-P_{22}, I]\begin{bmatrix} u \\ y \end{bmatrix} = 0.$$

(b) The equation

$$P_{21}w = y - P_{22}u$$

is consistent if and only if $\text{rank}[P_{21}, y - P_{22}u] = \text{rank } P_{21}$.

(c) It is obvious that (c) is equivalent to (b).

(d) Considering (a), again we only need to show that this condition is independent of the particular $\{1\}$-inverse. Let P_{21}^g be any other $\{1\}$-inverse of P_{21}, from the proof of (a), it follows that

$$(I - P_{21}P_{21}^g)[-P_{22}, I] = (I - P_{21}K)(I - P_{21}P_{21}^-)[-P_{22}, I],$$

for a certain rational matrix K. So

$$Ker(I - P_{21}P_{21}^-)[-P_{22}, I] \subseteq Ker(I - P_{21}P_{21}^g)[-P_{22}, I].$$

In a similar way, since P^- can be expressed parameterizedly in terms of P^g, one arrives at

$$Ker(I - P_{21}P_{21}^g)[-P_{22}, I] \subseteq Ker(I - P_{21}P_{21}^-)[-P_{22}, I].$$

Consequently

$$Ker(I - P_{21}P_{21}^g)[-P_{22}, I] = Ker(I - P_{21}P_{21}^-)[-P_{22}, I].$$

$\square$

Corollary 11.2.1. *The plant P is input-output consistent about w for arbitrary (u, y) if and only if P_{21} has full row rank.*

Proof. If P_{21} has full row rank, it follows that $I - P_{21}P_{21}^- = 0$, and so (a) is obviously satisfied for arbitrary (u, y). Contrarily, if the plant P is input-output consistent about w for arbitrary (u, y), then one can choose those input-output pairs $(u_i, y_i)(i = 1, \ldots, m_2)$ such that $y_i = P_{22}u_i + e_i$, where e_i is the identity vector. From (a), one yields

$$(I - P_{21}P_{21}^-)[-P_{22}, I] \begin{bmatrix} u_1 & \cdots & u_{m_2} \\ y_1 & \cdots & y_{m_2} \end{bmatrix} = (I - P_{21}P_{21}^-) [e_1, \ldots, e_{m_2}]$$

$$= (I - P_{21}P_{21}^-)I_{m_2} = I - P_{21}P_{21}^- = 0,$$

thus P_{21} has full row rank.

$\square$

A similar result holds for the plant to be consistent about u. From Corollary 11.2.1, the assumption of P_{21} having full row rank made in Kimura [22] is seen as the requirement that P be input-output consistent about w for arbitrary (u, y). The CSR in this case is well established by Kimura [22] by introducing the augmentation of plants. However, as far as the existence of the CSR is concerned this rank condition on P_{21} is stronger than necessary. In the general case, if the input-output pair satisfies one of the above conditions of consistency, then the CSRs are still available even though the matrices P_{21} and P_{12} are not of full rank. This is the subject of the following theorems.

Theorem 11.2.2. *If the plant P is consistent about w with respect to the input-output pair (u, y), then Eq. (11.1) can be written as*

$$\begin{bmatrix} z \\ w \end{bmatrix} = GCHAIN(P) \begin{bmatrix} u \\ y \\ h \end{bmatrix}, \tag{11.2}$$

where we denote the matrix

$$GCHAIN(P) := \left[\begin{array}{cc|c} P_{12} - P_{11}P_{21}^{-}P_{22} & P_{11}P_{21}^{-} & P_{11}(I - P_{21}^{-}P_{21}) \\ -P_{21}^{-}P_{22} & P_{21}^{-} & I - P_{21}^{-}P_{21} \end{array} \right], \quad (11.3)$$

where h is arbitrary rational vector. P_{21}^{-} is any {1}-inverse of P_{21}, i.e., any matrix satisfying $(\cdot)(\cdot)^{-}(\cdot) = (\cdot)$.

Proof. If one of the equivalent conditions (a), (b), (c), and (d) in Theorem 11.2.1 is satisfied, then using Lemma 11.1.1 in conjunction with the equation

$$P_{21}w = y - P_{22}u,$$

one obtains the general solution for w as

$$w = -P_{21}^{-}P_{22}u + P_{21}^{-}y + (I - P_{21}^{-}P_{21})h,$$
$$z = P_{11}w + P_{12}u$$
$$= (P_{12} - P_{11}P_{21}^{-}P_{22})u + P_{11}P_{21}^{-}y + P_{11}(I - P_{21}^{-}P_{21})h.$$

Hence

$$GCHAIN(P) = \left[\begin{array}{cc|c} P_{12} - P_{11}P_{21}^{-}P_{22} & P_{11}P_{21}^{-} & P_{11}(I - P_{21}^{-}P_{21}) \\ -P_{21}^{-}P_{22} & P_{21}^{-} & I - P_{21}^{-}P_{21} \end{array} \right].$$

$\square$

Definition 11.2.2. The relation (11.2) is called the GCSR of the plant P and any matrix $GCHAIN(P)$ is termed as a *GCSR matrix*.

The GCSR (Eq. 11.2) is schematically shown in Fig. 11.2.

Theorem 11.2.3. *If the plant P is consistent about u with respect to the input-output pair (z, w), then Eq. (11.1) can be written as*

$$\left[\begin{array}{c} u \\ y \end{array} \right] = GCHAIN(P) \left[\begin{array}{c} z \\ w \\ q \end{array} \right], \quad (11.4)$$

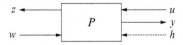

Fig. 11.2 Generalized chain-scattering representation.

where we denote the matrix as follows

$$GCHAIN(P) := \left[\begin{array}{cc|c} P_{12}^{-} & -P_{12}^{-}P_{11} & I - P_{12}^{-}P_{12} \\ P_{22}P_{12}^{-} & P_{21} - P_{22}P_{12}^{-}P_{11} & P_{22}(I - P_{12}^{-}P_{12}) \end{array} \right], \quad (11.5)$$

where q is arbitrary rational vector. P_{12}^{-} is any $\{1\}$-inverse of P_{12}, i.e., any matrix satisfying $(\cdot)(\cdot)^{-}(\cdot) = (\cdot)$.

Proof. The proof is similar to that of Theorem 11.2.2.   □

Definition 11.2.3. The relation (11.4) is called the DGCSR of the plant P and any matrix *DGCHAIN(P)* is termed as a *DGCSR matrix*.

The DGCSR (Eq. 11.4) is schematically shown in Fig. 11.3. It should be noted immediately that unlike the Kimura approach [22], the formulations of the GCSRs and the DGCSRs are not unique due to the fact that the $\{1\}$-inverse of a matrix are not unique. Obviously the set of all GCSR, DGCSR matrices of the plant P is composed of all the matrices in form (11.3), (11.5), where the $\{1\}$-inverse matrices P_{21}^{-}, P_{12}^{-} can in fact be parameterized by

$$P_{21}^{-} = P_{21}^{\dagger} + K - P_{21}^{\dagger}P_{21}KP_{21}P_{21}^{\dagger},$$
$$P_{12}^{-} = P_{12}^{\dagger} + Q - P_{12}^{\dagger}P_{12}QP_{12}P_{12}^{\dagger},$$

where $P_{21}^{\dagger}, P_{12}^{\dagger}$ are the Moore-Penrose inverse of P_{21}, P_{12}, K, and Q are the parameters. However, if input-output consistency is taken into account, the following two theorems parameterize all the formulations of the GCSRs and the DGCSRs in a much simplified manner.

Theorem 11.2.4. *Assume that the plant P is consistent about w with respect to the input-output pair (u, y). If the set of all GCSR matrices of the plant P is denoted by $\Gamma(P)$, then*

$$\Gamma(P) = \{GCHAIN(P; K) : K \text{ is rational matrix}\},$$

where we denote

$$GCHAIN(P; K) := \left[GCHAIN^{*}(P; K) | \Delta G(P; K) \right]$$
$$:= \left[\begin{array}{cc|c} P_{12} - P_{11}P_{21}^{G}P_{22} & P_{11}P_{21}^{G} & P_{11}(I - P_{21}^{G}P_{21}) \\ -P_{21}^{G}P_{22} & P_{21}^{G} & I - P_{21}^{G}P_{21} \end{array} \right], (11.6)$$

and

$$P_{21}^{G} := P_{21}^{\dagger} + (I - P_{21}^{\dagger}P_{21})K,$$

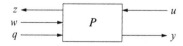

Fig. 11.3 Dual of generalized chain-scattering representation.

$P_{21}^{\dagger}$ is the Moore-Penrose inverse of P_{21}.

Proof. ($\subseteq$) This is to show that any GCSR matrix can be written into the form (11.6). The consistency of P about w suggests us that

$$(I - P_{21}P_{21}^{\dagger})(y - P_{22}u) = 0.$$

For any $\{1\}$-inverse P_{21}^{-} of P_{21}, Lemma 11.1.2 tells us that there is a rational matrix K such that

$$P_{21}^{-} = P_{21}^{\dagger} + K - P_{21}^{\dagger}P_{21}KP_{21}P_{21}^{\dagger}.$$

Following the proof of theorem 11.2.2, one obtains

$$
\begin{aligned}
w &= P_{21}^{-}(y - P_{22}u) + (I - P_{21}^{-}P_{21})h \\
&= (P_{21}^{\dagger} + K - P_{21}^{\dagger}P_{21}KP_{21}P_{21}^{\dagger})(y - P_{22}u) \\
&\quad + [I - (P_{21}^{\dagger} + K - P_{21}^{\dagger}P_{21}KP_{21}P_{21}^{\dagger})P_{21}]h \\
&= [P_{21}^{\dagger} + (I - P_{21}^{\dagger}P_{21})K](y - P_{22}u) + (I - P_{21}^{\dagger}P_{21})(I - KP_{21})h \\
&= -P_{21}^{G}P_{22}u + P_{21}^{G}y + (I - P_{21}^{G}P_{21})h, \\
z &= (P_{12} - P_{11}P_{21}^{-}P_{22})u + P_{11}P_{21}^{-}y + P_{11}(I - P_{21}^{-}P_{21})h \\
&= [P_{12} - P_{11}(P_{21}^{\dagger} + K - P_{21}^{\dagger}P_{21}KP_{21}P_{21}^{\dagger})P_{22}]u \\
&\quad + P_{11}(P_{21}^{\dagger} + K - P_{21}^{\dagger}P_{21}KP_{21}P_{21}^{\dagger})y + P_{11}(I - P_{21}^{\dagger}P_{21})(I - KP_{21})h \\
&= P_{12}u + P_{11}[P_{21}^{\dagger} + (I - P_{21}^{\dagger}P_{21})K](y - P_{22}u) + P_{11}(I - P_{21}^{\dagger}P_{21})(I - KP_{21})h \\
&= (P_{12} - P_{11}P_{21}^{G}P_{22})u + P_{11}P_{21}^{G}y + P_{11}(I - P_{21}^{G}P_{21})h,
\end{aligned}
$$

where

$$P_{21}^{G} := P_{21}^{\dagger} + (I - P_{21}^{\dagger}P_{21})K,$$

so

$$\begin{bmatrix} z \\ w \end{bmatrix} = GCHAIN(P; K) \begin{bmatrix} u \\ y \\ h \end{bmatrix},$$

where the GCSR matrix $GCHAIN(P; K)$ is just Eq. (11.6).

($\supseteq$) Conversely, referring to Eq. (11.6) and Theorem 7.2.2, we only need to show that any matrix P_{21}^{G} is one $\{1\}$-inverse of P_{21}. This is obvious from $P_{21}P_{21}^{G}P_{21} = P_{21}[P_{21}^{\dagger} + (I - P_{21}^{\dagger}P_{21})K]P_{21} = P_{21}$, for any rational matrix K. $\square$

Theorem 11.2.5. *Assume that the plant P is consistent about u with respect to the input-output pair (z, w). If the set of all DGCSR matrices of the plant P is denoted by $\Phi(P)$, then*

$$\Phi(P) = \{DGCHAIN(P; Q) : Q \text{ is rational matrix}\},$$

where

$$DGCHAIN(P; Q) := \left[DCHAIN^*(P; K) | \Delta DG(P; K) \right]$$

$$:= \left[\begin{array}{cc|cc} P_{12}^G & -P_{12}^G P_{11} & I - P_{12}^G P_{12} \\ P_{22} P_{12}^G & P_{21} - P_{22} P_{12}^G P_{11} & P_{22}(I - P_{12}^G P_{12}) \end{array} \right] (11.7)$$

and

$$P_{12}^G := P_{12}^\dagger + (I - P_{12}^\dagger P_{12})Q,$$

$P_{12}^\dagger$ *is the Moore-Penrose inverse of* P_{12}.

Proof. The proof is similar to that of Theorem 11.2.4. $\qquad\qquad\square$

Remark 11.2.1. These theorems clarify the essential structure of the GCSRs and the DGCSRs in a general setting. Even though the structure of the GCSR and DGCSR is very similar to that proposed by Kimura [22] and Ball et al. [25] via augmentation of the plant by introduction of fictitious observation output or control input for the cases where P_{21} is row full rank and P_{12} is column full rank, the GCSR and DGCSR proposed here are based on a relaxed assumption and do not require any such rank condition on P_{21} and P_{12}.

Remark 11.2.2. If in the plant P, the matrix P_{21} is invertible, then for any $GCHAIN(P; K) \in \Gamma(P)$, one has

$$GCHAIN(P; K) = [CHAIN(P) | 0],$$

where

$$CHAIN(P) = \left[\begin{array}{cc} P_{12} - P_{11} P_{21}^{-1} P_{22} & P_{11} P_{21}^{-1} \\ -P_{21}^{-1} P_{22} & P_{21}^{-1} \end{array} \right]$$

is the standard CSR of Kimura [22]. Similarly, if in the plant P, the matrix P_{12} is invertible, then for any $DGCHAIN(P; K) \in \Phi(P)$, one has

$$DGCHAIN(P; K) = [DCHAIN(P) | 0],$$

where

$$DCHAIN(P) = \left[\begin{array}{cc} P_{12}^{-1} & -P_{12}^{-1} P_{11} \\ P_{22} P_{12}^{-1} & P_{21} - P_{22} P_{12}^{-1} P_{11} \end{array} \right].$$

11.3 Algebraic system properties of GCSR and DGCSR

Not only can the CSR illuminate the fundamental structure of H^∞ control systems (see, e.g., [22–24]), but also it has many applications in circuit theory and signal processing (see, e.g., [113]). Two fundamental algebraic properties of the CSR, i.e., the cascade structure and the duality, are considered to be relevant to such control system design problems. In the following, these two issues are discussed in a more general setting.

Similar to that of CSR and DCSR, one observation is the following.

Theorem 11.3.1 (Duality). *(i) The set of GCSR matrices $\Gamma(P^T)$ is isomorphic to the set of DGCSR matrices $\Phi(P)$. Furthermore, the following identities, which describe the matching relationship between any pair in the two sets, displays the essential duality*

$$\begin{bmatrix} 0 & I \\ -I & 0 \end{bmatrix} GCHAIN^*(P^T; K) \begin{bmatrix} 0 & -I \\ I & 0 \end{bmatrix} = [DGCHAIN^*(P; K^T)]^T, \forall K,$$

(GD11)

$$\begin{bmatrix} 0 & -I \\ I & 0 \end{bmatrix} GCHAIN^*(P^T; K) \begin{bmatrix} 0 & I \\ -I & 0 \end{bmatrix} = [DGCHAIN^*(P; K^T)]^T, \forall K,$$

(GD12)

$$\begin{bmatrix} 0 & I \\ -I & 0 \end{bmatrix} [DGCHAIN^*(P; Q)]^T \begin{bmatrix} 0 & -I \\ I & 0 \end{bmatrix} = GCHAIN^*(P^T; Q^T), \forall Q,$$

(DG13)

$$\begin{bmatrix} 0 & -I \\ I & 0 \end{bmatrix} [DGCHAIN^*(P; Q)]^T \begin{bmatrix} 0 & I \\ -I & 0 \end{bmatrix} = GCHAIN^*(P^T; Q^T), \forall Q,$$

(DG14)

where the matrices $GCHAIN^*(\cdot\ ;\ \cdot)$ and the matrices $DGCHAIN^*(\cdot\ ;\ \cdot)$ are the block matrices in the GCSR and DGCSR formulations as denoted in Eqs. (11.6), (11.7), respectively.

(ii) Similarly, $\Phi(P^T)$ is isomorphic to $\Gamma(P)$, furthermore the following identities display the duality

$$\begin{bmatrix} 0 & I \\ -I & 0 \end{bmatrix} DGCHAIN^*(P^T; Q) \begin{bmatrix} 0 & -I \\ I & 0 \end{bmatrix} = [GCHAIN^*(P; Q^T)]^T, \forall Q,$$

(DG21)

$$\begin{bmatrix} 0 & -I \\ I & 0 \end{bmatrix} DGCHAIN^*(P^T; Q) \begin{bmatrix} 0 & I \\ -I & 0 \end{bmatrix} = [GCHAIN^*(P; Q^T)]^T, \forall Q,$$

(DG22)

$$\begin{bmatrix} 0 & I \\ -I & 0 \end{bmatrix} [GCHAIN^*(P; K)]^T \begin{bmatrix} 0 & -I \\ I & 0 \end{bmatrix} = DGCHAIN^*(P^T; K^T), \forall K,$$

(GD23)

$$\begin{bmatrix} 0 & -I \\ I & 0 \end{bmatrix} [GCHAIN^*(P; K)]^T \begin{bmatrix} 0 & I \\ -I & 0 \end{bmatrix} = DGCHAIN^*(P^T; K^T), \forall K,$$

(GD24)

Proof. We only need to prove (i). The proof of (ii) is similar. If the plant P is given by Eq. (11.1), then

$$P^T = \begin{bmatrix} P_{11}^T & P_{21}^T \\ P_{12}^T & P_{22}^T \end{bmatrix}.$$

$\forall GCHAIN(P^T; K) \in \Gamma(P^T)$ given by,

$$GCHAIN(P^T; K) = \left[GCHAIN^*(P^T; K) | \Delta G(P^T; K) \right]$$

$$= \left[\begin{array}{cc|c} P_{21}^T - P_{11}^T(P_{12}^T)^G P_{22}^T & P_{11}^T(P_{12}^T)^G & P_{11}^T(I - (P_{12}^T)^G P_{12}^T) \\ -(P_{12}^T)^G P_{22}^T & (P_{12}^T)^G & I - (P_{12}^T)^G P_{12}^T \end{array} \right],$$

one can find a rational matrix K such that $(P_{12}^T)^G = (P_{12}^T)^\dagger + [I - (P_{12}^T)^\dagger P_{12}^T]K$. Due to the consistency of the system, the operator $(P_{12}^T)^\dagger + [I - (P_{12}^T)^\dagger P_{12}^T]K$, as seen in the proof of theorem 11.2.4, is equal to $(P_{12}^T)^\dagger + K - (P_{12}^T)^\dagger P_{12}^T K P_{12}^T (P_{12}^T)^\dagger$, i.e., for this rational matrix

$$(P_{12}^T)^G = (P_{12}^T)^\dagger + K - (P_{12}^T)^\dagger P_{12}^T K P_{12}^T (P_{12}^T)^\dagger.$$

Thus one finds a rational matrix $Q = K^T$, such that

$$(P_{12})^G = ((P_{12}^T)^T)^G = ((P_{12}^T)^G)^T = [(P_{12}^T)^\dagger + K - (P_{12}^T)^\dagger P_{12}^T K P_{12}^T (P_{12}^T)^\dagger]^T$$

$$= P_{12}^\dagger + Q - P_{12}^\dagger P_{12} Q P_{12}^\dagger P_{12} = P_{12}^\dagger + (I - P_{12}^\dagger P_{12})Q,$$

consequently, the corresponding DGSR matrix $DCHAIN(P; Q) \in \Phi(P)$ is given by Eq. (11.7), therefore

$$[DGCHAIN(P; Q)]^T = \left[\frac{(DGCHAIN^*(P; Q))^T}{(\Delta DG(P; Q))^T} \right]$$

$$= \left[\begin{array}{cc} (P_{12}^T)^G & (P_{12}^T)^G P_{22}^T \\ -P_{11}^T(P_{12}^T)^G & P_{21}^T - P_{11}^T(P_{12}^T)^G P_{22}^T \\ \hline I - P_{12}^T(P_{12}^T)^G & (I - P_{12}^T(P_{12}^T)^G)P_{22}^T \end{array} \right].$$

The statements (GD11) and (GD12) thus follow from a direct calculation. By repeating the similar augment as above, one can prove the dual statements (DG13) and (DG14). The statement (GD11) (or (GD12)) together with the statement (DG13) (or (DG14)) suggests us that $\Gamma(P^T)$ is isomorphic to $\Phi(P)$. □

For the full GCSR matrices and the DGCSR matrices, the basic result concerning duality is as follows.

Theorem 11.3.2 (Duality). *The equality*

$$\left[\begin{array}{cc} 0 & I \\ -I & 0 \\ \hline 0 & 0 \end{array} \right] GCHAIN(P^T; K) \left[\begin{array}{cc} 0 & -I \\ I & 0 \\ \hline 0 & 0 \end{array} \right] = [DGCHAIN(P; K^T)]^T, \forall K \,(11.8)$$

holds if and only if P_{12} is of full column rank. The equality

$$\begin{bmatrix} 0 & I \\ -I & 0 \\ \hline 0 & 0 \end{bmatrix} DGCHAIN(P^T; Q) \begin{bmatrix} 0 & -I \\ I & 0 \\ \hline 0 & 0 \end{bmatrix} = [GCHAIN(P; Q^T)]^T, \forall Q \quad (11.9)$$

holds if and only if P_{21} is of full column rank.

Proof. If P_{12} is of full column rank, then $I - P_{12}^G P_{12} = 0$, thus $I - P_{12}^T (P_{12}^T)^G = 0$, refer to the proof of Theorem 11.3.1, Eq. (11.8) holds. On the other hand, if Eq. (11.8) holds, then from Eqs. (11.6), (11.7) one follows that $I - P_{12}^T (P_{12}^T)^G = 0$, this is true iff P_{12} is of full column rank. The second result follows in a similar way.   □

Now we consider the cascade properties of the GCSR and DGCSR. The following theorems concerning this subject are independent of the specific formulation of the GCSR matrix, so for concreteness we shall use a general notation $GCHAIN(P)$ in the following. Let a series of plants $P^{(i)} (i \in \mathbf{r} := 1, 2, \ldots, r)$ represented by

$$\begin{bmatrix} z_i \\ y_i \end{bmatrix} = P^{(i)} \begin{bmatrix} w_i \\ u_i \end{bmatrix} = \begin{bmatrix} P_{11}^{(i)} & P_{12}^{(i)} \\ P_{21}^{(i)} & P_{22}^{(i)} \end{bmatrix} \begin{bmatrix} w_i \\ u_i \end{bmatrix}, \quad (11.10)$$

$$i = 1, 2, \ldots, r \quad (11.11)$$

be interconnected (Fig. 11.4) simply according to

$$u_{i+1} = z_i,$$
$$y_{i+1} = w_i, \quad i = 1, 2, \ldots, r - 1. \quad (11.12)$$

This interconnection was denoted [88] by

$$P^r := \bigwedge_{i=1}^{r} P^{(i)}.$$

Definition 11.3.1. An input-output pair (u_1, y_1) is called *consistent* about w to the interconnection plant P^r if the resulting input-output pairs (u_i, y_i) $(i = 2, 3, \ldots, r)$ are all consistent about w_i to $P^{(i)} (i \in \mathbf{r})$.

An immediate observation from the above definitions is the following Proposition.

Proposition 11.3.1. $\forall r_1, r_1 \in \mathbf{r}$, an input-output pair (u_1, y_1) is consistent about w to P^r if and only if (u_1, y_1) is consistent about w to $P^{r_1-1} := \bigwedge_{i=1}^{r_1-1} P^{(i)}$ and the resulting pair (u_{r_1}, y_{r_1}) is consistent about w to $\overline{P^{r_1-1}} := \bigwedge_{i=r_1}^{r} P^{(i)}$.

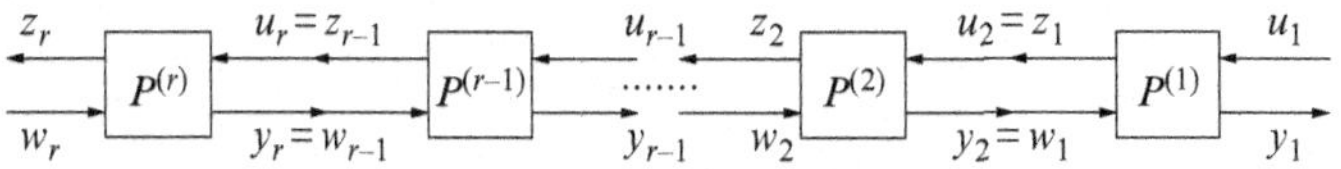

Fig. 11.4 Interconnections of plants.

Theorem 11.3.3. *An input-output pair (u_1, y_1) is consistent about w to P^r if and only if the following conditions hold*

$$\left[I - P_{21}^{(1)}(P_{21}^{(1)})^{-}\right]\left[-P_{22}^{(1)}, I\right]\left[\begin{array}{c} u_1 \\ y_1 \end{array}\right] = 0, \tag{E_1}$$

$$\left[I - P_{21}^{(2)}(P_{21}^{(2)})^{-}\right]\left[-P_{22}^{(2)}, I\right] GCHAIN(P^{(1)})\left[\begin{array}{c} u_1 \\ y_1 \\ h_1 \end{array}\right] = 0, \tag{E_2}$$

$$\vdots \qquad\qquad \vdots$$

$$\left[I - P_{21}^{(r)}(P_{21}^{(r)})^{-}\right]\left[-P_{22}^{(r)}, I\right] GCHAIN(P^{(r-1)})$$

$$\times \left[\begin{array}{c} GCHAIN(P^{(r-2)}) \left[\begin{array}{c} \cdots \left[\begin{array}{c} GCHAIN(P^{(1)})\left[\begin{array}{c} u_1 \\ y_1 \\ h_1 \end{array}\right] \\ h_2 \\ \cdots \end{array}\right] \cdots \end{array}\right] \\ h_{r-1} \end{array}\right] = 0. \tag{E_r}$$

Proof. We proceed by induction. When $r = 1$, from Definition 11.2.1, an input-output pair (u_1, y_1) is consistent about w to $P^1 = P^{(1)}$ if and only if (E_1) is satisfied. Now assume (u_1, y_1) is consistent about w to P^k if and only if the first kth equations are satisfied. If (u_1, y_1) is consistent about w to P^{k+1}, from Proposition 11.3.1, it follows that (u_{k+1}, y_{k+1}) is consistent about w to $P^{(k+1)}$, thus the $K+1$th equation follows from the assumption and Definition 11.3.1. □

From Definition 11.3.1 and Theorem 11.3.3, one immediately notes the following Corollary.

Corollary 11.3.1. *The interconnection P^r is consistent about w to arbitrary an input-output pair (u_1, y_1), if and only if $P_{21}^{(i)}(i \in r)$ and all of full row rank.*

The main reason for using the CSR in classical circuit theory lies in the remarkable feature that enables cascade connection to be represented in a much simplified manner, i.e., the *port characteristics* of connection that essentially reflect the physical structure of the system can be precisely represented by the CSR [21, 22]. The following result details this property of GCSRs in a general setting.

Theorem 11.3.4. *If an input-output pair (u_1, y_1) is consistent about w to the interconnection P^r, then*

$$\left[\begin{array}{c} z_r \\ w_r \end{array}\right] = GCHAIN(P^{(r)})\left[\begin{array}{c} GCHAIN(P^{(r-1)})\left[\begin{array}{c} \cdots \left[\begin{array}{c} GCHAIN(P^{(1)})\left[\begin{array}{c} u_1 \\ y_1 \\ h_1 \end{array}\right] \\ h_2 \\ \cdots \end{array}\right] \cdots \end{array}\right] \\ h_r \end{array}\right],$$

where $h_1, \ldots, h_r$ are arbitrary rational vectors.

Proof. If (u_1, y_1) is consistent about w to P^r, then $GCHAIN(P^{(1)}), \ldots, GCHAIN(P^{(r)})$ all exist. Thus $\begin{bmatrix} z_r \\ w_r \end{bmatrix}$ can be formulated recursively from $\begin{bmatrix} z_{r-1} \\ w_{r-1} \end{bmatrix}$. $\square$

The above theorem provides a useful way of determining the last port from the first port in terms of the GCSR matrix in an interconnection that also displays the cascade connection in an input-output format, the next result states it in a more explicit form. By denoting,

$$
GCHAIN(P^{(i)}) = \left[\begin{array}{cc|c} P_{12}^{(i)} - P_{11}^{(i)}(P_{21}^{(i)})^- P_{22}^{(i)} & P_{11}^{(i)}(P_{21}^{(i)})^- & P_{11}^{(i)}(I - (P_{21}^{(i)})^- P_{21}^{(i)}) \\ -(P_{21}^{(i)})^- P_{22}^{(i)} & (P_{21}^{(i)})^- & I - (P_{21}^{(i)})^- P_{21}^{(i)} \end{array} \right]
$$
$$
:= \left[\begin{array}{c|c} GCHAIN^*(P^{(i)}) & \Delta G(P^{(i)}) \end{array} \right], \quad i \in \mathbf{r},
$$

one establishes the following theorem.

Theorem 11.3.5. *If an input-output pair (u_1, y_1) is consistent about w to the interconnection P^r, then*

$$
\begin{bmatrix} z_r \\ w_r \end{bmatrix} = \prod_{i=1}^{r} GCHAIN^*(P^{(i)}) \begin{bmatrix} u_1 \\ y_1 \end{bmatrix} + \prod_{i=2}^{r} GCHAIN^*(P^{(i)}) \Delta G(P^{(1)}) h_1
$$
$$
+ \prod_{i=3}^{r} GCHAIN^*(P^{(i)}) \Delta G(P^{(2)}) h_2 + \cdots
$$
$$
+ GCHAIN^*(P^{(r)}) \Delta G(P^{(r-1)}) h_{r-1} + \Delta G(P^{(r)}) h_r. \tag{11.13}
$$

Proof. We proceed by induction. When $r = 1$, if (u_1, y_1) is consistent about w to $P^1 = P^{(1)}$, then from Theorem 11.2.2, it follows that

$$
\begin{bmatrix} z_1 \\ w_1 \end{bmatrix} = GCHAIN(P^{(1)}) \begin{bmatrix} u_1 \\ y_1 \\ h_1 \end{bmatrix} = GCHAIN^*(P^{(1)}) \begin{bmatrix} u_1 \\ y_1 \end{bmatrix} + \Delta G(P^{(1)}) h_1,
$$

for any rational vector h_1. Now assuming that (u_1, y_1) is consistent about w to P^k, and

$$
\begin{bmatrix} z_k \\ w_k \end{bmatrix} = \prod_{i=1}^{k} GCHAIN^*(P^{(i)}) \begin{bmatrix} u_1 \\ y_1 \end{bmatrix} + \prod_{i=2}^{k} GCHAIN^*(P^{(i)}) \Delta G(P^{(1)}) h_1
$$
$$
+ \prod_{i=3}^{k} GCHAIN^*(P^{(i)}) \Delta G(P^{(2)}) h_2 + \cdots
$$
$$
+ GCHAIN^*(P^{(k)}) \Delta G(P^{(k-1)}) h_{k-1} + \Delta G(P^{(k)}) h_k,
$$

for any rational vector $h_1, \ldots, h_k$, by Proposition 11.3.1, $\begin{bmatrix} u_{k+1} \\ y_{k+1} \end{bmatrix} = \begin{bmatrix} z_k \\ w_k \end{bmatrix}$ is consistent about w to $P^{(k+1)}$, form Theorem 11.2.2, one follows that

$$\left[\begin{array}{c} z_{k+1} \\ w_{k+1} \end{array} \right] = GCHAIN^*(P^{(k+1)}) \left[\begin{array}{c} u_{k+1} \\ y_{k+1} \\ h_{k+1} \end{array} \right]$$

$$= GCHAIN^*(P^{(k+1)}) \left[\begin{array}{c} u_{k+1} \\ y_{k+1} \end{array} \right] + \Delta G(P^{(k+1)}) h_{k+1}$$

$$= \prod_{i=1}^{k+1} GCHAIN^*(P^{(i)}) \left[\begin{array}{c} u_1 \\ y_1 \end{array} \right] + \prod_{i=2}^{k+1} GCHAIN^*(P^{(i)}) \Delta G(P^{(1)}) h_1$$

$$+ \prod_{i=3}^{k+1} GCHAIN^*(P^{(i)}) \Delta G(P^{(2)}) h_2 + \cdots$$

$$+ GCHAIN^*(P^{(k+1)}) \Delta G(P^{(k)}) h_k + \Delta G(P^{(k+1)}) h_{k+1},$$

for any rational vector $h_1, \ldots, h_{k+1}$.   □

Corollary 11.3.2. *If an input-output pair (u_1, y_1) is consistent about w to the interconnection P^r, and if for all $i \in r, P_{21}^{(i)}$ is of full column rank, then*

$$\left[\begin{array}{c} z_r \\ w_r \end{array} \right] = \prod_{i=1}^{r} GCHAIN^*(P^{(i)}) \left[\begin{array}{c} u_1 \\ y_1 \end{array} \right]. \tag{11.14}$$

Proof. For this statement, if for all $i \in \mathbf{r}, P_{21}^{(i)}$ is of full column rank, then

$$\Delta G(P^{(i)}) = 0, \quad i \in \mathbf{r},$$

Eq. (11.14) follows from Eq. (11.13) directly.   □

Remark 11.3.1. Theorem 11.3.4, Theorem 11.3.5, and Corollary 11.3.2 are extensions of the classical result, which was termed *star product* and was given by Redheffer [91] from the regular case to the general case. They are also extensions to the regular cascade connection, which is one of the remarkable properties of the chain matrix [21, 22] in circuit theory. The use of GCSR makes it possible to formulate the general cascade connection explicitly.

11.4   Realizations of GCSR and DGCSR

The main purpose of this section is to determine the state space realizations of GCSR and DGCSR.

If $P(s)$ is given in state-space form by

$$P(s) = \left[\begin{array}{c|cc} A & B_1 & B_2 \\ \hline C_1 & D_{11} & D_{12} \\ C_2 & D_{21} & D_{22} \end{array} \right],$$

i.e., the input-output relations are governed by

$$\dot{x} = Ax + B_1 w + B_2 u \tag{11.15}$$

$$z = C_1 x + D_{11} w + D_{12} u \tag{11.16}$$

$$y = C_2 x + D_{21} w + D_{22} u, \tag{11.17}$$

satisfying

$$(I - D_{21} D_{21}^-)(y - C_2 x - D_{22} u) = 0,$$

Eq. (11.17) is written as

$$w = D_{21}^-(y - C_2 x - D_{22} u) + (I - D_{21}^- D_{21})h,$$

where h is any function vector and D_{21}^- is any {1}-inverse of D_{21}. Substituting this relation into Eqs. (11.15), (11.16) yields

$$\dot{x} = (A - B_1 D_{21}^- C_2)x + (B_2 - B_1 D_{21}^- D_{22})u + B_1 D_{21}^- y + B_1(I - D_{21}^- D_{21})h,$$

$$z = (C_1 - D_{11} D_{21}^- C_2)x + (D_{12} - D_{11} D_{21}^- D_{22})u + D_{11} D_{21}^- y + D_{11}(I - D_{21}^- D_{21})h.$$

Thus we have the following state-space realization of GCSR of $\sum$,

$$\begin{bmatrix} \dot{x} \\ z \\ w \end{bmatrix} = GCHAIN(P) \begin{bmatrix} x \\ u \\ y \\ h \end{bmatrix},$$

where

$$GCHAIN(P) = \left[\begin{array}{cc|cc||c} A - B_1 D_{21}^- C_2 & B_2 - B_1 D_{21}^- D_{22} & B_1 D_{21}^- & B_1(I - D_{21}^- D_{21}) \\ C_1 - D_{11} D_{21}^- C_2 & D_{12} - D_{11} D_{21}^- D_{22} & D_{11} D_{21}^- & D_{11}(I - D_{21}^- D_{21}) \\ -D_{21}^- C_2 & -D_{21}^- D_{22} & D_{21}^- & (I - D_{21}^- D_{21}) \end{array} \right].$$

If the plant P satisfies

$$(I - D_{12} D_{12}^-)(z - C_1 x - D_{11} w) = 0,$$

by repeating a similar procedure as above, one obtains the realization for DGCSR of $\sum$ as follows

$$GCHAIN(P) = \left[\begin{array}{cc|cc||c} A - B_2 D_{12}^- C_1 & B_2 D_{12}^- & B_1 - B_2 D_{12}^- D_{11} & B_2(I - D_{12}^- D_{12}) \\ -D_{12}^- C_1 & D_{12}^- & -D_{12}^- D_{11} & (I - D_{12}^- D_{12}) \\ C_2 - D_{22} D_{12}^- C_1 & D_{22} D_{12}^- & D_{21} - D_{22} D_{12}^- D_{11} & D_{22}(I - D_{12}^- D_{12}) \end{array} \right].$$

11.5 Conclusions

In this chapter, the CSR proposed by Kimura for the regular case is extended to the general case where no rank condition is assumed on P_{21} and P_{12}. The conditions under which the GCSR and the DGCSR exist are developed from the point of view of input-output consistency. The generalized chain-scattering matrices are formulated into parameterized form by using the generalized inverse of matrices. The essential structure of them is clarified in a general setting. Some algebraic system properties of these general matrices which are relevant to control system design requirements, especially the H^{∞} control problem, are developed. The application of this approach to analysis and synthesis problems in general case is a subject of further research.

Realization of behavior **12**

12.1 Introduction

In classical network theory, a circuit representation called the *chain matrix* [21] was widely used to deal with the cascade connection of circuits arising in analysis and synthesis problems. Based on this, Kimura [22] developed the *chain-scattering representation* (CSR), which was subsequently used to provide a unified framework for H^∞ control theory. Kimura's approach is, however, only available to the special cases where the matrices P_{21} and P_{12} (refer to Eq. 11.1) satisfy some assumptions of full rank. Recently, in [31] this approach was extended to the general case in which such conditions are essentially relaxed. From an input-output consistency point of view, the *generalized chain-scattering representation* (GCSR) and the *dual generalized chain-scattering representation* (DGCSR) emerge and are there successfully used to characterize the cascade structure property and the symmetry of general plants in a general setting.

Latterly *behavioral theory* (see, e.g., [26, 27]) has received broad acceptance as an approach for modeling dynamical systems. One of the main features of the behavioral approach is that it does not use the conventional input-output structure in describing systems. Instead, a mathematical model is used to represent the systems in which the collection of time trajectories of the relevant variables are viewed as the *behavior* of the dynamical systems. This approach has been shown [26, 28] to be powerful in system modeling and analysis. In contrast to this, the classical theories such as Kalman's state-space description and Rosenbrock's polynomial matrix description (PMD) take the input-output representation as their starting point. In many control contexts it has proven to be very convenient to adopt the classical input/state/output framework. It is often found that the system models, in many situations, can easily be formulated into the input/state/output models such as the state space descriptions and the PMDs. Based on such input/state/output representations, the action of the controller can usually be explained in a very natural manner and the control aims can usually be attained very effectively.

As far as the issue of system modeling is concerned the computer-aided procedure, i.e., the automated modeling technology, has been well developed as a practical approach over recent years. If the *physical description* of a system is known, the automated modeling approach can be applied to find a set of equation to describe the *dynamical behavior* of the given system. It is seen that, in many cases, such as in an electrical circuit or, more generally, in an interconnection of blocks, such a

A Generalized Framework of Linear Multivariable Control. http://dx.doi.org/10.1016/B978-0-08-101946-7.00012-3

physical description is more conveniently specified through the *frequency behavior* of the system. Consequently in these cases, a more general theoretic problem arises, i.e., if the *frequency behavior description* of a system is known, what is the corresponding dynamical behavior? In other words, what is the input/output or input/state/output structure in the time domain that generates the given frequency domain description. It turns out that this question can be interpreted through the notion of *realization of behavior*, which we shall introduce in this chapter.

In fact, as we shall see later, realization of behavior, in many cases, amounts to introduction of latent variables in the time domain. From this point of view, realization of behavior can be understood as a converse procedure of the latent variable elimination theorem [27] in one way or another. It should be emphasized that realization of behavior also generalizes the notion of transfer function matrix realization in the classical control theory framework, as the behavior equation is a more general description than the transfer function matrix. As a special case, the behavior equations determine a transfer matrix of the system if they represent an input-output system [27, 29], i.e., when the matrices describing the system satisfy a full rank condition.

Recently in [30], a realization approach was suggested that reduces high-order linear differential equations to the first-order system representations by using the method of "linearization." From the point of view of *realization* in a physical sense one is, however, forced to start from the system frequency behavior description into which system behavior is generally described rather than from the high-order linear differential equations in time domain. Also a constructive procedure for autoregressive (AR) equation realization has been introduced in [114].

One of the main aims of this chapter is to present a new notion of *realization of behavior*. Further to the results of [31], the input-output structure of the GCSRs and the DGCSRs are thus clarified by using this approach. Subsequently the corresponding autoregressive-moving-average (ARMA) representations are proposed and are proved to be realizations of behavior for any GCSR and for any DGCSR. Once these ARMA representations are proposed, one can further find the corresponding first-order system representations by using the method of [30] or other well-developed realization approaches such as Kailath [32].

These results are interesting in that they provide a good insight into the natural relationship between the (frequency) behavior of any GCSR and any DGCSR and the (dynamical) behavior of the corresponding ARMA representations. Since no numerical computation is involved in this approach, realization of behavior is particularly accessible for situations in which the coefficient are symbolic rather than numerical. Based on the realization of behavior, one can further find the corresponding first-order system representations by using the classical realization approaches [30, 32]. Also a constructive procedure for AR equation realization suggested by Vafiadis and Karcanias [114] can be used to obtain the required first-order system representations.

The following result can be easily proposed and can be easily proved by using the definition of 1-inverse of a polynomial matrix. It is presented here for later use.

Proposition 12.1.1. *The ring of the polynomial matrices in denoted by* $\mathfrak{R}[s]$. *Let* $A[s] \in \mathfrak{R}[s]^{n \times m}$, $\mathrm{rank}_{\mathfrak{R}[s]} A[s] = r \leq \min\{m, n\}$, *there exist nonsingular polynomial matrices* $Q(s) \in \mathfrak{R}[s]^{n \times n}$, $P(s) \in \mathfrak{R}[s]^{m \times m}$, *such that*

$$Q(s)A(s)P(s) = \begin{bmatrix} B(s) & 0 \\ 0 & 0 \end{bmatrix},$$

where the polynomial matrix $B[s] \in \mathfrak{R}[s]^{r \times r}$ *is invertible, then for any polynomial matrix* $L[s] \in \mathfrak{R}[s]^{(n-r) \times (m-r)}$, *the following matrix* $X(s)$ *is a {1}-inverse of* $A(s)$. $X(s)$ *is given by*

$$X(s) = P(s) \begin{bmatrix} B^{-1}(s) & 0 \\ 0 & L(s) \end{bmatrix} Q(s).$$

12.2 Behavior realization

This section introduce the concept of *realization of behavior*. Recall that in the behavioral framework the (dynamical) behavioral equations of an ARMA representation [27] are

$$R_1(\rho)u_1(t) + R_2(\rho)y_1(t) = S(\rho)\xi(t), \tag{12.1}$$

where $w' := \left[(y_1(t))^T, (u_1(t))^T \right]^T$ stands for the external variables representing the dynamical behavior of the underlying dynamical system, $\xi(t)$ which are called latent variables corresponding to auxiliary variables resulting from the modeling procedure. $R_1(\rho)$, $R_2(\rho)$ and $S(\rho)$ are polynomial matrices containing the differential operator $\rho = d/dt$. In order to distinguish them from the existing notation y, u of Eq. (11.1), the external variables are denoted by y_1 and u_1. When $S(\rho) = 0$, Eq. (12.1) is termed [27] an AR representation.

In the following approach we are interested in the external behavior of the system (12.1), where we choose the underlying function space to be $C^\infty := C^\infty(\mathcal{R}, \mathfrak{R})$, this function space consists of all the infinitely differentiable functions that are defined for all time $\mathcal{R} := [0, +\infty)$ and take values in the real number field $\mathfrak{R}$. For brevity, we write $C_k^\infty := C^\infty(\mathcal{R}, \mathfrak{R}^k)$. Then the dynamical external behavior of Eq. (12.1) is given by

$$\mathfrak{B}_d(R_1, R_2; S) = \left\{ \begin{bmatrix} y_1(t) \\ u_1(t) \end{bmatrix} \in C_{m+p}^\infty \mid \exists\, \xi(t) \in C_n^\infty \text{ so that}(8.1) \text{ is valid} \right\}$$

$$= \left\{ \begin{bmatrix} y_1(t) \\ u_1(t) \end{bmatrix} \in C_{m+p}^\infty \left[R_2(\rho), R_1(\rho) \right] \begin{bmatrix} y_1(t) \\ u_1(t) \end{bmatrix} \in ImS(\rho) \right\}. \tag{12.2}$$

From the above, to avoid the trivial case that the external behavior is empty, i.e., to ensure that there exist latent variables, every pair $(y_1(t), u_1(t))$ in the external behavior must be consistent, that is

$$(I - S(\rho)S^-(\rho))(R_1(\rho)u_1(t) + R_2(\rho)y_1(t)) = 0, \quad \forall \begin{bmatrix} y_1(t) \\ u_1(t) \end{bmatrix} \in \mathfrak{B}_d(R_1, R_2; S),$$

where $\{1\}$-inverse is arbitrary. It is immediately noted that, when $S(\rho)$ is invertible or more specially when $S(\rho) = I$, the above condition is automatically satisfied.

In many real cases (for example, electrical circuits), however, system behavior is usually described (see Example 12.2.1) in the frequency domain as

$$A(s)u^*(s) + B(s)y^*(s) = C(s)\eta(s), \tag{12.3}$$

where $A(s) \in \Re[s]^{q \times p}$, $B(s) \in \Re[s]^{q \times m}$, and $C(s) \in \Re[s]^{q \times n}$, as the following example suggests, are not polynomial but rational matrices. The vector-valued signals $u^*(s)$, $y^*(s)$, and $\eta(s)$ live in the square (Lebesgue) integrable functional spaces L_2^p, L_2^m, and L_2^n respectively.

Example 12.2.1. Consider the system. The loop equations of Fig. 12.1 are written as

$$\left(R + \frac{2}{sC}\right)I_1(s) - \frac{1}{sC}I_2(s) = E_{in}(s) \tag{12.4}$$

$$-\frac{1}{sC}I_1(s) + \left(R + \frac{1}{sC}\right)I_2(s) = 0. \tag{12.5}$$

In order to obtain the transfer function $Z_t(s) = \frac{E_0(s)}{E_{in}(s)}$, one needs to solve for $I_2(s)$ to obtain

$$I_2(s) = \frac{E_{in}(s)Cs}{s^2R^2C^2 + 3sRC + 1}$$

and then note that the output voltage is

$$E_0(s) = RI_2(s).$$

This suggests that $I_1(s)$ can be regarded as latent variable, the original loop equations thus can be written into

$$\begin{bmatrix} 1 \\ 0 \end{bmatrix} E_{in}(s) + \begin{bmatrix} \frac{1}{sC} \\ R + \frac{1}{sC} \end{bmatrix} I_2(s) = \begin{bmatrix} R + \frac{2}{sC} \\ \frac{1}{sC} \end{bmatrix} I_1(s), \tag{12.6}$$

which is just in the form of Eq. (12.3). The coefficient matrices of the external variables and the latent variables are seen to be rational but not necessarily polynomial.

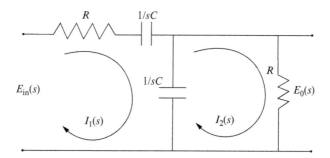

Fig. 12.1 An RC circuit.

As a special case when $C(s) = 0$ and $B(s)$ is invertible, Eq. (12.3) determines a transfer function $G(s) = -B^{-1}(s)A(s)$.

The frequency behavior of Eq. (12.3) is given by

$$\mathcal{B}_f(A, B; C) = \left\{ \begin{bmatrix} y^*(s) \\ u^*(s) \end{bmatrix} \in \mathbf{L}_2^{m+p} \mid \exists\, \eta(s) \in \mathbf{L}_2^n \text{ so that (8.3) is valid} \right\}$$

$$= \left\{ \begin{bmatrix} y^*(s) \\ u^*(s) \end{bmatrix} \in \mathbf{L}_2^{m+p} \,\middle|\, [B(s), A(s)] \begin{bmatrix} y^*(s) \\ u^*(s) \end{bmatrix} \in ImC(s) \right\}. \quad (12.7)$$

In here the matrix $B(s)$ is not necessarily invertible. It will be seen later that this is the reason why the realization of behavior comes out to be a generalization of realization of transfer function. Instead of this condition, we only make a relaxed assumption that every u^* in the behavior is consistent, i.e.,

$$(I - B(s)B^-(s))A(s)u^*(s) = 0, \quad \forall u^*(s) \in \mathcal{B}_f(A, B),$$

where $B^-(s)$ is any $\{1\}$-inverse of $B(s)$. Again it should be noted that, when $B(s)$ is invertible or more specially when $B(s) = I$, the above condition is automatically satisfied. By denoting

$$\mathcal{L}(\mathcal{B}_d)(R_1, R_2; S) = \left\{ \begin{bmatrix} \hat{y}_1(s) \\ \hat{u}_1(s) \end{bmatrix} \,\middle|\, \begin{bmatrix} \hat{y}_1(s) \\ \hat{u}_1(s) \end{bmatrix} = \mathcal{L}\left(\begin{bmatrix} y_1(t) \\ u_1(t) \end{bmatrix} \right),$$

$$\begin{bmatrix} y_1(t) \\ u_1(t) \end{bmatrix} \in \mathcal{B}_d(R_1, R_2; S),\ \rho^i y_1(0) = 0,\ \rho^i u_1(0) = 0,\ \rho^i \xi(0) = 0, i = 0, 1, \dots \right\},$$

where $\mathcal{L}(f(t))$ denotes the Laplace transformation of $f(t)$, the definition of realization of behavior follows.

Definition 12.2.1. Given a frequency behavior description (Eq. 12.3), if there exists an ARMA representation (Eq. 12.1), i.e., there exist polynomial matrices $R_1(\rho)$, $R_2(\rho)$, and $S(\rho)$ such that

$$\mathcal{L}(\mathcal{B}_d)(R_1, R_2; S) = \mathcal{B}_f(A, B; C),$$

then the ARMA representation (Eq. 12.1) is said to be a *realization of behavior* for Eq. (12.3).

Remark 12.2.1. It should be noted that, in the frequency behavior description (Eq. 12.3), the matrices $A(s)$ and $B(s)$ are assumed to be rational, not necessarily polynomial. If they are polynomial, it is obvious that the following AR representation [27]

$$A(\rho)u_1(t) + B(\rho)y_1(t) = 0$$

is a realization of behavior for Eq. (12.3). In this case, one does not need to introduce any latent variable.

Remark 12.2.2. The above concept is a generalization to the classical notion of realization of transfer function matrix.

To see this, let us consider the special case, when $B(s)$ is invertible and $C(s) = 0$. Then Eq. (12.3) determines a transfer matrix and can be written as

$$y^*(s) = -B^{-1}(s)A(s)u^*(s). \tag{12.8}$$

If there exist polynomial matrices $T(\rho), U(\rho), V(\rho)$, and $W(\rho)$ of appropriate dimensions such that $T(\rho)$ is invertible and

$$- B^{-1}(s)A(s) = V(s)T^{-1}(s)U(s) + W(s), \tag{12.9}$$

then by Definition 12.2.1, it is easy to verify that the following ARMA representation

$$\begin{bmatrix} I \\ 0 \end{bmatrix} y_1(t) + \begin{bmatrix} -W(\rho) \\ U(\rho) \end{bmatrix} u_1(t) = \begin{bmatrix} V(\rho) \\ T(\rho) \end{bmatrix} x(t) \tag{12.10}$$

is a realization of behavior for the frequency behavior description (Eq. 12.3). It is noted that Eq. (12.10) is nothing but the Rosenbrock PMD

$$\begin{cases} T(\rho)x(t) = U(\rho)u_1(t) \\ \quad y_1(t) = V(\rho)x(t) + W(\rho)u_1(t). \end{cases}$$

The condition of consistency is seen to be satisfied because of the invertibility of $T(\rho)$.

When $T(\rho) = \rho E - A$ with E singular, the above PMD is termed a singular system, while when $T(\rho) = \rho I - A$, the above description is known as the conventional state space system. It is clearly seen that in the above special cases, realization of behavior is equivalent to realization of transfer function in the classical sense.

12.3   Realization of behavior for GCSRs and DGCSRs

Before developing the realization of behavior for GCSRs, we will establish a realization of behavior for the general plant. Given the general plant described by

$$\begin{bmatrix} z(s) \\ y(s) \end{bmatrix} = P(s) \begin{bmatrix} w(s) \\ u(s) \end{bmatrix} = \begin{bmatrix} P_{11}(s) & P_{12}(s) \\ P_{21}(s) & P_{22}(s) \end{bmatrix} \begin{bmatrix} w(s) \\ u(s) \end{bmatrix}, \tag{12.11}$$

where $P_{ij}(s)$ ($i = 1, 2; j = 1, 2$) are all rational matrices with dimensions $m_i \times k_j$ ($i = 1, 2; j = 1, 2$), consider the rational matrix $P(s) \in \Re[s]^{(m_1+m_2)\times(k_1+k_2)}$. It is well-known [4] that there always exist nonunique polynomial pairs $P_1(s) \in \Re[s]^{(m_1+m_2)\times(m_1+m_2)}$ and $P_2(s) \in \Re[s]^{(m_1+m_2)\times(k_1+k_2)}$ such that

$$P(s) = P_1^{-1}(s)P_2(s). \tag{12.12}$$

It should be noted in here $P_1(s)$ and $P_2(s)$ do not need to be coprime. In this way, the following result is obtained.

Theorem 12.3.1. *The following AR representation*

$$- P_2(\rho)u_1(t) + P_1(\rho)y_1(t) = 0 \tag{12.13}$$

is a realization of behavior for the general plant (Eq. 12.11), where the external variables are denoted by

$$\begin{bmatrix} y_1(t) \\ u_1(t) \end{bmatrix} := \begin{bmatrix} z(t) \\ y(t) \\ \hline w(t) \\ u(t) \end{bmatrix},$$

and the polynomial matrices $P_1(\rho), P_2(\rho)$ satisfy Eq. (12.12).

Proof. Under the decomposition (Eq. 12.12), Eq. (12.11) can be written into

$$P_1(s) \begin{bmatrix} z(s) \\ y(s) \end{bmatrix} = P_2(s) \begin{bmatrix} w(s) \\ u(s) \end{bmatrix},$$

the above frequency behavior is seen to be $\mathcal{B}_f = Ker([P_1(s), -P_2(s)])$, while the dynamical external behavior of the AR representation (Eq. 12.13) is

$$\mathcal{B}_d = \left\{ \begin{bmatrix} y_1(t) \\ u_1(t) \end{bmatrix} \in C^\infty_{m_1+m_2+k_1+k_2} \middle| [P_1(s), -P_2(s)] \begin{bmatrix} y_1(t) \\ u_1(t) \end{bmatrix} = 0 \right\}.$$

Now let

$$P_1(\rho) = P_1^0 + P_1^1\rho + \cdots + P_1^{q_1}\rho^{q_1}, \quad P_1^j \in \Re[s]^{(m_1+m_2)\times(m_1+m_2)}, \quad j = 0, 1, \ldots, q_1$$

$$P_2(\rho) = P_2^0 + P_2^1\rho + \cdots + P_2^{q_2}\rho^{q_2}, \quad P_2^j \in \Re[s]^{(m_1+m_2)\times(k_1+k_2)}, \quad j = 0, 1, \ldots, q_2.$$

The Laplace transformation of Eq. (12.13) with zero initial condition yields

$$- P_2(s)\hat{u}_1(s) + P_1(s)\hat{y}_1(s) = 0, \tag{12.14}$$

where $\hat{u}_1(s) := \int_0^{+\infty} u_1(t)e^{-st}dt$, $\hat{y}_1(s) := \int_0^{+\infty} y_1(t)e^{-st}dt$. Thus Eq. (8.14) gives

$$\mathcal{L}(\mathcal{B}_d) = \mathcal{B}_f.$$

Hence the theorem follows from Definition 12.2.1. □

Remark 12.3.1. The above realizations are not unique due to the fact that the decompositions (Eq. 12.12) are not unique. Of course there exists a minimal realization among all the realizations. The main difference between the minimal realization and a nonminimal one is if the realization description contains redundant information. In some situations, it is important to have a description of behavior structure in terms of the original data, and the system may depend on parameters, redundant descriptions are subsequently preferable to minimal description because it allows more freedom to incorporate the dependence on the parameters in a nice way. Usually one is forced to deal with the issue of redundancy in actual system analysis. In the context of the state space systems and the generalized state space systems, such importance has already been stressed in Rosenbrock [5] and Luenberger [115]. For properties of first order realizations of AR equations, one can refer to [114].

Recall now Theorem 11.2.2. If the input-output pair $(u(s), y(s))$ is consistent about w to the plant P, then the GCSR is represented by

$$\left[\begin{array}{c} z(s) \\ w(s) \end{array}\right] = GCHAIN(P; P_{21}^-) \left[\begin{array}{c} u(s) \\ y(s) \\ h(s) \end{array}\right]$$

$$= GCHAIN^*(P; P_{21}^-) \left[\begin{array}{c} u(s) \\ y(s) \end{array}\right] + \Delta GCHAIN(P; P_{21}^-)h(s), \qquad (12.15)$$

where we denote the GCSR matrix

$$GCHAIN(P) := \left[GCHAIN^*(P; P_{21}^-) \,\middle|\, \Delta GCHAIN(P; P_{21}^-) \right]$$

$$:= \left[\begin{array}{cc|c} P_{12} - P_{11}P_{21}^-P_{22} & P_{11}P_{21}^- & P_{11}(I - P_{21}^-P_{21}) \\ -P_{21}^-P_{22} & P_{21}^- & I - P_{21}^-P_{21} \end{array}\right].$$

The above GCSR gives rise to the frequency behavior

$$\mathcal{B}_f(I, -GCHAIN(P; P_{21}^-))$$

$$:= \left\{ \left[\begin{array}{c} y^*(s) \\ u^*(s) \end{array}\right] := \left[\begin{array}{c} z(s) \\ w(s) \\ \hline u(s) \\ y(s) \end{array}\right] \,\middle|\, \begin{array}{l} y^*(s) = GCHAIN^*(P; P_{21}^-)u^*(s) \\ \\ +\Delta GCHAIN(P; P_{21}^-)h(s), \\ \quad h(s) \text{ is arbitrary}, \\ (u(s), y(s)) \text{ is consistent to } P \end{array} \right\}.$$

It can be proved that every GCSR gives rise to the same frequency behavior, in other words, the frequency behavior of GCSRs is independent of the particular {1}-inverse. The following theorem establishes this observation.

Theorem 12.3.2. *Given any two GCSEs* $GCHAIN(P; P_{21}^-)$, $GCHAIN(P; P_{21}^g)$, *which are formulated in terms of two {1}-inverse of* P_{21} *respectively, one has*

$$\mathcal{B}_f(I, -GCHAIN(P; P_{21}^-)) = \mathcal{B}_f(I, -GCHAIN(P; P_{21}^g)).$$

Proof. One only needs to prove that

$$\mathcal{B}_f(I, -GCHAIN(P; P_{21}^-)) \subseteq \mathcal{B}_f(I, -GCHAIN(P; P_{21}^g)).$$

The converse statement

$$\mathcal{B}_f(I, -GCHAIN(P; P_{21}^g)) \subseteq \mathcal{B}_f(I, -GCHAIN(P; P_{21}^-))$$

can be proved similarly.   □

From Lemma 11.1.2, there exists a matrix $K(s)$ such that

$$P_{21}^-(s) = P_{21}^g(s) + K(s) - P_{21}^g(s)P_{21}(s)K(s)P_{21}(s)P_{21}^g(s). \qquad (12.16)$$

Also the input-output pair $(u(s), y(s))$, if being consistent to the plant P, must satisfy (Theorem 11.2.1)

$$(I - P_{21}(s)P_{21}^g(s))[-P_{22}(s), I] \left[\begin{array}{c} u(s) \\ y(s) \end{array}\right] = 0.$$

This follows that

$$y(s) - P_{22}(s)u(s) = P_{21}(s)P_{21}^g(s)(y(s) - P_{22}(s)u(s)). \tag{12.17}$$

Given any

$$\begin{bmatrix} y^*(s) \\ u^*(s) \end{bmatrix} \in \mathcal{B}_f(I, -GCHAIN(P; P_{21}^-)),$$

there should be a rational vector $h_1(s)$ such that

$$\begin{bmatrix} z(s) \\ w(s) \end{bmatrix} = y^*(s) = GCHAIN^*(P; P_{21}^-)u^*(s) + \Delta GCHAIN(P; P_{21}^-)h_1(s)$$

$$= \begin{bmatrix} P_{12}(s) - P_{11}(s)P_{21}^-(s)P_{22}(s) & P_{11}(s)P_{21}^-(s) \\ -P_{21}^-(s)P_{22}(s) & P_{21}^-(s) \end{bmatrix} \begin{bmatrix} u(s) \\ y(s) \end{bmatrix}$$

$$+ \begin{bmatrix} P_{11}(s)(I - P_{21}^-(s)P_{21}(s)) \\ I - P_{21}^-(s)P_{21}(s) \end{bmatrix} h_1(s). \tag{12.18}$$

By substituting Eqs. (12.16), (12.17) into Eq. (12.18), one yields

$$z(s) = (P_{12}(s) - P_{11}(s)P_{21}^-(s)P_{22}(s))u(s) + P_{11}(s)P_{21}^-(s)y(s)$$

$$+ P_{11}(s)(I - P_{21}^-(s)P_{21}(s))h_1(s)$$

$$= [P_{12}(s) - P_{11}(s)(P_{21}^g(s) + K(s) - P_{21}^g(s)P_{21}(s)K(s)P_{21}(s)P_{21}^g(s))P_{22}(s)]u(s)$$

$$+ P_{11}(s)(P_{21}^g(s) + K(s) - P_{21}^g(s)P_{21}(s)K(s)P_{21}(s)P_{21}^g(s))y(s)$$

$$+ P_{11}(s)[I - (P_{21}^g(s) + K(s) - P_{21}^g(s)P_{21}(s)K(s)P_{21}(s)P_{21}^g(s))P_{21}(s)]h_1(s)$$

$$= (P_{12}(s) - P_{11}(s)P_{21}^g(s)P_{22}(s))u(s) + P_{11}(s)P_{21}^g(s)y(s)$$

$$+ P_{11}(s)(I - P_{21}^g(s)P_{21}(s))[K(s)(y(s) - P_{22}(s)u(s))$$

$$+ (I - K(s)P_{21}(s))h_1(s)];$$

$$w(s) = -P_{21}^-(s)P_{22}(s)u(s) + P_{21}^-(s)y(s) + (I - P_{21}^-(s)P_{21}(s))h_1(s)$$

$$= -(P_{21}^g(s) + K(s) - P_{21}^g(s)P_{21}(s)K(s)P_{21}(s)P_{21}^g(s))P_{22}(s)u(s)$$

$$+ (P_{21}^g(s) + K(s) - P_{21}^g(s)P_{21}(s)K(s)P_{21}(s)P_{21}^g(s))y(s)$$

$$+ [I - (P_{21}^g(s) + K(s) - P_{21}^g(s)P_{21}(s)K(s)P_{21}(s)P_{21}^g(s))P_{21}(s)]h_1(s)$$

$$= -P_{21}^g(s)P_{22}(s)u(s) + P_{21}^g(s)y(s)$$

$$+ (I - P_{21}^g(s)P_{21}(s))[K(s)(y(s) - P_{22}(s)u(s)) + (I - K(s)P_{21}(s))h_1(s)].$$

By letting

$$h_2(s) := K(s)(y(s) - P_{22}(s)u(s)) + (I - K(s)P_{21}(s))h_1(s),$$

the above formulations about $z(s)$ and $w(s)$ can be written into the following matrix form

$$\begin{bmatrix} z(s) \\ w(s) \end{bmatrix} = GCHAIN^*(P; P_{21}^g)u^*(s) + \Delta GCHAIN(P; P_{21}^g)h_2(s),$$

which displays the fact that

$$\begin{bmatrix} y^*(s) \\ u^*(s) \end{bmatrix} \in \mathcal{B}_f(I, -GCHAIN(P; P_{21}^g)).$$

Subsequently

$$\mathcal{B}_f(I, -GCHAIN(P; P_{21}^-)) \subseteq \mathcal{B}_f(I, -GCHAIN(P; P_{21}^g)).$$

This finishes the proof.

By the virtue of the above theorem, the frequency behavior of any GCSR thus can be simply denoted by $\mathcal{B}_f(I, -GCHAIN(P))$.

One of the remaining aims of this section is to show how the frequency behavior of GCSRs can be realized as the dynamical behavior of an ARMA representation through the approach of realization of behavior. To this end, the general plant (Eq. 12.11) is rewritten into

$$\begin{bmatrix} z(s) \\ y(s) \end{bmatrix} = \begin{bmatrix} \frac{P_{11}^*(s)}{g(s)} & \frac{P_{12}^*(s)}{g(s)} \\ \frac{P_{21}^*(s)}{g(s)} & \frac{P_{22}^*(s)}{g(s)} \end{bmatrix} \begin{bmatrix} w(s) \\ u(s) \end{bmatrix}, \tag{12.19}$$

where $g(s)$ is the least common (monic) multiple of the denominator polynomials of all the entries in $P(s)$, and $P_{ij}^*(s)/g(s) = P_{ij}(s)$, $i = 1, 2$; $j = 1, 2$. It is immediately noted that the above decomposition of P is a special case of Eq. (12.12). By letting

$$\begin{bmatrix} z_c(s) \\ y_c(s) \end{bmatrix} = g(s)I_{m_1+m_2} \begin{bmatrix} z(s) \\ y(s) \end{bmatrix}, \tag{12.20}$$

$$P^*(s) = \begin{bmatrix} P_{11}^*(s) & P_{12}^*(s) \\ P_{21}^*(s) & P_{22}^*(s) \end{bmatrix}, \tag{12.21}$$

where $I_{m_1+m_2}$ is the identity matrix with dimension $m_1 + m_2$. Eq. (12.19) thus takes the form

$$\begin{bmatrix} z_c(s) \\ y_c(s) \end{bmatrix} = P^*(s) \begin{bmatrix} w(s) \\ u(s) \end{bmatrix}. \tag{12.22}$$

It should be noted that in here $P^*(s)$ is a polynomial matrix and $g(s)$ is a polynomial.

As is shown before, the realization of behavior for a general plant is rather straightforward, while the realization of behavior for GCSR is more obscure, as in this case the introduction of latent variables is necessary. To propose a realization of behavior for GCSRs, consider the following ARMA representation

$$\begin{bmatrix} P_{22}^*(\rho) & -g(\rho)I \\ P_{12}^*(\rho) & 0 \\ 0 & 0 \end{bmatrix} \begin{bmatrix} u(t) \\ y(t) \end{bmatrix} + \begin{bmatrix} 0 & 0 \\ -g(\rho)I & 0 \\ 0 & -I \end{bmatrix} \begin{bmatrix} z(t) \\ w(t) \end{bmatrix}$$

$$= \begin{bmatrix} P_{21}^*(\rho) \\ P_{11}^*(\rho) \\ I \end{bmatrix} x(t), \quad t \geq 0. \tag{12.23}$$

The above ARMA representation is in fact

$$
\begin{cases}
P_{21}^*(\rho)x(t) = [P_{22}^*(\rho), -g(\rho)I] \begin{bmatrix} u(t) \\ y(t) \end{bmatrix} \\
\begin{bmatrix} g(\rho)I & 0 \\ 0 & I \end{bmatrix} \begin{bmatrix} z(t) \\ w(t) \end{bmatrix} = \begin{bmatrix} -P_{11}^*(\rho) \\ -I \end{bmatrix} x(t) + \begin{bmatrix} P_{12}^*(\rho) & 0 \\ 0 & 0 \end{bmatrix} \begin{bmatrix} u(t) \\ y(t) \end{bmatrix},
\end{cases}
$$
$$(12.24)$$

where all the identity matrices and all the zero block matrices are of appropriate dimensions. $x(t) \in C_{k_1}^\infty$ are the latent variables. It is noted that every condition pair

$$
\begin{bmatrix} y_1(t) \\ u_1(s) \end{bmatrix} := \begin{bmatrix} z(t) \\ w(t) \\ \hline u(t) \\ y(t) \end{bmatrix} \in \mathcal{B}_d
$$

is consistent about the latent variables $x(t)$ is equivalent that every input-output pair $(u(s), y(s))$ is consistent about w.

The following technical result will be used in the proof of Theorem 12.3.3. It is stated as a lemma.

Lemma 12.3.1. *Let all the initial conditions and their derivatives of $u(\cdot)$ and $y(\cdot)$ be zero. The following condition of consistency in the time domain*

$$
[I - P_{21}^*(\rho)(P_{21}^*(\rho))^-][P_{22}^*(\rho), -g(\rho)I] \begin{bmatrix} u(t) \\ y(t) \end{bmatrix} = 0
$$

is equivalent to the following consistency condition in the frequency domain

$$
[I - P_{21}^*(s)(P_{21}^*(s))^-][P_{22}^*(s), -g(s)I] \begin{bmatrix} \hat{u}(s) \\ \hat{y}(s) \end{bmatrix} = 0,
$$

i.e.,

$$
[I - P_{21}(s)(P_{21}(s))^-][P_{22}(s), -I] \begin{bmatrix} \hat{u}(s) \\ \hat{y}(s) \end{bmatrix} = 0,
$$

where $\hat{u}(s)$, $\hat{y}(s)$ denote the Laplace transformation of $u(t)$ and $y(t)$ respectively.

Proof. Viewing the Laplace transformation and inverse Laplace transformation properties, one only needs to note that the matrix $I - P_{21}^*(\cdot)(P_{21}^*(\cdot))^-$ is a polynomial matrix for any {1}-inverse of P_{21}^*. This can be done easily by using Proposition 12.1.1. □

Now we are ready to state and prove the following result.

Theorem 12.3.3. *The ARMA representation (Eq. 12.23) is a realization of behavior for any GCSR GCHAIN$(P; P_{21}^-)$.*

Proof. Given any GCSR, its frequency behavior is $\mathcal{B}_f(I, -GCHAIN(P))$. The dynamical external behavior of the ARMA representation (Eq. 12.23) is

$$
\mathcal{B}_d = \left\{ \begin{bmatrix} y_1(t) \\ u_1(t) \end{bmatrix} = \begin{bmatrix} z(t) \\ w(t) \\ \hline u(t) \\ y(t) \end{bmatrix} \in C_{m_1+m_2+k_1+k_2}^\infty \;\middle|\; \begin{array}{l} \exists x(t) \in C_{k_1}^\infty \\ \\ \text{so that (8.23) is valid} \end{array} \right\} (12.25)
$$

$\forall \begin{bmatrix} y_1(t) \\ u_1(t) \end{bmatrix} \in \mathcal{B}_d$, to ensure that there exists $x(t)$ such that Eq. (12.23), i.e., Eq. (12.24) is valid, $u_1(t)$ must be consistent to Eq. (12.24). This suggests that

$$[I - P_{21}^*(\rho)(P_{21}^*(\rho))^-][P_{22}^*(\rho), -g(\rho)I] \begin{bmatrix} u(t) \\ y(t) \end{bmatrix} = 0.$$

It is easily seen to be equivalent to

$$\begin{bmatrix} \hat{u}(s) \\ \hat{y}(s) \end{bmatrix} \in Ker\{[I - P_{21}(s)P_{21}^-(s)^-][P_{22}(s), -I]\}.$$

The Laplace transformation of Eq. (12.24) (with zero initial conditions) yields

$$P_{21}^*(s)\hat{x}(s) = [P_{22}^*(s), -g(s)I] \begin{bmatrix} \hat{u}(s) \\ \hat{y}(s) \end{bmatrix}, \tag{12.26}$$

and

$$\begin{bmatrix} g(s)I & 0 \\ 0 & I \end{bmatrix} \begin{bmatrix} \hat{z}(s) \\ \hat{w}(s) \end{bmatrix} = \begin{bmatrix} -P_{11}^*(s) \\ -I \end{bmatrix} \hat{x}(s) + \begin{bmatrix} P_{12}^*(s) & 0 \\ 0 & 0 \end{bmatrix} \begin{bmatrix} \hat{u}(s) \\ \hat{y}(s) \end{bmatrix}, \tag{12.27}$$

where $\hat{f}(s) := \int_0^{+\infty} f(t)e^{-st} dt$. Due to the reason stated in the proof of Theorem 12.3.1 all the Laplace transformed initial vectors that are associated with the initial values of the variables and their derivatives are zeros. $\qquad\square$

Due to the consistency of $\begin{bmatrix} \hat{u}(s) \\ \hat{y}(s) \end{bmatrix}$, Eq. (12.26) determines the latent variables $\hat{x}(s)$. By solving the latent variables $\hat{x}$ in Eq. (12.26) and then substituting into Eq. (12.27), one obtains

$$\begin{bmatrix} g(s)I & 0 \\ 0 & I \end{bmatrix} \begin{bmatrix} \hat{z}(s) \\ \hat{w}(s) \end{bmatrix}$$
$$= \begin{bmatrix} P_{12}^*(s) - P_{11}^*(s)(P_{21}^*)^-(s)P_{22}^*(s) & g(s)P_{11}^*(s)(P_{21}^*)^-(s) \\ -(P_{21}^*)^-(s)P_{22}^*(s) & g(s)(P_{21}^*)^-(s) \end{bmatrix} \begin{bmatrix} \hat{u}(s) \\ \hat{y}(s) \end{bmatrix}$$
$$+ \begin{bmatrix} P_{11}^*(s)[I - (P_{21}^*)^-(s)P_{21}^*(s)] \\ I - (P_{21}^*)^-(s)P_{21}^*(s) \end{bmatrix} h(s), \tag{12.28}$$

where $h(s)$ is any rational vector. By noticing that $P_{ij}^*(s) = g(s)P_{ij}(s)$, $i = 1, 2; j = 1, 2$, and that $P_{21}^-(s) = g(s)(P_{21}^*)^-(s)$, Eq. (12.28) can also be written into

$$\begin{bmatrix} \hat{z}(s) \\ \hat{w}(s) \end{bmatrix} = \begin{bmatrix} P_{12}(s) - P_{11}(s)P_{21}^-(s)P_{22}(s) & P_{11}(s)P_{21}^-(s) \\ -P_{21}^-(s)P_{22}(s) & P_{21}^-(s) \end{bmatrix} \begin{bmatrix} \hat{u}(s) \\ \hat{y}(s) \end{bmatrix}$$
$$+ \begin{bmatrix} P_{11}(s)[I - P_{21}^-(s)P_{21}(s)] \\ I - P_{21}^-(s)P_{21}(s) \end{bmatrix} h(s). \tag{12.29}$$

It is thus seen that

$$
\begin{bmatrix} y^*(s) \\ u^*(s) \end{bmatrix} = \begin{bmatrix} \hat{z}(s) \\ \hat{w}(s) \\ \hline \hat{u}(s) \\ \hat{y}(s) \end{bmatrix} \in \mathcal{B}_f(I, -GCHAIN(P)).
$$

So far it has been proved that

$$
\mathcal{L}(\mathcal{B}_d) \subseteq \mathcal{B}_f(I, -GCHAIN(P)).
$$

To prove the statement

$$
\mathcal{B}_f(I, -GCHAIN(P)) \subseteq \mathcal{L}(\mathcal{B}_d),
$$

let

$$
\begin{bmatrix} y^*(s) \\ u^*(s) \end{bmatrix} \in \mathcal{B}_f(I, -GCHAIN(P)),
$$

there should be a rational vector $h_1(s)$ such that

$$
y^*(s) = GCHAIN^*(P; P_{21}^-)u^*(s) + \Delta GCHAIN(P; P_{21}^-)h_1(s), \tag{12.30}
$$

for any {1}-inverse of P_{21}. Furthermore, the input-output pair $(u(s), y(s))$ must be consistent to the plant P. Now let

$$
x(s) = (P_{21}^*)^-(s)[P_{22}^*(s), -g(s)I]u^*(s) + [I - (P_{21}^*)^-(s)P_{21}^*(s)]h_1(s),
$$

using the consistency of $(u(s), y(s))$, it is easy to verify that the variables y^*, u^*, and x satisfy Eqs. (12.26), (12.27), this is to say that

$$
\begin{bmatrix} y^*(s) \\ u^*(s) \end{bmatrix} \in \mathcal{L}(\mathcal{B}_d).
$$

On noticing Eq. (12.20), one can write the ARMA representation (Eq. 12.23) into the following Rosenbrock PMD

$$
\begin{cases} P_{21}^*(\rho)x(t) = [P_{22}^*(\rho), -I] \begin{bmatrix} u(t) \\ y_c(t) \end{bmatrix} \\ \begin{bmatrix} z_c(t) \\ w(t) \end{bmatrix} = \begin{bmatrix} -P_{11}^*(\rho) \\ -I \end{bmatrix} x(t) + \begin{bmatrix} P_{12}^*(\rho) & 0 \\ 0 & 0 \end{bmatrix} \begin{bmatrix} u(t) \\ y_c(t) \end{bmatrix}. \end{cases} \tag{12.31}
$$

It is easily seen that

$$
P_{21}^-(s) = g(s)(P_{21}^*)^-(s).
$$

By substituting the above and

$$
P_{ij}(s) = P_{ij}^*(s)/g(s), \quad i = 1, 2; \ j = 1, 2,
$$

into the GCSR

$$
\begin{bmatrix} z(s) \\ w(s) \end{bmatrix}
$$

$$
= \begin{bmatrix} P_{12}(s) - P_{11}(s)P_{21}^-(s)P_{22}(s) & P_{11}(s)P_{21}^-(s) & P_{11}(s)[I - P_{21}^-(s)P_{21}(s)] \\ -P_{21}^-(s)P_{22}(s) & P_{21}^-(s) & I - P_{21}^-(s)P_{21}(s) \end{bmatrix} \begin{bmatrix} u(s) \\ y(s) \\ h(s) \end{bmatrix},
$$

using the notations of $z_c(s) = g(s)z(s)$ and $y_c(s) = g(s)y(s)$, one can find that

$$
\begin{bmatrix} z_c(s) \\ w(s) \end{bmatrix} = GCHAIN(P^*; (P_{21}^*)^-) \begin{bmatrix} u(s) \\ y_c(s) \\ h(s) \end{bmatrix},
$$

where we denote the matrix

$$
GCHAIN(P^*; (P_{21}^*)^-)
$$
$$
:= \begin{bmatrix} P_{12}^*(s) - P_{11}^*(s)(P_{21}^*)^-(s)P_{22}^*(s) & P_{11}^*(s)(P_{21}^*)^-(s) & I - P_{21}^-(s)P_{21}(s) \\ -(P_{21}^*)^-(s)P_{22}^*(s) & P_{21}^-(s) & I - (P_{21}^*)^-(s)P_{21}^*(s) \end{bmatrix}.
$$

A further result concerning the realization of behavior for any GCSR $GCHAIN(P^*; (P_{21}^*)^-)$ is the following theorem.

Theorem 12.3.4. *The Rosenbrock PMD (Eq. 12.31) is a realization of behavior for any GCSR GCHAIN$(P^*; (P_{21}^*)^-)$.*

Proof. This follows readily from Theorem 12.3.3 on noting that

$$
\begin{bmatrix} z_c(s) \\ w(s) \end{bmatrix} = GCHAIN(P^*; (P_{21}^*)^-) \begin{bmatrix} u(s) \\ y_c(s) \\ h(s) \end{bmatrix},
$$

where $h(s)$ is arbitrary rational vector, and that in the matrix P^* the least common (monic) multiple of the denominator polynomials of all the entries is 1. $\qquad\square$

The realization of behavior for DGCSRs of the plant P can be proposed in a completely analogous manner. To this end, consider the following ARMA representation

$$
\begin{bmatrix} -g(\rho)I & P_{11}^*(\rho) & 0 & 0 \\ 0 & 0 & -I & 0 \\ 0 & P_{21}^*(\rho) & 0 & -g(\rho)I \end{bmatrix} \begin{bmatrix} z(t) \\ w(t) \\ u(t) \\ y(t) \end{bmatrix} = \begin{bmatrix} P_{12}^*(\rho) \\ I \\ P_{22}^*(\rho) \end{bmatrix} x(t), \quad t \geq 0.
$$

$$(12.32)$$

The above ARMA representation is in fact

$$
\begin{cases} P_{12}^*(\rho)x(t) = [-g(\rho)I, P_{11}^*(\rho)] \begin{bmatrix} z(t) \\ w(t) \end{bmatrix} \\ \begin{bmatrix} I & 0 \\ 0 & g(\rho)I \end{bmatrix} \begin{bmatrix} u(t) \\ y(t) \end{bmatrix} = \begin{bmatrix} -I \\ -P_{22}^*(\rho) \end{bmatrix} x(t) + \begin{bmatrix} 0 & 0 \\ 0 & P_{21}^*(\rho) \end{bmatrix} \begin{bmatrix} z(t) \\ w(t) \end{bmatrix}. \end{cases}
$$

$$(12.33)$$

Theorem 12.3.5. *The ARMA representation (Eq. 12.32) is a realization of behavior for any DGCSR DGCHAIN$(P; P_{12}^-)$.*

Proof. The proof is similar to that of Theorem 12.3.3.   □

Also if Eq. (12.32) is written into the following Rosenbrock PMD

$$
\begin{cases}
P_{12}^*(\rho)x(t) = [-I, P_{11}^*(\rho)] \begin{bmatrix} z_c(t) \\ w(t) \end{bmatrix} \\[2mm]
\begin{bmatrix} u(t) \\ y_c(t) \end{bmatrix} = \begin{bmatrix} -I \\ -P_{22}^*(\rho) \end{bmatrix} x(t) + \begin{bmatrix} 0 & 0 \\ 0 & P_{21}^*(\rho) \end{bmatrix} \begin{bmatrix} z_c(t) \\ w(t) \end{bmatrix},
\end{cases}
\tag{12.34}
$$

one proposes a realization of behavior for any DGCSR $DGCHAIN(P^*; (P_{21}^*)^-)$, where any $DGCHAIN(P^*; (P_{12}^*)^-)$ is given by

$$
DGCHAIN(P^*; (P_{12}^*)^-)
$$
$$
:= \left[\begin{array}{cc|c} (P_{12}^*)^-(s) & -(P_{12}^*)^-(s)P_{11}^*(s) & I - (P_{12}^*)^-(s)P_{12}^*(s) \\ P_{22}^*(s)(P_{12}^*)^-(s) & P_{21}^*(s) - P_{22}^*(s)(P_{12}^*)^-(s)P_{11}^*(s) & P_{22}^*(s)[I - (P_{12}^*)^-(s)P_{12}^*(s)] \end{array} \right],
$$

and satisfies

$$
\begin{bmatrix} u(s) \\ y_c(s) \end{bmatrix} = DGCHAIN(P^*; (P_{12}^*)^-) \begin{bmatrix} z_c(s) \\ w(s) \\ q(s) \end{bmatrix},
$$

where $q(s)$ is arbitrary rational vector. $(P_{12}^*)^-(s)$ is any {1}-inverse of $P_{12}^*(s)$. This result is stated in the following theorem.

Theorem 12.3.6. *The Rosenbrock PMD (Eq. 12.34) is a realization of behavior for any DGCSR DGCHAIN$(P^*; (P_{12}^*)^-)$.*

Proof. The proof follows from a direct application of Theorem 8.3.5 on noticing the special formulation of DGCSR $DGCHAIN(P^*; (P_{12}^*)^-)$.   □

Remark 12.3.2. The above theorems are interesting, not least for the way in which they clarify the input-output structure of GCSRs and that of DGCSRs. More importantly than this, however, is the observation that frequency behavior of any GCSR, or any DGCSR, can be completely recovered in a precise way, by introducing latent variables to the dynamical behavior of the ARMA representations via the approach of realization of behavior.

12.4 Conclusions

This chapter has presented a new notion of realization of behavior. It has been shown that realization of behavior generalizes the classical concept of realization of transfer function matrix by virtue of the input consistency assumption essentially relaxing the condition of full rank, which is put on the relevant matrix to ensure the existence of the transfer function. The basic idea in this approach is to find an ARMA representation for a given frequency behavior description such that the known frequency behavior is completely recovered from the corresponding dynamical behavior. From this point of view, realization of behavior is seen to be a converse procedure to the latent variable elimination process that was studied by Willems [27]. Such a realization approach is believed to be highly significant in modeling dynamical systems in some real cases where the system behavior is conveniently described in the frequency domain. Since no numerical computation is needed, the realization of behavior is believed to be particularly suitable for situations in which the coefficients are symbolic rather than numerical.

From an input/output viewpoint, a CSR is in fact an alternative form of system description. It is well-known and is widely used in classical circuit theory. This approach when combined with the theory of J-spectral factorization provides the most compact theory for $\mathcal{H}_\infty$ control. CSR is undoubtedly a potential research area due to the fact that it finds many applications in circuit theory, $\mathcal{H}_\infty$ control, behavior control, and infinite-dimension system theory. This chapter has investigated the input/output structure of the GCSR and has made contributions regarding this fundamental issue. Based on the approach of realization of behavior, the behavior structures of the GCSRs and the DGCSRs have been clarified. It has been shown that any GCSR or any DGCSR develops the same (frequency) behavior. Subsequently the corresponding ARMA representations are proposed and are proved to be realizations of behavior for any GCSR and for any DGCSR. More specifically, two Rosenbrock PMDs are found to be the realizations of behavior for any GCSR $GCHAIN(P^*; (P_{21}^*)^-)$ and any DGCSR $DGCHAIN(P^*; (P_{12}^*)^-)$. Once these ARMA representations are proposed, one can further find the corresponding first-order system representations by using the method of [30] or other well-developed realization approaches such as Kailath [32]. Also a constructive procedure for AR equation realization suggested by Vafiadis and Karcanias [114] can be used to obtain the required first-order system representations. The results are thus interesting in that they provide a natural linkage between the new chain-scattering approach and the well-developed Rosenbrock PMD theory or the developing behavior theory.

Related extensions to system well-posedness and internal stability

13.1 Introduction

Consider the standard feedback configuration given in Fig. 13.1, where P is the plant and K is the *controller*. The plant accounts for all system components except for the controller. The signal w contains all external inputs, including disturbances, sensor noise, and commands. In some circumstances, the input signal w belongs to a set that can be characterized to some degree [39]. The output z is an error signal, y is the vector of measured variables, and u is the control input. All these signals z, y, w, and u are *real rational* vectors in the complex variable $s = \lambda + j\omega$. The equations that describe the feedback configuration of Fig. 13.1 are

$$\begin{bmatrix} z \\ y \end{bmatrix} = P \begin{bmatrix} w \\ u \end{bmatrix},$$
$$u = Ky, \tag{13.1}$$

where P and K are *real rational* transfer matrices in the complex variable $s = \lambda + j\omega$. This means that only finite-dimensional linear time-invariant systems are considered. We write

$$P = \begin{bmatrix} P_{11} & P_{12} \\ P_{21} & P_{22} \end{bmatrix}, \tag{13.2}$$

where the partitioning is consistent with the dimensions of z, y and w, u. These will be denoted, respectively, by l_1, l_2 and m_1, m_2.

Roughly speaking, the $\mathcal{H}_\infty$ control problem consists of finding a controller that makes the closed-loop system well-posed, internally stable and minimizes the $\mathcal{H}_\infty$ norm of the closed-loop transfer function. The basic requirements in this problem are *well-posedness* (WP) and *internal stability* (IS) [36, 39, 40]. The so-called WP requirement is to require the matrix $(I - P_{22}K)$ (or $(I - KP_{22})$) to be invertible, while IS guarantees bounded signals z, u, y for all bounded input signals w.

The $\mathcal{H}_\infty$ or $\mathcal{H}_2$ control problem in fact amounts to the constrained minimization problems. The constraints come from a WP requirement and an IS requirement and the object we seek to minimize is the $\mathcal{H}_\infty$ norm or $\mathcal{H}_2$ norm of the closed-loop transfer function. If these two requirements are relaxed, it will definitely enable us to search the appropriate controllers over a wider class, the resulting controllers might thus be more flexible in accomplishing the control aims. This need is indeed seen from the

A Generalized Framework of Linear Multivariable Control. http://dx.doi.org/10.1016/B978-0-08-101946-7.00013-5

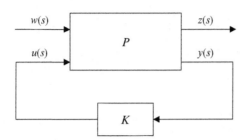

Fig. 13.1 Standard control system.

following situations. The fact that P_{12} and P_{21} do not need to have full rank at ∞, as usually required, allows [41] one to study the celebrated minimum-sensitivity problem [35]. Also by relaxing the usual requirement that the entire feedback system be stable, the relaxed requirement that K merely stabilize P_{22} allows one to include unstable weighting filters (including ones that have poles on the imaginary axis) in the mixed sensitivity problem [41].

For some plants, there may be no controller that can make the closed-loop systems internally stable. There exist, however, controllers to make the input-output transfer function $T_{z,w}$ (from w to z) stable. To such plants, one may still be interested in designing a controller to achieve this input-output stability and also to accomplish other control aims.

The main aim of this chapter is to present a certain generalization to the classical concepts of WP and IS. The input consistency and output uniqueness [116] of the closed-loop system in the standard control feedback configurations are investigated. Based on this, a number of notions are introduced, such as *fully internal well-posedness* (FIWP), *externally internal well-posedness* (EIWP), and *externally internal stability* (EIS), which characterize the rich input-output and stability features of the general control systems in a general setting. It is shown that FIWP is equivalent to the classical assumption WP, which has been widely adopted in control system designs, while EIWP and EIS generalize the notions of WP and IS. Some conditions are established to verify EIWP by using the approach of matrix generalized inverses. Natural links between EIWP and WP, EIS and IS are pointed out. This approach also leads to a generalization of the *linear fractional transformation* (LFT). The *generalized linear fractional transformations* (GLFTs) are formulated into certain general parameterized forms by using the generalized inverse of matrices. Finally on the basis of these notions of EIWP, EIS, and GLFT, the extended $\mathcal{H}_\infty$ control problem is defined in a general setting.

The Hardy 2-space $\mathcal{H}_2$ consists of functions of a complex variable that are analytic in the open right-half of the complex plane and such that the norm

$$\|f\|_2 = \left\{ \sup_{\lambda > 0} \frac{1}{2\pi} \int_{-\infty}^{\infty} f^*(\lambda + j\omega) f(\lambda + j\omega) d\omega \right\}^{1/2}$$

is finite, where $f^* := f(-s)^T$. That is

$$\mathcal{H}_2 = \left\{ f : f \text{ is analytic in } Re > 0 \text{ and } \|f\|_2 < \infty \right\}.$$

The $\mathcal{H}_2$ space, which consists of the functional vectors with size $m \times 1$, is denoted by $\mathcal{H}_2^m$. The class of systems for which the transfer function G is analytic in the open-right-half plane and this supremum, which is finite, is known as $\mathcal{H}_\infty$

$$\mathcal{H}_\infty = \{ G : G \text{ is analytic in } Re > 0 \text{ and } \|G\|_\infty < \infty \},$$

in which

$$\|G\|_\infty = \sup_{\lambda > 0} \left\{ \sup_\omega \hat{\sigma} (G(\lambda + j\omega)) \right\},$$

where $\hat{\sigma}(G)$ denotes the *maximum singular value* of G. A transfer function matrix G defines a stable system if and only if $G \in \mathcal{H}_\infty$. The $\mathcal{H}_\infty$ space, which consists of the $r \times m$ matrices, is denoted by $\mathcal{H}_\infty^{r \times m}$.

13.2 Input consistency, output uniqueness, fully internal well-posedness, and externally internal well-posedness

In connection with generalizing the concepts of WP and IS we will present two main issues, *input consistency* and *output uniqueness*, in what follows.

Recall that a standard general feedback configuration in terms of which many problems of interest in control theory can be posed is given in Fig. 13.1. Fig. 13.1 represents the algebraic equations (13.1) and (13.2), i.e.,

$$\begin{aligned} z &= P_{11}w + P_{12}u \\ y &= P_{21}w + P_{22}un \\ u &= Ky, \end{aligned} \tag{13.3}$$

where $w \in \mathcal{W}$, and according to Doyle et al. [39] $\mathcal{W}$ is assumed to be a known subspace of $\mathcal{H}_2^{m_1}$.

To define [36] what it means for K to stabilize P, two additional inputs v_1 and v_2 are introduced, as shown in Fig. 13.2. Fig. 13.1 can be deemed as a special case of Fig. 13.2 by simply letting $v_1 = v_2 = 0$.

The equation relating the three inputs w, v_1, v_2 and the three signals z, u, y can be written in the form

$$\Sigma : Y_1(P, K)Z_O + Y_2(P)W_I = 0, \tag{13.4}$$

where

$$Y_1(P, K) := \begin{bmatrix} -I & P_{12} & 0 \\ 0 & P_{22} & -I \\ 0 & I & -K \end{bmatrix}, \quad Z_O := \begin{bmatrix} z \\ u \\ y \end{bmatrix}$$

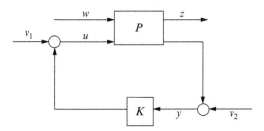

Fig. 13.2 Diagram for stability definition.

$$Y_2(P) := \begin{bmatrix} P_{11} & 0 & 0 \\ P_{21} & 0 & I \\ 0 & -I & 0 \end{bmatrix}, \quad W_I := \begin{bmatrix} w \\ v_1 \\ v_2 \end{bmatrix},$$

where 0 denotes the zero matrix, and I denotes the identity matrix in appropriate dimensions, $W_I \in \mathcal{W}_I$, $\mathcal{W}_I$ is a subspace of $\mathcal{H}_2^{m_1+m_2+l_2}$. As a special case of Eq. (13.4), Eq. (13.3) which describes the closed-loop system shown in Fig. 13.1 can subsequently be written into

$$\Sigma' : Y_1(P,K)Z_O + Y_2(P)W_I' = 0, \tag{13.5}$$

where $W_I' := [w^T, 0^T, 0^T]^T \in \mathcal{W}_I'$, $\mathcal{W}_I' := \{W_I', w \in \mathcal{W}\} \subseteq \mathcal{H}_2^{m_1+m_2+l_2}$.

The general control design theories such as $\mathcal{H}_\infty$ control theory and $\mathcal{H}_2$ control theory impose two requirements on the above closed-loop system. One is the WP requirement, which requires the matrix $(I - P_{22}K)$(or $(I - KP_{22})$) to be invertible for every K. This is to guarantee that the coefficient matrix $Y_1(P,K)$ in Eq. (13.4) or Eq. (13.5) has a real-rational inverse for every K, which will be seen clearly later.

The case where the controllers result in a singular matrix $(I - P_{22}K)$(or $(I - KP_{22})$) is said to be *ill-posed* in the literature (see, e.g., [117]). To our knowledge, such controllers have not been considered in the $\mathcal{H}_\infty$ theory, and neither have they been applied to control engineering yet. The other is the IS requirement, which requires the *nine transfer function* from w, v_1, v_2 to z, u, y to be stable. If the assumption WP holds, this nine transfer function can be written into [25]

$$\mathcal{T}_{OI} = -Y_1^{-1}(P,K)Y_2(P)$$

$$:= \begin{bmatrix} T_{z,w} & T_{z,v_1} & T_{z,v_2} \\ T_{u,w} & T_{u,v_1} & T_{u,v_2} \\ T_{y,w} & T_{y,v_1} & T_{y,v_2} \end{bmatrix}$$

$$= \begin{bmatrix} P_{11} + P_{12}K\Delta^{-1}P_{21} & P_{12} + P_{12}K\Delta^{-1}P_{22} & P_{12}K\Delta^{-1} \\ K\Delta^{-1}P_{21} & I + K\Delta^{-1}P_{22} & K\Delta^{-1} \\ \Delta^{-1}P_{21} & \Delta^{-1}P_{22} & \Delta^{-1} \end{bmatrix},$$

where $\Delta := I - P_{22}K$. The transfer function $T_{z,w} = P_{11} + P_{12}K(I - P_{22})^{-1}P_{21}$ is known as the LFT [91], which has been widely used in circuit and system theory (see, e.g., [118]).

For some plants, there is no controller that can make the closed-loop systems be internal stable. There exist, however, controllers to make the input-output transfer function $T_{z,w}$ (from w to z) be stable. An obvious example is, in the plant $P, P_{22} = 0, P_{21}$ is unstable. The transfer function $T_{y,w} = P_{21}$ is thus unstable for any K, the input-output transfer function $T_{z,w} = P_{11} + P_{12}KP_{21}$ is, however, possibly stable for some K. For such plants, one may still be interested in designing a controller to achieve this input-output stability and also to accomplish other control aims. In this situation, one thus has to relax the IS requirement.

We do not make the above assumptions of WP and IS in the following. Instead, it is found that, if the input signal information W_I or W_I' is taken into account, it will result in a new interpretation of the issue of WP in this more general setting. Based on this, the stability of the closed-loop system Σ (or Σ') can thus be characterized in the input-output setting by using the generalized matrix inverse.

A primary issue that plays a key role in this approach is termed *input consistency*, i.e., the conditions under which the input is *acceptable* to the closed-loop system. A second important issue is, given a consistent input, whether the closed-loop system Σ is able to give rise to a unique output Z_O or z. These two issues are central to the existence of the relevant *generalized transfer function* [116]. In *behavior theory* (see, e.g., [27]), when an input is consistent for the system, it is said that the input is *free* to the system, while if any free input produces a unique output it is said that the output *possesses* the input. These two issues are central to guaranteeing the existence of the generalized transfer function [116] in polynomial matrix description (PMD) theory. Once the input consistency and the output uniqueness are determined, it is natural to characterize the stability of the system in terms of the resulting generalized transfer function.

Definition 13.2.1. An input signal $W_I(W_I')$ is be said to be consistent for the closed-loop system $\Sigma(\Sigma')$ if there exists at least one Z_O satisfying Eqs. (13.4), (13.5).

Lemma 11.1.1 leads to the following immediate characterization of the input consistency.

Theorem 13.2.1.

(c1) *The closed-loop system Σ is input consistent with respect to the input signal $W_I \in \mathcal{W}_I$ if and only if*

$$(I - Y_1(P, K)Y_1^-(P, K))Y_2(P)W_I = 0, \tag{13.6}$$

(c2) *The closed-loop system Σ' is input consistent with respect to the input signal $W_I' \in \mathcal{W}_I'$ if and only if*

$$(I - Y_1(P, K)Y_1^-(P, K))Y_2(P)W_I' = 0, \tag{13.7}$$

where $Y_1^-(P, K)$ denotes any {1}-inverse of $Y_1(P, K)$.

Proof. The proof follows directly from Lemma 11.1.1.   □

Obviously, the signals z, u, and y in the standard general feedback configuration Fig. 13.1 or Fig. 13.2 play different roles and have different implications due to the fact that the signal z is indeed the true *external output* signal of the closed-loop system. It is thus of main concern in circumstances when the input-output feature of the closed-loop system is considered, since the signals u and y can be deemed as *internal output* signals of the closed-loop system. Thus there are two remaining interesting question to ask. First, given a consistent input signal $W_I(W_I')$ in $\Sigma(\Sigma')$, does the closed-loop system $\Sigma(\Sigma')$ have a unique output signal Z_O? Second, given a consistent input signal $W_I(W_I')$ in $\Sigma(\Sigma')$, does the closed-loop system $\Sigma(\Sigma')$ have a unique output signal z? These two questions naturally lead to the following definitions.

Definition 13.2.2. The closed-loop system $\Sigma(\Sigma')$ is said to be *fully internal well-posed* if for any input signal $W_I \in \mathcal{W}_I(W_I' \in \mathcal{W}_I')$ the closed-loop system $\Sigma(\Sigma')$ is input consistent and the closed-loop system $\Sigma(\Sigma')$ has a unique output signal Z_O for any consistent input $W_I \in \mathcal{W}_I(W_I' \in \mathcal{W}_I')$.

In view of Lemma 11.1.1, it is observed that FIWP is essentially equivalent to WP. The following theorem gives this statement.

Theorem 13.2.2. The closed-loop system $\Sigma(\Sigma')$ is fully internal well-posed if and only if it is well-posed.

Proof. We only need to prove this statement holds for Σ, the statement about Σ' can be proved similarly.   □

(if) If the closed-loop system Σ is well-posed for a feedback K, then $(I - KP_{22})$ is invertible for this K. From

$$
Y_1(P,K) = \begin{bmatrix} -I & P_{12} & 0 \\ 0 & P_{22} & -I \\ 0 & I & -K \end{bmatrix}
$$
$$
= \begin{bmatrix} I & 0 & 0 \\ 0 & I & 0 \\ 0 & K & I \end{bmatrix} \begin{bmatrix} -I & 0 & 0 \\ 0 & -I & 0 \\ 0 & 0 & I - KP_{22} \end{bmatrix} \begin{bmatrix} I & P_{12} & 0 \\ 0 & -P_{22} & I \\ 0 & I & 0 \end{bmatrix} \quad (13.8)
$$

it thus follows that $Y_1(P,K)$ is subsequently invertible. Hence Eq. (13.6) is satisfied automatically, and for any $W_I \in \mathcal{W}_I$ the closed-loop system Σ have a unique output signal Z_O.

(only if) For any $W_I \in \mathcal{W}_I$ if it is input consistent, then Eq. (13.6) is satisfied. By Lemma 11.1.1, one can write the solution for Z_O (in Eq. 13.4) as

$$
Z_O = -Y_1^-(P,K)Y_2(P)W_I + (I - Y_1^-(P,K)Y_1(P,K))h, \quad (13.9)
$$

where I is the identity matrix with a dimension of $l_1 + l_2 + m_2$, $h \in \mathcal{H}_2$ is any rational vector also with a dimension of $l_1 + l_2 + m_2$. The definition of FIWP requires that Z_O be unique, thus

$$
I - Y_1^-(P,K)Y_1(P,K) = 0,
$$

which tell us that $Y_1(P, K)$ is of full column rank. By the virtue of the fact that $Y_1(P, K)$ is in fact a square matrix, the invertibility of $I - KP_{22}$ thus follows from Eq. (13.8).

Definition 13.2.3. The closed-loop system $\Sigma(\Sigma')$ is said to be *externally internal well-posed* if for any input signal $W_I \in \mathcal{W}_I (W_I' \in \mathcal{W}_I')$ the closed-loop system $\Sigma(\Sigma')$ is input consistent and the closed-loop system $\Sigma(\Sigma')$ has a unique output signal z for any consistent input $W_I \in \mathcal{W}_I (W_I' \in \mathcal{W}_I')$.

Let us partition the matrix $I - Y_1^-(P, K)Y_1(P, K)$ as

$$I - Y_1^-(P, K)Y_1(P, K) = \begin{bmatrix} (I - Y_1^-(P, K)Y_1(P, K))_1 \\ (I - Y_1^-(P, K)Y_1(P, K))_2 \\ (I - Y_1^-(P, K)Y_1(P, K))_3 \end{bmatrix}, \tag{13.10}$$

where $(I - Y_1^-(P, K)Y_1(P, K))_1$, $(I - Y_1^-(P, K)Y_1(P, K))_2$, and $(I - Y_1^-(P, K)Y_1(P, K))_3$ have the dimensions of $l_1 \times (l_1 + l_2 + m_2)$, $l_2 \times (l_1 + l_2 + m_2)$, and $m_2 \times (l_1 + l_2 + m_2)$ respectively.

Theorem 13.2.3. *The closed-loop system Σ is* externally internal well-posed *if and only if*

(c1) $(I - Y_1(P, K)Y_1^-(P, K))Y_2(P)W_I = 0, \ \forall \, W_I \in \mathcal{W}_I$
(c2) $(I - Y_1^-(P, K)Y_1(P, K))_1 = 0;$

The closed-loop system Σ' is externally internal well-posed *if and only if*

(c1') $(I - Y_1(P, K)Y_1^-(P, K))Y_2(P)W_I' = 0, \ \forall \, W_I' \in \mathcal{W}_I'$
(c2') $(I - Y_1^-(P, K)Y_1(P, K))_1 = 0.$

Proof. By noting Eqs. (13.9), (13.10), the proof follows from Definition 13.2.3, Lemma 11.1.1, and Theorem 13.2.1 directly. □

It is noted that EIWP is a relaxation on FIWP, also it is a generalization to WP.

Corollary 13.2.1. *If the closed-loop system $\Sigma(\Sigma')$ is fully internal well-posed, then it is externally internal well-posed. If the closed-loop system $\Sigma(\Sigma')$ is well-posed, then it is externally internal well-posed.*

Proof. One only needs to check the statement about Σ, the other one about Σ' can be checked similarly. From Definition 13.2.2, the FIWP of the closed-loop system Σ suggests that

$$(I - Y_1(P, K)Y_1^-(P, K))Y_2(P)W_I = 0, \ \forall \, W_I \in \mathcal{W}_I,$$

and $I - Y_1^-(P, K)Y_1(P, K) = 0$, which in turn means that $(I - Y_1^-(P, K)Y_1(P, K))_1 = 0$. Thus the closed-loop system Σ is externally internal well-posed. □

If the closed-loop system Σ is well-posed, i.e., $I - KP_{22}$ is invertible, from Eq. (13.8) it follows that $Y_1(P, K)$ is invertible, and so the conditions in Theorem 13.2.3 are thus satisfied automatically.

The above corollary suggests that EIWP is indeed a generalization to WP, this will also be seen clearly in the next section.

13.3 Further characterizations of externally internal well-posedness

In the above section, it is found that the concept of WP can be generalized to the EIWP from the point of view of input consistency and output uniqueness. This generalized concept of EIWP is important not only in that it is suggested to take into account the input signal information in control system design, but also in that it only requires the external input signal z to be unique. The closed-loop system stability can thus be defined according to the generalized transfer function from w to z in Σ' (from (w, v_1, v_2) to z in Σ) in the sense of BIBO (*bounded-input bound-output*) stability [33], which is the subject of the notion of EIS we will present in the next section.

The main aim of this section is to derive further explicit characterizations of EIWP in terms of P and K by using the approach of the generalized inverse of matrix. For the sake of conciseness, we denote

$$F(P, K) := \begin{bmatrix} P_{22} & -I \\ I & -K \end{bmatrix}.$$

The following observation enables us to obtain the generalized inverse matrix $Y_1^-(P, K)$, which is very important in the sequel.

Lemma 13.3.1. *The following decompositions about $F(P, K)$ hold*

$$(a)\ F(P, K) = \begin{bmatrix} I & 0 \\ K & I \end{bmatrix} \begin{bmatrix} -I & 0 \\ 0 & I - KP_{22} \end{bmatrix} \begin{bmatrix} -P_{22} & I \\ I & 0 \end{bmatrix},$$

$$(b)\ F(P, K) = \begin{bmatrix} P_{22} & I \\ I & 0 \end{bmatrix} \begin{bmatrix} I & 0 \\ 0 & -I + P_{22}K \end{bmatrix} \begin{bmatrix} I & -K \\ 0 & I \end{bmatrix}.$$

Proof. The proof follows from a direct calculation. □

We can now state the following theorems that give further explicit characterizations of EIWP in terms of P and K.

Theorem 13.3.1. *The closed-loop system Σ is* externally internal well-posed *if and only if*

(c1) $[I - (I - KP_{22})(I - KP_{22})^-](K(P_{21}w + v_2) + v_1) = 0, \quad \forall\, W_I \in \mathcal{W}_I$
(c2) $P_{12}[I - (KP_{22})^-(I - KP_{22})] = 0;$

The closed-loop system Σ' is externally internal well-posed *if and only if*

(c1') $[I - (I - KP_{22})(I - KP_{22})^-]KP_{21}w = 0, \quad \forall\, w \in \mathcal{W}$
(c2') $P_{12}[I - (KP_{22})^-(I - KP_{22})] = 0;$

where the {1}-inverse is arbitrary.
Proof. The statement about Σ' follows immediately from that about Σ simply by letting $v_1 = v_2 = 0$. So we only need to prove the statement about Σ. □

From the decomposition (a) of $F(P, K)$ in Lemma 13.3.1, one finds that any {1}-inverse of $F(P, K)$ can be formulated as

$$F^-(P,K) = \begin{bmatrix} -P_{22} & I \\ I & 0 \end{bmatrix}^{-1} \begin{bmatrix} -I & 0 \\ 0 & (I-KP_{22})^- \end{bmatrix} \begin{bmatrix} I & 0 \\ K & I \end{bmatrix}^{-1}$$

$$= \begin{bmatrix} 0 & I \\ I & P_{22} \end{bmatrix} \begin{bmatrix} -I & 0 \\ 0 & (I-KP_{22})^- \end{bmatrix} \begin{bmatrix} I & 0 \\ -K & I \end{bmatrix}, \quad (13.11)$$

subsequently

$$F(P,K)F^-(P,K) = \begin{bmatrix} I & 0 \\ K & I \end{bmatrix} \begin{bmatrix} -I & 0 \\ 0 & I-KP_{22} \end{bmatrix} \begin{bmatrix} -P_{22} & I \\ I & 0 \end{bmatrix}$$

$$\times \begin{bmatrix} -P_{22} & I \\ I & 0 \end{bmatrix}^{-1} \begin{bmatrix} -I & 0 \\ 0 & (I-KP_{22})^- \end{bmatrix} \begin{bmatrix} I & 0 \\ K & I \end{bmatrix}^{-1}$$

$$= \begin{bmatrix} I & 0 \\ K & I \end{bmatrix} \begin{bmatrix} I & 0 \\ 0 & (I-KP_{22})(I-KP_{22})^- \end{bmatrix} \begin{bmatrix} I & 0 \\ K & I \end{bmatrix}^{-1}$$

$$= \begin{bmatrix} I & 0 \\ K-(I-KP_{22})(I-KP_{22})^-K & (I-KP_{22})(I-KP_{22})^- \end{bmatrix}, \quad (13.12)$$

similarly

$$F^-(P,K)F(P,K) = \begin{bmatrix} -P_{22} & I \\ I & 0 \end{bmatrix}^{-1} \begin{bmatrix} -I & 0 \\ 0 & (I-KP_{22})^- \end{bmatrix} \begin{bmatrix} I & 0 \\ K & I \end{bmatrix}^{-1}$$

$$\times \begin{bmatrix} I & 0 \\ K & I \end{bmatrix} \begin{bmatrix} -I & 0 \\ 0 & I-KP_{22} \end{bmatrix} \begin{bmatrix} -P_{22} & I \\ I & 0 \end{bmatrix}$$

$$= \begin{bmatrix} -P_{22} & I \\ I & 0 \end{bmatrix}^{-1} \begin{bmatrix} I & 0 \\ 0 & (I-KP_{22})^-(I-KP_{22}) \end{bmatrix} \begin{bmatrix} -P_{22} & I \\ I & 0 \end{bmatrix}^{-1}$$

$$= \begin{bmatrix} (I-KP_{22})^-(I-KP_{22}) & 0 \\ -P_{22}[I-(I-KP_{22})^-(I-KP_{22})] & I \end{bmatrix}. \quad (13.13)$$

From

$$Y_1(P,K) = \left[\begin{array}{c|c} -I & [P_{12},0] \\ \hline 0 & F(P,K) \end{array} \right], \quad (13.14)$$

it follows immediately that

$$Y_1^-(P,K) = \left[\begin{array}{c|c} -I & [P_{12},0]F^-(P,K) \\ \hline 0 & F^-(P,K) \end{array} \right]. \quad (13.15)$$

Thus some direct calculations yield

$$I - Y_1(P,K)Y_1^-(P,K) = \begin{bmatrix} I & 0 \\ 0 & I \end{bmatrix} - \begin{bmatrix} -I & [P_{12},0] \\ \hline 0 & F(P,K) \end{bmatrix} \begin{bmatrix} -I & [P_{12},0]F^-(P,K) \\ \hline 0 & F^-(P,K) \end{bmatrix}$$

$$= \begin{bmatrix} 0 & 0 \\ 0 & I - F(P,K)F^-(P,K) \end{bmatrix}$$

$$= \begin{bmatrix} 0 & 0 & 0 \\ 0 & 0 & 0 \\ 0 & -[I - (I - KP_{22})(I - KP_{22})^-]K & I - (I - KP_{22})(I - KP_{22})^- \end{bmatrix},$$

$$\tag{13.16}$$

and

$$I - Y_1^-(P,K)Y_1(P,K) = \begin{bmatrix} I & 0 \\ 0 & I \end{bmatrix} - \begin{bmatrix} -I & [P_{12},0]F^-(P,K) \\ \hline 0 & F^-(P,K) \end{bmatrix} \begin{bmatrix} -I & [P_{12},0] \\ \hline 0 & F(P,K) \end{bmatrix}$$

$$= \begin{bmatrix} 0 & P_{12}[I - (I - KP_{22})^-(I - KP_{22})] & 0 \\ 0 & I - (I - KP_{22})^-(I - KP_{22}) & 0 \\ 0 & P_{22}[I - (I - KP_{22})^-(I - KP_{22})] & 0 \end{bmatrix}. \tag{13.17}$$

Based on Theorem 13.2.3, the above formulae finally lead to the required statement.

Now we need to show that all these conditions are independent of the particular {1}-inverse used. We only check (c2) is independent of the particular {1}-inverse used, the remaining conditions can be checked similarly. Let $(I - KP_{22})^g$ be any other {1}-inverse of $I - KP_{22}$. By Lemma 11.1.2, it follows that

$$(I - KP_{22})^g = (I - KP_{22})^- + Q - (I - KP_{22})^-(I - KP_{22})Q(I - KP_{22})(I - KP_{22})^-$$

for a suitable choice of the arbitrary rational matrix Q. Hence

$$\begin{aligned} & P_{12}[I - (I - KP_{22})^g(I - KP_{22})] \\ &= P_{12}\{I - [(I - KP_{22})^- + Q - (I - KP_{22})^- \\ &\quad - (I - KP_{22})Q(I - KP_{22})(I - KP_{22})^-](I - KP_{22})\} \\ &= P_{12}[I - (I - KP_{22})^-(I - KP_{22})][I - Q(I - KP_{22})] \\ &= 0, \end{aligned}$$

whenever

$$P_{12}[I - (I - KP_{22})^-(I - KP_{22})] = 0.$$

This finishes the proof.

Based on the decomposition (b) of $F(P,K)$ given by Lemma 9.3.1, the following result can be proposed dually.

Theorem 13.3.2. *The closed-loop system Σ is externally internal well-posed if and only if*

(dc1) $[I - (I - P_{22}K)(I - P_{22}K)^-](P_{21}w + v_2 + P_{22}v_1) = 0, \quad \forall W_I \in \mathcal{W}_I$

(dc2) $P_{12}K[I - (I - P_{22}K)^-(I - P_{22}K)] = 0;$

The closed-loop system Σ' *is externally internal well-posed if and only if*

(dc1') $[I - (I - P_{22}K)(I - P_{22}K)^-]P_{21}w = 0, \quad \forall w \in \mathcal{W}$

(dc2') $P_{12}K[I - (I - P_{22}K)^-(I - P_{22}K)] = 0;$

where the {1}-inverse is arbitrary.

Proof. By using the decomposition (b) of $F(P, K)$ given by Lemma 13.3.1, following the same procedures as the proof of Theorem 13.3.1, one obtains this proof. $\quad\square$

Remark 13.3.1. It is noted again that the classical assumption of WP is a special case of the above conditions of EIWP, due to the fact that, when $I - KP_{22}$ is invertible, $(I - KP_{22})^- = (I - KP_{22})^{-1}$, and the conditions in Theorem 13.3.1 are automatically satisfied. When $I - P_{22}K$ is invertible, i.e., $(I - P_{22}K)^- = (I - P_{22}K)^{-1}$, the conditions in Theorem 13.3.2 are automatically satisfied. The extra generality of EIWP is achieved by its facility for allowing the input signal information to be related, for allowing the internal output signals u and y not necessarily to be determined by W_I uniquely, and for allowing the matrix P_{21} not necessarily to be of full row rank and the matrix P_{12} not necessarily to be of full column rank.

A basic observation arising from Theorems 13.3.1 and 13.3.2 is as follows.

Corollary 13.3.1. *If* $\mathcal{W}_I = \mathcal{H}_2^{m_1+m_2+l_2}$ *or if* P_{12} *has full column rank, then the closed-loop system* Σ *is externally internal well-posed if and only if it is well-posed. if* $\mathcal{W} = \mathcal{H}_2^{m_1}$, *the matrix* P_{21} *has full row rank or if* P_{12} *has full column rank, then the closed-loop system* Σ' *is externally internal well-posed if and only if it is well-posed.*

Proof. We only need to prove the first statement. The proof of the second one can be established in a similar manner. For the first statement, by the virtue of Corollary 13.2.1, one only needs to prove the necessity. When $\mathcal{W}_I = \mathcal{H}_2^{m_1+m_2+l_2}$, it is also seen that the matrix $[P_{21}, I, P_{22}]$ always has full row rank, the condition (dc1) in Theorem 13.3.2 tells us that

$$I - (I - P_{22}K)(I - P_{22}K)^- = 0,$$

which suggests that the matrix $I - P_{22}K$ is of full row rank, it is thus invertible due to the fact that it is indeed a square matrix. If P_{12} has full column rank, (c2) in Theorem 13.3.1 leads to

$$I - (I - KP_{22})^-(I - KP_{22}) = 0,$$

which shows that the matrix $I - KP_{22}$ is of full column rank, it is thus invertible due to the fact that it is indeed a square matrix. $\quad\square$

Remark 13.3.2. The above corollary shows that, under the conditions that P_{21}(or $[P_{21}, I, P_{22}]$ in Σ) or P_{12} satisfies the standard rank assumption, EIWP is equivalent to WP. It is thus seen again that EIWP generalizes WP.

13.4 Generalized linear fractional transformations, externally internal stability, and their characterizations

In terms of a state-space model, there exists controller K for the system P such that the closed-loop system is internally stabilizable if and only if the unstable modes of P are *controllable* from u (*stabilizability*) and *observable* from y (*detectability*) [36]. If either of the above conditions fails, the system is *internally unstabilizable*. In this case, one is still frequently concerned with the input-output properties of the closed-loop system [119]. In the approach of Doyle et al. [37], for those systems that can be internally stabilized, it was established that, under the assumption that P_{12} has full column rank and P_{21} has full row rank, the IS requirement is equivalent to the input-output stability, which requires the transfer function $T_{z,w}$ (from w to z), i.e., the LFT of the closed-loop system Σ' to be stable. In some circumstances, the system is, however, impossible to be internally stabilized. In other circumstances, such as in the celebrated minimum-sensitivity problem Zames [35], the assumptions of full rank on P_{12} and P_{21} are simply not satisfied [38] either.

The main aims of this section are firstly to obtain the GLFT and then to parameterize them by using the generalized inverse of matrices. We then present the notion of EIS and give characterizations of it in a matrix generalized inverse approach. Finally on the basis of the notions of EIWP, EIS, and GLFT the extended $\mathcal{H}_\infty$ control problem is defined in a general setting. The concept of EIS will generalize the notion of IS as it will include the above nonregular case where the system is internally unstabilizable. It will also deal with the issue of stability of the closed-loop system in the case where the general rank assumptions on P_{21} and P_{12} do not necessarily hold, which is in turn seen to be a generalization to that of Doyle et al. [37]. The assumption of EIWP ensures that the generalized transfer functions $GT_{z,w}$ for Σ' and GT_{z,W_l} for Σ exist. To find these generalized transfer functions, the following result will be useful, it is also of independent interest.

Theorem 13.4.1. *Consider Eq. (13.4). If one denotes* $G := -Y_1^-(P,K)Y_2(P)$, *then*

$$
G = \begin{bmatrix}
P_{11} + P_{12}(I - KP_{22})^- KP_{21} & P_{12}(I - KP_{22})^- & P_{12}(I - KP_{22})^- K \\
(I - KP_{22})^- KP_{21} & (I - KP_{22})^- & (I - KP_{22})^- K \\
P_{21} + P_{22}(I - KP_{22})^- KP_{21} & P_{22}(I - KP_{22})^- & I + P_{22}(I - KP_{22})^- K
\end{bmatrix},
\tag{13.18}
$$

dually G can also be formulated as following

$$
G = \begin{bmatrix}
P_{11} + P_{12}K(I - P_{22}K)^- P_{21} & P_{12} + P_{12}K(I - P_{22}K)^- KP_{22} & P_{12}K(I - P_{22}K)^- \\
K(I - P_{22}K)^- P_{21} & I + K(I - P_{22}K)^- P_{22} & K(I - P_{22}K)^- \\
(I - P_{22}K)^- P_{21} & (I - P_{22}K)^- P_{22} & (I - P_{22}K)^-
\end{bmatrix}.
\tag{13.19}
$$

The above {1}-inverses are arbitrary.

Proof. We are going to prove Eq. (13.18) by using the decomposition (a) of $F(P, K)$ in Lemma 13.3.1. The proof of Eq. (13.19) is omitted in here, which can be processed in a similar manner by using the decomposition (b) of $F(P, K)$ in Lemma 13.3.1. ☐

From (a) in Lemma 13.3.1, one finds that

$$
\begin{aligned}
F^-(P, K) &= \begin{bmatrix} -P_{22} & I \\ I & 0 \end{bmatrix} \begin{bmatrix} -I & 0 \\ 0 & (I - KP_{22})^- \end{bmatrix} \begin{bmatrix} I & 0 \\ K & I \end{bmatrix}^{-1} \\
&= \begin{bmatrix} 0 & I \\ I & P_{22} \end{bmatrix} \begin{bmatrix} -I & 0 \\ 0 & (I - KP_{22})^- \end{bmatrix} \begin{bmatrix} I & 0 \\ -K & I \end{bmatrix} \\
&= \begin{bmatrix} -(I - KP_{22})^- K & (I - KP_{22})^- \\ -I - P_{22}(I - KP_{22})^- K & P_{22}(I - KP_{22})^- \end{bmatrix}.
\end{aligned}
\tag{13.20}
$$

Now any $\{1\}$-inverse of $Y_1(P, K)$ is calculated to be

$$
Y_1^-(P, K) = \begin{bmatrix} -I & [P_{12}, 0]F^-(P, K) \\ 0 & F^-(P, K) \end{bmatrix}.
\tag{13.21}
$$

By substituting Eq. (13.20) into Eq. (13.21), one can seen that

$$
Y_1^-(P, K) = \begin{bmatrix} -I & -P_{12}(I - KP_{22})^- K & P_{12}(I - KP_{22})^- \\ 0 & -(I - KP_{22})^- K & (I - KP_{22})^- \\ 0 & -I - P_{22}(I - KP_{22})^- K & P_{22}(I - KP_{22})^- \end{bmatrix}.
\tag{13.22}
$$

Therefore, one has

$$
\begin{aligned}
G &= -Y_1^-(P, K)Y_2(P) \\
&= -\begin{bmatrix} -I & -P_{12}(I - KP_{22})^- K & P_{12}(I - KP_{22})^- \\ 0 & -(I - KP_{22})^- K & (I - KP_{22})^- \\ 0 & -I - P_{22}(I - KP_{22})^- K & P_{22}(I - KP_{22})^- \end{bmatrix} \begin{bmatrix} P_{11} & 0 & 0 \\ P_{12} & 0 & I \\ 0 & -I & 0 \end{bmatrix} \\
&= \begin{bmatrix} P_{11} + P_{12}(I - KP_{22})^- KP_{21} & P_{12}(I - KP_{22})^- & P_{12}(I - KP_{22})^- K \\ (I - KP_{22})^- KP_{21} & (I - KP_{22})^- & (I - KP_{22})^- K \\ P_{21} + P_{22}(I - KP_{22})^- KP_{21} & P_{22}(I - KP_{22})^- & I + P_{22}(I - KP_{22})^- K \end{bmatrix}.
\end{aligned}
\tag{13.23}
$$

This finishes the proof.

Theorems 13.4.1 and 13.2.3 immediately lead to the following results.

Theorem 13.4.2. *If the closed-loop system Σ (shown in Fig. 13.2) is externally internal well-posed, then the generalized input-output transfer functions GT_{z,W_I} (from W_I to z) exists. Moreover*

$$
GT_{z,W_I} = \begin{bmatrix} P_{11} + P_{12}(I - KP_{22})^- KP_{21} & P_{12}(I - KP_{22})^- & P_{12}(I - KP_{22})^- K \end{bmatrix},
\tag{13.24}
$$

dually GT_{z,W_I} can be formulated as

$$GT_{z,W_I} = \begin{bmatrix} P_{11}+P_{12}K(I-P_{22}K)^-P_{21} & P_{12}+P_{12}K(I-P_{22}K)^-P_{22} & P_{12}K(I-P_{22}K)^- \end{bmatrix},$$
$$(13.25)$$

where the {1}-inverses are arbitrary.

Proof. Recall that the closed-loop system Σ has been written into

$$Y_1(P,K)Z_O + Y_2(P)W_I = 0.$$

EIWP guarantees that the system is input consistent, by Lemma 11.1.1, the solution Z_O in the above equation is thus worked out to be

$$Z_O = GW_I + (I - Y_1^- Y_1(P,K))h,$$

where h is any rational vector with appropriate dimension, $G := -Y_1^-(P,K)Y_2(P)$. The partitioning G and $I - Y_1^-(P,K)Y_1(P,K)$ according to Z_O follows that

$$z = GT_{z,W_I}W_I + (I - Y_1^-(P,K)Y_1(P,K))_1 h.$$

By Theorem 13.2.3, EIWP suggests that

$$(I - Y_1^-(P,K)Y_1(P,K))_1 = 0.$$

Therefore, the generalized input-output transfer function from W_I to z exists, which is just GT_{z,W_I}. Theorem 13.4.1 shows in turn the explicit formulations (13.24) and (13.25). □

Theorem 13.4.3. *If the closed-loop system Σ' (show in Fig. 13.1) is externally internal well-posed, then the generalized input-output transfer functions $GT_{z,w}$ (from w to z) exists. Moreover*

$$GT_{z,w} = P_{11} + P_{12}(I - KP_{22})^-P_{21}, \qquad (13.26)$$

dually $GT_{z,w}$ can also be formulated as

$$GT_{z,w} = P_{11} + P_{12}K(I - P_{22}K)^-P_{21}, \qquad (13.27)$$

where the {1}-inverses are arbitrary.

Proof. This proof follows from Theorem 13.4.2 directly. □

Remark 13.4.1. The above theorems clarify the essential structures of the relevant generalized input-output transfer functions of the closed-loop system Σ and Σ' in a general setting, which makes it possible for one to study the input-output properties, such as the stability of the systems in this general setting. In fact, Theorem 13.4.3 generalizes the regular LFT from the WP case to the more general setting EIWP. Let us give this generalization an explicit definition.

Definition 13.4.1 (GLFT). Assume that the standard control system Σ' (Fig. 13.1) is *externally internal well-posed*. Any generalized input-output transfer function from w to z either in the form

$$GT_{z,w} = P_{11} + P_{12}(I - KP_{22})^-KP_{21}$$

or in the form

$$GT_{z,w} = P_{11} + P_{12}K(I - P_{22}K)^- P_{21}$$

is called a GLFT.

It should be noted immediately that, unlike the regular LFT, the formulations of the above GLFTs are not unique due to the fact the $\{1\}$-inverses of a matrix are not unique. The following theorem parameterizes all the formulations of the GLFTs.

Theorem 13.4.4. *Assume that the standard control system* Σ' *(Fig. 13.1) is externally internal well-posed. If the set of all GLFT matrices corresponding to a specific plant P and a controller K is denoted by* $\Gamma(P, K)$, *then*

$$\Gamma(P, K) = \{GT_{z,w}(P, K; Q_1) : Q_1 \text{ is rational matrix}\},$$

where we denote

$$GT_{z,w}(P, K; Q_1) := P_{11} + P_{12}(I - KP_{22})^G KP_{21},$$

$$(I - KP_{22})^G := (I - KP_{22})^\dagger + Q_1 - (I - KP_{22})^\dagger (I - KP_{22}) Q_1 (I - KP_{22})(I - KP_{22})^\dagger,$$

$(I - KP_{22})^\dagger$ *is the Moore-Penrose inverse of* $(I - KP_{22})$.

Alternatively

$$\Gamma(P, K) = \{GT_{z,w}(P, K; Q_2) : Q_2 \text{ is rational matrix}\},$$

where we denote

$$GT_{z,w}(P, K; Q_2) := P_{11} + P_{12}K(I - P_{22}K)^G P_{21},$$

$$(I - P_{22}K)^G := (I - P_{22}K)^\dagger + Q_2 - (I - P_{22}K)^\dagger (I - P_{22}K) Q_2 (I - P_{22}K)(I - P_{22}K)^\dagger,$$

$(I - P_{22}K)^\dagger$ *is the Moore-Penrose inverse of* $(I - P_{22}K)$.

Proof. This proof is a direct application of Lemma 11.1.2. □

It is apparent that the generalized input-output transfer function GT_{z,W_I} of the closed-loop system Σ can also be parameterized in a much similar manner. The details are omitted in here.

Specifically, in the following, $\mathcal{RH}_\infty$ denotes the space consists of those real-rational matrices that are stable and proper, $\mathcal{RH}_2$ denotes the space consists of those real-rational vector-valued strictly proper stable functions. Based on the notion of EIWP, the following definition of EIS is suggested.

Definition 13.4.2. Assume that the closed-loop system Σ (or Σ') is *externally internal well-posed*. The system Σ is called *externally internal stable* if $z(s) \in \mathcal{RH}_2^{l_1}$ whenever $W_I \in \mathcal{W}_I \subseteq \mathcal{RH}_2^{m_1+m_2+l_2}$; the system Σ' is called *externally internal stable* if $z(s) \in \mathcal{RH}_2^{l_1}$ whenever $W_I' \in \mathcal{W}_I' \subseteq \mathcal{RH}_2^{m_1+m_2+l_2}$.

Now we are ready to state the following important result, which can be deemed as a generalization to that of Doyle et al. [37].

Theorem 13.4.5. *Assume that the closed-loop system* Σ' *is externally internal well-posed and* $\mathcal{W} = \mathcal{RH}_2^{m_1}$. *The system* Σ' *is externally internal stable if for any GLFT*

$GT_{z,w}$, one has $GT_{z,w} \in \mathcal{RH}_{\infty}^{l_1 \times m_1}$. *Assume that the closed-loop system Σ is externally internal well-posed and* $\mathcal{W}_l = \mathcal{H}_2^{m_1+m_2+l_2}$. *The system Σ is externally internal stable if for any generalized transfer function GT_{z,W_l}, one has $GT_{z,W_l} \in \mathcal{RH}_{\infty}^{l_1 \times (m_1+m_2+l_2)}$.*

Proof. Only the first statement is proved in here. The proof of the second one is similar. $\qquad\qquad\qquad\qquad\qquad\qquad\qquad\qquad\qquad\qquad\qquad\qquad\qquad\qquad$ $\square$

By the BIBO stability theory (see, e.g., [33]) it is sufficient to prove that, for any two GLFTs

$$GT_{z,w}(P, K; Q_1) := P_{11} + P_{12}(I - KP_{22})^G KP_{21}$$

and

$$GT_{z,w}(P, K; Q_2) := P_{11} + P_{12}(I - KP_{22})^{G'} KP_{21},$$

where the only distinction between these two GLFTs is the $\{1\}$-inverse of $(I - KP_{22})$ used, one has

$$\left\| GT_{z,w}(P, K; Q_1) \right\|_{\infty} = \left\| GT_{z,w}(P, K; Q_2) \right\|_{\infty}. \tag{13.28}$$

To prove Eq. (13.28), Lemma 11.1.2 suggests that there exists a rational matrix Q such that

$$(I - KP_{22})^G = (I - KP_{22})^{G'} + Q - (I - KP_{22})^{G'}(I - KP_{22})Q(I - KP_{22})(I - KP_{22})^{G'}.$$

For any $w \in \mathcal{RH}_2^{m_1}$, it follows that

$$
\begin{aligned}
& GT_{z,w}(P, K; Q_1)w \\
&= [P_{11} + P_{12}(I - KP_{22})^G KP_{21}]w \\
&= \{P_{11} + P_{12}[(I - KP_{22})^{G'} + Q - (I - KP_{22})^{G'}(I - KP_{22})Q \times \\
&\quad \times (I - KP_{22})(I - KP_{22})^{G'}]KP_{21}\}w \\
&= P_{11}w + P_{12}[(I - KP_{22})^{G'}KP_{21}w + QKP_{21}w - (I - KP_{22})^{G'}(I - KP_{22})Q \times \\
&\quad \times (I - KP_{22})(I - KP_{22})^{G'}KP_{21}w] \\
&= P_{11}w + P_{12}[(I - KP_{22})^{G'}KP_{21}w + QKP_{21}w - (I - KP_{22})^{G'}(I - KP_{22})QKP_{21}w] \\
&= P_{11}w + P_{12}[(I - KP_{22})^{G'}KP_{21}w + (I - (I - KP_{22})^{G'}(I - KP_{22}))QKP_{21}w] \\
&= [P_{11} + P_{12}(I - KP_{22})^{G'}KP_{21}]w \\
&= GT_{z,w}(P, K; Q_2)w,
\end{aligned}
$$

where the conditions (c1$'$) and (c2$'$) of system EIWP in Theorem 13.3.1 were used. Thus

$$
\begin{aligned}
\hat{\sigma}(GT_{z,w}(P, K; Q_1)) &= \max_{\|w\|=1} \left\| GT_{z,w}(P, K; Q_1)w \right\| \\
&= \max_{\|w\|=1} \left\| GT_{z,w}(P, K; Q_2)w \right\| \\
&= \hat{\sigma}(GT_{z,w}(P, K; Q_2)),
\end{aligned}
$$

where $\hat{\sigma}(f)$ denotes the maximum singular value of f and the vector norm is the Euclidean norm. Therefore

$$
\begin{aligned}
\left\| GT_{z,w}(P,K;Q_1) \right\|_\infty &= \sup_{\lambda>0}\{\sup_\omega \hat{\sigma}(GT_{z,w}(P,K;Q_1)(\lambda+j\omega))\} \\
&= \sup_{\lambda>0}\{\sup_\omega \hat{\sigma}(GT_{z,w}(P,K;Q_2)(\lambda+j\omega))\} \\
&= \left\| GT_{z,w}(P,K;Q_2) \right\|_\infty,
\end{aligned}
$$

which finishes the proof.

Remark 13.4.2. It is not difficult to see that, if the given system can be internally stabilized and P_{12}, P_{21} satisfy the corresponding rank assumption, EIS is equivalent to IS due to Corollary 13.3.1 and Lemma 9.3.1 in Francis [36]. From this point of view, Theorem 13.4.5 can be deemed as a generalization to that of Doyle et al. [37].

On the basis of these notions of EIWP, EIS, and GLFT, we can formulate the *extended $\mathcal{H}_\infty$ control problem*, which is: find a real-rational matrix K to minimize the $\mathcal{H}_\infty$ norm of any GLFT under the constraints that the resulting closed-loop system Σ' is *externally internal well-posed* and it is *externally internal stable*. This extended problem is general enough to include the case where the system is internally unstabilized, the case where the system does not satisfy the relevant rank assumptions on P_{12} and P_{21}, and the nonregular case where the resulting closed-loop system is "ill-posed."

13.5 Conclusions

It is remarkable that the $\mathcal{H}_\infty$ control scheme is quite rich in input-output structure and in stability features. Looking into this scheme in detail, we presented certain generalizations to the classical concepts of WP and IS, which are termed as EIWS and EIS. The main significance of these generalizations lies in that the issue of WP can be unifiedly clarified from a general setting, the *input-output stability* of the closed-loop system can be characterized in terms of the generalized input-output transfer functions. Such generalizations also throw a new light on the impact of the input signal information on system control design.

It was shown that EIWP and EIS generalize the classical notions of WP and IS respectively. The generality of EIWP was seen from the fact that, the resulting closed-loop system can be "ill-posed" provided that the related input signals are appropriate, the matrices P_{21} and P_{12} do not necessarily satisfy the full rank assumptions. The generality of EIS is achieved by its facility for allowing the system to be internally unstabilized, for allowing the system not necessarily to satisfy the relevant rank assumptions. Especially when these restrictions on the system are satisfied, system EIWP is essentially equivalent to system WP and EIS is equivalent to IS.

Our approach was exposed by using the matrix generalized inverses, which in turn led to a generalization of LFT and an extended definition of $\mathcal{H}_\infty$ control problem. The *extended $\mathcal{H}_\infty$ control problem* was cast in such a general setting that it allows one to cancel two of the restrictions that the known approaches in $\mathcal{H}_\infty$ control theories usually impose on the system. One is that the considered system must be internally stabilized, the other is that the system must satisfy the relevant rank assumptions on P_{21} and P_{12}. The solution set for this extended problem is thus enlarged from that of the regular $\mathcal{H}_\infty$ control problem, which enables one to search the appropriate controllers in a wider area; the resulting controllers might be subsequently more flexible in achieving the control aims.

In the next chapter we investigate such "ill-posed" controller structures along the line of the generalized chain-scattering approach and propose the extended $\mathcal{H}_\infty$ control problem in a general setting.

Nonstandard H_∞ control problem: A generalized chain-scattering representation approach

14

The nonstandard H_∞ control problem is the case where the direct feedthroughs from the input to the error and from the exogenous signal to the output are not necessarily of full rank. This problem is reformulated based on the generalized chain-scattering representation (GCSR). The GCSR approach leads naturally to a generalization of the homographic transformation. The state-space realization for this generalized homographic transformation and a number of fundamental cascade structures of the H_∞ control systems are further studied in a unified framework of GCSR. Certain sufficient conditions for the solvability of the nonstandard H_∞ control problem are therefore established via a (J, J')-lossless factorization of GCSR. These results present extensions to Kimura's results on the chain-scattering representation (CSR) approach to the H_∞ control in the standard case.

14.1 Introduction

Consider the plant

$$
\begin{bmatrix} z \\ y \end{bmatrix} = P \begin{bmatrix} w \\ u \end{bmatrix} = \begin{bmatrix} P_{11} & P_{12} \\ P_{21} & P_{22} \end{bmatrix} \begin{bmatrix} w \\ u \end{bmatrix},
\tag{14.1}
$$

where $z \in \Re(s)^m$, $y \in \Re(s)^q$, $w \in \Re(s)^r$, and $u \in \Re(s)^p$ are the controlled error, the observation output, the exogenous input, and the control input, respectively. The H_∞ control problem is to find a controller given by

$$
u = Ky,
\tag{14.2}
$$

which internally stabilizes the closed-loop system and satisfies $\|\Phi\|_\infty < 1$, where Φ is the closed-loop transfer function from w to z, which is generally called a *linear fractional transformation* in the control literature and is denoted [22] by $LF(P; K)$, i.e.,

$$
\Phi := LF(P; K) = P_{11} + P_{12}K(I - P_{22}K)^{-1}P_{21}.
\tag{14.3}
$$

Concerning the above problem, one of the assumptions is the following

(A)   rank $P_{21} = q$,   rank $P_{12} = p$.

A Generalized Framework of Linear Multivariable Control. http://dx.doi.org/10.1016/B978-0-08-101946-7.00014-7

When the plant satisfies the above assumption, the H_∞ control problem is called [37] the *regular* or *standard* case, while the case where the above assumption does not hold is referred [120] to a *singular* H_∞ control problem. We do not make the above assumption, the considered problem is thus nonstandard, which includes both the regular case and the singular case. From a system point of view, since no measurement is error free and no control action is costless [117] it is physically reasonable and general (see, e.g., [90]) to assume that the dimension of z is at least that of u, while the dimension of w must be at least that of y, without loss of generality only the right noninvertibility of P_{21} is thus considered in the following development. It should also be noted that, in Assumption (A) rank $P_{21} = q$ means that, rank $P_{21}(j\omega) = q$, $\forall \omega \in \mathbf{R}$ and rank $P_{21}(\infty) = q$.

Several approaches (e.g., [25, 36, 37, 121], and references therein) have already been proposed to solve the regular H_∞ control problem both in the state-space framework and in the input-output operator-theoretic techniques. The singular H_∞ control problem has attracted considerable research interests in the last few years [120, 122]. For a monograph on this subject, one can refer to [122–124] considered controller design for the singular H_∞ control problem based on the linear matrix inequality (LMI) approach. One basic observation to the aforementioned contributions on this theme is, however, that the problem is generally attacked in the state-space framework. Although state-space techniques are powerful for computing solutions, and are growing even more so with the advent of efficient numerical techniques, they do not always give physical insight into the nature of problems and bear fundamental limitations. Such insight can be obtained much more effectively using input-output techniques that allow solutions to be revealed and understood in their most transparent forms, and without the often obscuring one details that occur in specific problems.

Recently Kimura [22, 90] has developed the CSR, and it was successfully used there to provide a unified framework of H_∞ control theory. In this new framework, the H_∞ control problem is essentially reduced to a J-lossless factorization of the CSR of the plant. The J-lossless conjugation then provides a powerful tool for computing the required factorization. Kimura's CSR approach seems to be the most compact theory and the simplest method to the regular H_∞ control problem. The known CSR approach is, however, not applicable to the singular H_∞ control problem. This situation may improve with the advent of [125]. In [125] Kimura's results on CSR has been extended to the general case in which the condition (A) is essentially relaxed. From an input-output consistency point of view, the GCSR emerges and is successfully used there to characterize the cascade structure property and the symmetry of general plants in a general setting.

The main motivation of our current work presented herein is to extend Kimura's results [22, 90] on CSR approach to the H_∞ control from the standard case to the nonstandard case by using the GCSR approach [125]. To this end, the nonstandard H_∞ control problem is first reformulated based on the GCSR. The GCSR approach then leads naturally to a generalization of the homographic transformation. The state-space realization for this generalized homographic transformation and a number of fundamental cascade structures of the H_∞ control systems are further studied in a unified framework of GCSR. Certain sufficient conditions for the solvability of the nonstandard H_∞ control problem are thus established via a $(J; J')$-lossless factorization of GCSR.

Definition 14.1.1 ([5]). For every rational matrix $A \in \Re(S)^{n \times m}$ a unique matrix $A^\dagger \in \Re(S)^{m \times n}$, which is called the *Moore-Penrose inverse*, exists satisfying

$$AA^\dagger A = A,$$
$$A^\dagger AA^\dagger = A^\dagger,$$
$$(AA^\dagger)^T = AA^\dagger,$$
$$(A^\dagger A)^T = A^\dagger A,$$

where $A^\dagger$ denotes the conjugate transpose of A. In the special case that A is a square nonsingular matrix, the Moore-Penrose inverse of A is simply its inverse, i.e., $A^\dagger = A^{-1}$. In case a matrix A^{-1} satisfies only the first condition, it is called a {1}-inverse. {1}-inverses are not unique but play an important role in the following development.

Notions

$RL_{m \times r}^\infty$: the set of all $m \times r$ rational proper matrices without pole on the $j\omega$-axis

$RH_{m \times r}^\infty$: the set of all $m \times r$ rational stable proper matrices

$BH_{p \times q}^\infty$: $\{F(s) \in \Re(s)^{p \times q} : stable, \|F\|_\infty < 1\}$

$$\begin{bmatrix} A & B \\ C & D \end{bmatrix} := C(sI - A)^{-1} + D.$$

14.2 Reformulation of the nonstandard H_∞ control problem via generalized chain-scattering representation

The main reason for using the CSR lies in its ability of representing the feedback connection as a cascade one [22, 90], in this context any feedback of an input-output system is subsequently equivalent to a termination of the corresponding CSR. The regular H_∞ control problem, when described [22, 90] through CSR, is thus greatly simplified for the commonly used linear fraction transformation has been replaced by a much simpler form of homographic transformation. Motivated by a similar reason, the main aim of this section is to give a reformulation of the nonstandard H_∞ control problem via GCSR on the basis of a generalization to homographic transformation.

It is seen that, concerning the existence of CSR, if the input-output pair (u, y) satisfies the condition of consistency given in [125], i.e.,

$$(I - P_{21} P_{21}^-)[-P_{22}, I] \begin{bmatrix} u \\ y \end{bmatrix} = 0$$

then the CSRs are still available for the nonstandard case. It should be noted that such a condition of consistency essentially relaxes [125] the condition (A). These representations, termed as the GCSRs therein, are not unique. The set of them has, however, been parameterized in [125]. To facilitate exposition, only the special GCSR form, which is formed in terms of the Moore-Penrose inverse of P_{21}, has been chosen for use in the following development. Let us recall this result from [125].

Theorem 14.2.1 ([125]). *If the plant P is consistent about w with respect to the input-output pair* $(u; y)$*, then the original plant (1) can be written into the GCSR form*

$$\begin{bmatrix} z \\ w \end{bmatrix} = GCHAIN(P) \begin{bmatrix} u \\ y \\ h \end{bmatrix}, \tag{14.4}$$

where we denote the GCSR matrix

$$GCHAIN(P) := \begin{bmatrix} P_{12} - P_{11}P_{21}^{\dagger}P_{22} & P_{11}P_{21}^{\dagger} & P_{11}(I - P_{21}^{\dagger}P_{21}) \\ -P_{21}^{\dagger}P_{22} & P_{21}^{\dagger} & I - P_{21}^{\dagger}P_{21} \end{bmatrix} \in \Re(s)^{(m+r)\times(p+q+r)} \tag{14.5}$$

h is an arbitrary rational vector and $P_{21}^{\dagger}$ *is the Moore-Penrose inverse of* P_{21}*.*

If P_{21} is invertible, then the CSR [22, 90] exists for the plant P, Eq. (14.4) becomes

$$\begin{bmatrix} z \\ w \end{bmatrix} = CHAIN(P) \begin{bmatrix} u \\ y \end{bmatrix},$$

where the CSR matrix is

$$CHAIN(P) = \begin{bmatrix} P_{12} - P_{11}P_{21}^{-1}P_{22} & P_{11}P_{21}^{-1} \\ -P_{21}^{-1}P_{22} & P_{21}^{-1} \end{bmatrix}.$$

Further denote in Eq. (14.5)

$$G^* := GCHAIN(P) := \begin{bmatrix} G_{11} & G_{12} & G_{13} \\ G_{21} & G_{22} & G_{23} \end{bmatrix}, \tag{14.6}$$

where the sizes of the block matrices G_{11}, G_{12}, G_{13}, G_{21}, G_{22}, G_{23} are $m \times p$, $m \times q$, $m \times r$, $r \times p$, $r \times q$, $r \times r$, respectively.

Now consider the terminated cascade connection (14.2) and (14.4) (refer to Fig. 14.1), by denoting $\Sigma = [G_{21}K + G_{22}G_{23}]$, one can easily obtain a $\{1\}$-inverse of Σ, which is important in the sequel.

Proposition 14.2.1. *One {1}-inverse of* Σ *is*

$$\Sigma^- = \begin{bmatrix} (I - P_{22}K)^{-1}P_{21} \\ I - P_{21}^{\dagger}P_{21} \end{bmatrix}.$$

When considering the terminated cascade connection (14.2) and (14.4) (refer to Fig. 14.1), two primary issues that play a key role in our approach need to be

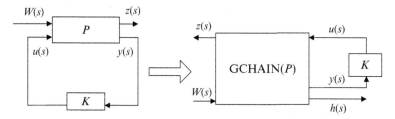

Fig. 14.1 H_∞ control scheme is reduced to a terminated cascade connection of the generalized chain-scattering representation.

considered. One is that the representation (14.4) together with the relation (14.2) can determine $[y^T, h^T]^T$. It will be seen clearly that this is always the case provided that the closed-loop system is *well-posed*.

The other issue is termed as *output uniqueness*, i.e., the condition under which the closed-loop system is able to give rise to a unique output z. These issues are central to the existence of the relevant *generalized transfer function* [116]. Interestingly, the determined generalized transfer function not only produces a unique output z, being a functional operator, but it is also exactly equal to the linear fractional transformation, once the condition of output uniqueness is satisfied. The following theorems establish the above interesting observations.

Theorem 14.2.2. *For the terminated cascade connection (14.2) and (14.4) of the GCSRs, the generalized transfer functions from w to z exist if and only if the condition of output uniqueness*

$$P_{12}K(I - P_{22}K)^{-1}(I - P_{21}P_{21}^\dagger) = 0 \tag{14.7}$$

is satisfied. The above condition is equivalent to

$$[G_{11}K + G_{12} \quad G_{13}](I - \Sigma^- \Sigma) = 0. \tag{14.8}$$

Proof. Considering the terminated cascade connection (14.2) and (14.4) of the GCSRs, which is described by $[z^T, w^T]^T = G^*[u^T, y^T, h^T]^T$, $u = Ky$, where G^* is given by Eq. (14.5), (14.6), one has

$$w = \Sigma[y^T, h^T]^T, \tag{14.9}$$

$$z = [G_{11}K + G_{12} \quad G_{13}][y^T, h^T]^T. \tag{14.10}$$

Well-posedness of the closed-loop system means the matrix $[P_{21}^*(I - P_{22}K) \quad I]$ is invertible, from the relation

$$\Sigma = [G_{21}K + G_{22} \quad G_{23}] = [P_{21}^\dagger(I - P_{22}K) \quad I - P_{21}^\dagger P_{21}]$$

$$= [P_{21}^\dagger(I - P_{22}K) \quad I] \begin{bmatrix} I & -(I - P_{22}K)^{-1}P_{21} \\ 0 & I \end{bmatrix}$$

noting that the matrix $[P_{21}^{\dagger}(I - P_{22}K) \; I]$ is of full row rank, one thus concludes that Σ is of full row rank. Therefore, one can always solve Eq. (14.9) for $[y^T, h^T]^T$ and obtain

$$[y^T, h^T]^T = \Sigma^- w + (I - \Sigma^- \Sigma)q, \tag{14.11}$$

where q is any rational vector. Substituting Eq. (14.11) into Eq. (14.10), Eq. (14.10) reads

$$z = [G_{11}K + G_{12} \; G_{13}]\Sigma^- w + [G_{11}K + G_{12} \; G_{13}](I - \Sigma^- \Sigma)q. \tag{14.12}$$

Hence for Eq. (14.12) to determine the generalized transfer function from w to z, it is sufficient and necessary that the following condition of output uniqueness is satisfied

$$[G_{11}K + G_{12} \; G_{13}](I - \Sigma^- \Sigma) = 0 \tag{14.13}$$

due to the fact that q is arbitrary. One can further verify that the output uniqueness condition (14.13) is equivalent to Eq. (14.7). From Proposition 14.2.1 it follows that

$$
I - \Sigma^- \Sigma = I - \begin{bmatrix} (I - P_{22}K)^{-1}P_{21} \\ I - P_{21}^{\dagger}P_{21} \end{bmatrix} [P_{21}^{\dagger}(I - P_{22}K) \; I - P_{21}^{\dagger}P_{21}]
$$

$$
= \begin{bmatrix} I - (I - P_{22}K)^{-1}P_{21}P_{21}^{\dagger}(I - P_{22}K) & -(I - P_{22}K)^{-1}P_{21}(I - P_{21}^{\dagger}P_{21}) \\ -(I - P_{21}^{\dagger}P_{21})P_{21}^{\dagger}(I - P_{22}K) & I - (I - P_{21}^{\dagger}P_{21}) \end{bmatrix}
$$

$$
= \begin{bmatrix} (I - P_{22}K)^{-1}(I - P_{21}P_{21}^{\dagger})(I - P_{22}K) & 0 \\ 0 & P_{21}^{\dagger}P_{21} \end{bmatrix},
$$

where the relations (i) and (ii) in Definition 14.1.1 are used. Now

$$
[G_{11}K + G_{12} \; G_{13}](I - \Sigma^- \Sigma) = [P_{12}K + P_{11}P_{21}^{\dagger}(I - P_{22}K) \; P_{11}(I - P_{21}^{\dagger}P_{21})]
$$

$$
\times \begin{bmatrix} (I - P_{22}K)^{-1}(I - P_{21}P_{21}^{\dagger})(I - P_{22}K) & 0 \\ 0 & P_{21}^{\dagger}P_{21} \end{bmatrix}
$$

$$
= [P_{12}K(I - P_{22}K)^{-1}(I - P_{21}P_{21}^{\dagger})(I - P_{22}K) \; 0],
$$

where we have used

$$P_{21}^{\dagger}(I - P_{21}P_{21}^{\dagger}) = 0$$

and

$$(I - P_{21}^{\dagger}P_{21})P_{21}^{\dagger} = 0.$$

From this we can see that the output uniqueness condition (14.13) is equivalent to Eq. (14.7) for the specially chosen $\{1\}$-inverse of Σ given by Proposition 14.2.1. This finishes the proof. $\qquad \square$

Unlike Kimura's approach [22, 90] of augmentation in the regular case, in the GCSR matrix the block matrix G_{23} is of $r \times r$, the matrix $[G_{21}K + G_{22} \ G_{23}]$ is subsequently of $r \times (q + r)$, which is nonsquare and is not invertible. This is one of the reasons that we have to utilize the matrix generalized inverses and where the difficulties arise from the present approach. It is noted that, if the condition of output uniqueness is satisfied, Eq. (14.12) determines a generalized transfer function from w to z, which is given by

$$GHM(G^*; [K \ \ 0]) = [G_{11}K + G_{12} \ G_{13}]\Sigma^-,$$

where Σ^- is given by Proposition 14.2.1. As a matter of fact this generalized transfer function is an extension to the homographic transformation that is used [22, 90] in dealing with the regular H_∞ control problem. We therefore term it as *the generalized homographic transformation* and denote it by $GHM(G^*; [K \ \ 0])$ following the notation adopted in [22, 90].

It is of further interest to look at the relationship between the above proposed generalized homographic transformation and the linear fractional transformation. It is interesting that, being a functional operator, the generalized homographic transformation is equal to the linear fractional transformation. This observation is stated as the following theorem.

Theorem 14.2.3. *For the terminated cascade connection (14.2) and (14.4), if the output uniqueness s condition is satisfied, then*

$$GHM(G^*; [K \ \ 0]) = LF(P; K). \tag{14.14}$$

Proof. Considering Eqs. (14.5), (14.6) and Proposition 14.2.1, we have

$$
\begin{aligned}
GHM(G^*; [K \ \ 0]) &= [G_{11}K + G_{12} \ G_{13}]\Sigma^- \\
&= [P_{12}K + P_{11}P_{21}^\dagger(I - P_{22}K) \ \ P_{11}(I - P_{21}^\dagger P_{21})] \\
&\quad \times \begin{bmatrix} (I - P_{22}K)^{-1}P_{21} \\ I - P_{21}^\dagger P_{21} \end{bmatrix} \\
&= \{[P_{12}K + P_{11}P_{21}^\dagger(I - P_{22}K)\}(I - P_{22}K)^- P_{21} + \\
&\quad + P_{11}(I - P_{21}^\dagger P_{21})(I - P_{21}^\dagger P_{21}) \\
&= P_{11} + P_{12}K(I - P_{22}K)^{-1}P_{21} = LF\{P; K\}.
\end{aligned}
$$

$\square$

The above theorems are interesting, not least for the way in which they reduce the original closed-loop system (14.1) and (14.2) into a wave scatter $GCHAIN(P)$ terminated by a load K. This situation is illustrated in Fig. 14.1. More important than this, however, is that the implied mechanism enables us to pose the nonstandard H_∞ control problem in the framework of GCSRs in the following manner.

Nonstandard H_∞ problem reformulation: Find a controller (load) K such that the terminated cascade connection (14.2) and (14.4) (refer to Fig. 14.1) is well-posed, internally stable, the output uniqueness condition is satisfied and

$$\| GHM(G^*; [K \quad 0]) \|_\infty < 1. \tag{14.15}$$

One further observation arising from Theorem 14.2.3 is that the H_∞ norm of the closed-loop system can be given equivalently in terms of the generalized homographic transformation. It will also be seen clearly in the next section that the internal stability of the terminated cascade connection (14.2) and (14.4) can be characterized in terms of the A-matrix in the corresponding state-space realization to the generalized homographic transformation. These two observations are exactly the rationale behind the above reformulation of the nonstandard H_∞ control problem.

14.3   Solvability of nonstandard H_∞ control problem

This section mainly studies the state-space realization for the generalized homographic transformation and some fundamental cascade structures of the H_∞ control systems in the framework of GCSR. Certain sufficient conditions for the solvability of the nonstandard H_∞ control problem are then established via (J, J')-lossless factorization of GCSR.

If the state-space realizations for the GCSR and the controller K are given by

$$G^* = \begin{bmatrix} A & B_1 & B_2 & B_3 \\ C_1 & D_{11} & D_{12} & D_{13} \\ C_2 & D_{21} & D_{22} & D_{23} \end{bmatrix} \tag{14.16}$$

and

$$K = \begin{bmatrix} A_K & B_K \\ C_K & D_K \end{bmatrix} \tag{14.17}$$

respectively. The above relations (14.16) and (14.17) are then written as

$$\left. \begin{aligned} \dot{x} &= Ax + B_1 u + B_2 y + B_3 h \\ z &= C_1 x + D_{11} u + D_{12} y + D_{13} h \\ z &= C_2 x + D_{21} u + D_{22} y + D_{23} h \end{aligned} \right\} \tag{14.18}$$

and

$$\left. \begin{aligned} \dot{\xi} &= A_K \xi + B_K y \\ u &= C_K \xi + D_K y \end{aligned} \right\}. \tag{14.19}$$

Elimination of u from the above relations yields

$$\left. \begin{aligned} \begin{bmatrix} \dot{x} \\ \dot{\xi} \end{bmatrix} &= \begin{bmatrix} A & B_1 C_K \\ 0 & A_K \end{bmatrix} \begin{bmatrix} x \\ \xi \end{bmatrix} + \begin{bmatrix} \hat{B} \\ \hat{B}_K \end{bmatrix} \begin{bmatrix} y \\ h \end{bmatrix} \\ \begin{bmatrix} z \\ w \end{bmatrix} &= \begin{bmatrix} C_1 & D_{11} C_K \\ C_2 & D_{21} C_K \end{bmatrix} \begin{bmatrix} x \\ \xi \end{bmatrix} + \begin{bmatrix} D_1 \\ D_2 \end{bmatrix} \begin{bmatrix} y \\ h \end{bmatrix} \end{aligned} \right\} \tag{14.20}$$

where we denote

$$\begin{bmatrix} \hat{B} \\ D_1 \\ D_2 \end{bmatrix} = \begin{bmatrix} \begin{bmatrix} B_1 & B_2 \\ D_{11} & D_{12} \\ D_{21} & D_{22} \end{bmatrix} \begin{bmatrix} D_K \\ I \end{bmatrix} & \begin{bmatrix} B_3 \\ D_{13} \\ D_{23} \end{bmatrix} \end{bmatrix} \tag{14.21}$$

$$\hat{B}_K = [B_k \quad 0].$$

In a similar manner described in [90] and considering the output uniqueness condition, from the above relations one obtains the state-space realization of the generalized homographic transformation $GHM(G^*; [K \quad 0])$ given by

$$GHM(G^*; [K \quad 0]) = \begin{bmatrix} A(G^*; [K \quad 0]) & B(G^*; [K \quad 0]) \\ C(G^*; [K \quad 0]) & D(G^*; [K \quad 0]) \end{bmatrix}, \tag{14.22}$$

where

$$\left.\begin{aligned} A(G^*; [K \quad 0]) &= \begin{bmatrix} A & B_1 C_K \\ 0 & A_K \end{bmatrix} - \begin{bmatrix} \hat{B} \\ \hat{B}_K \end{bmatrix} \times D_2[C_2 \quad D_{21} C_K] \\ B(G^*; [K \quad 0]) &= \begin{bmatrix} \hat{B} \\ \hat{B}_K \end{bmatrix} D_2^- \\ C(G^*; [K \quad 0]) &= [C_1 - D(G^*; [K \quad 0])C_2, (D_{11} - D(G^*; [K \quad 0])C_k] \\ D(G^*; [K \quad 0]) &= D_1 D_2^- \end{aligned}\right\} \tag{14.23}$$

where D_2^- is a $\{1\}$-inverse of D_2 given by

$$D_2^- = \begin{bmatrix} (D_{21}D_K + D_{22})^{-1}(I - D_{23}D_{23}^\dagger) \\ D_{23}^\dagger \end{bmatrix}.$$

Now we are ready to introduce the following definition.

Definition 14.3.1. A terminated system (14.2) and (14.4) is said to be internally stable with respect to the generalized homographic transformation $GHM(G^*; [K \quad 0])$ if the corresponding A-matrix $A(G^*; [K \quad 0])$ given in Eq. (14.23) is stable.

For a closed-loop system (14.1) and (14.2), if the CSR [22, 90] exists and is given by

$$\begin{bmatrix} z \\ w \end{bmatrix} = CHAIN(P) \begin{bmatrix} u \\ y \end{bmatrix} := \begin{bmatrix} \Theta_{11} & \Theta_{12} \\ \Theta_{21} & \Theta_{22} \end{bmatrix} \begin{bmatrix} u \\ y \end{bmatrix}$$

then being the closed-loop transfer function from w to z the usually used linear fractional transformation can be represented [22, 90] as the homographic transformation, i.e.,

$$LF(P; K) = HM(CHAIN(P); K) := (\Theta_{11}K + \Theta_{12})(\Theta_{21}K + \Theta_{22})^{-1}.$$

It is not difficult to see that, in the case that the CSR exists for the plant P, any generalized homographic transformation comes out to be the homographic transformation, i.e.,

$$GHM(CHAIN(P); [K \quad 0]) = HM(CHAIN(P); K).$$

To derive the solvability of the nonstandard H_∞ control problem, the following results are crucial.

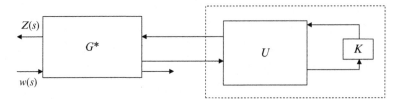

Fig. 14.2 Terminated cascade connection of a generalized chain-scattering representation and a chain-scattering representation.

Theorem 14.3.1. *For the terminated cascade connection (Fig. 14.2) of a GCSR* $G^* = GCHAIN(P_1)$ *and a CSR* $U = CHAIN(P_2)$, *if the condition of output uniqueness is satisfied, then we have*

$$GHM(G^*; [HM(U;K) \ 0]) = GHM\left(G^*\begin{bmatrix} U & 0 \\ 0 & I \end{bmatrix}; [K \ 0]\right). \tag{14.24}$$

Proof. Let

$$U = \begin{bmatrix} U_{11} & U_{12} \\ U_{21} & U_{22} \end{bmatrix}$$

and G^* be given by Eq. (14.6), subsequently

$$HM(U;K) = (U_{11}K + U_{12})(U_{21}K + U_{22})^{-1}. \tag{14.25}$$

From Theorem 14.2.2, the output uniqueness condition for the left-hand side of relation (14.24) is

$$[G_{11}HM(U;K) + G_{12} \quad G_{13}] \tag{14.26}$$
$$\times (I - [G_{21}HM(U;K) + G_{22} \quad G_{23}]^- \tag{14.27}$$
$$\times [G_{21}HM(U;K) + G_{22} \quad G_{23}]) = 0. \tag{14.28}$$

By substituting Eq. (14.25) into Eq. (14.26) and noting

$$G^*\begin{bmatrix} U & 0 \\ 0 & I \end{bmatrix} = \begin{bmatrix} G_{11}U_{11} + G_{12}U_{21} & G_{11}U_{12} + G_{12}U_{22} & G_{13} \\ G_{21}U_{11} + G_{22}U_{21} & G_{21}U_{12} + G_{22}U_{22} & G_{23} \end{bmatrix} \tag{14.29}$$
$$:= \begin{bmatrix} GU_{11} & GU_{12} & G_{13} \\ GU_{21} & GU_{22} & G_{23} \end{bmatrix}$$

one can see that Eq. (14.26) is equivalent to

$$[GU_{11}K + GU_{12} \quad G_{13}] \times (I - [GU_{21}K + GU_{22} \quad G_{23}]^-[GU_{21}K + GU_{22} \quad G_{23}]) = 0,$$

which is exactly the output uniqueness condition for the right-hand side of relation (14.24). One therefore has actually proved that the existence of $GHM(G^*; [HM(U;K) \ 0])$ implies the existency of

$$GHM\left(G^*\begin{bmatrix} U & 0 \\ 0 & I \end{bmatrix}; [K \ 0]\right)$$

and vice versa. The assertion that they are equal follows directly from the fact

$$GHM(G^*; [HM(U;K) \quad 0]) = [G_{11}HM(U;K) + G_{12} \quad G_{13}][G_{21}HM(U;K) + G_{22} \quad G_{23}]^-$$
$$= [G_{11}(U_{11}K + U_{12}) + G_{12}(U_{21}K + U_{22}) \quad G_{13}(U_{21}K + U_{22})]$$
$$\times [G_{21}(U_{11}K + U_{12}) + G_{22}(U_{21}K + U_{22}) \quad G_{23}(U_{21}K + U_{22})]^-$$
$$= [GU_{11}K + GU_{12} \quad G_{13}][GU_{21}K + GU_{22} \quad G_{23}]^-$$
$$= GHM\left(G^* \begin{bmatrix} U & 0 \\ 0 & I \end{bmatrix}; [K \quad 0]\right).$$

This completes the proof.   □

Theorem 14.3.2. *For the terminated cascade connection of a GCSR G^*, if the condition of output uniqueness is satisfied, then we have*

$$GHM\left(G^* \begin{bmatrix} I & 0 \\ R & T \end{bmatrix}; [K \quad 0]\right) = GHM(G^*; [K \quad 0]), \tag{14.30}$$

where R and T are any rational matrix with appropriate dimensions and T is nonsingular.

Proof. Let G^* be given by Eq. (14.6), and according to the partition of G^*

$$\begin{bmatrix} I & 0 \\ R & T \end{bmatrix} := \begin{bmatrix} I & 0 & 0 \\ 0 & I & 0 \\ R_1 & R_2 & T \end{bmatrix} \tag{14.31}$$

by using Theorem 14.2.2 and in a similar manner to the proof of Theorem 14.3.3, we can see that the output uniqueness condition for the left-hand side of relation (14.30) is equivalent to that for the right-hand side. By noting that

$$G^* \begin{bmatrix} I & 0 \\ R & T \end{bmatrix} = \begin{bmatrix} G_{11} + G_{13}R_1 & G_{12} + G_{13}R_2 & G_{13}T \\ G_{21} + G_{23}R_1 & G_{22} + G_{23}R_2 & G_{23}T \end{bmatrix}$$

we find that

$$GHM\left(G^* \begin{bmatrix} I & 0 \\ R & T \end{bmatrix}; [K \quad 0]\right) = [(G_{11} + G_{13}R_1)K + (G_{12} + G_{13}R_2) \quad G_{13}T]$$
$$\times [(G_{21} + G_{23}R_1)K + (G_{22} + G_{23}R_2)G_{23}T]^-$$
$$= [G_{11}K + G_{12} \quad G_{13}] \begin{bmatrix} I & 0 \\ R_1K + R_2 & T \end{bmatrix}$$
$$\times \left\{ \Sigma \begin{bmatrix} I & 0 \\ R_1K + R_2 & T \end{bmatrix} \right\}^-$$
$$= [G_{11}K + G_{12} \quad G_{13}]$$
$$\times \begin{bmatrix} I & 0 \\ R_1K + R_2 & T \end{bmatrix} \begin{bmatrix} I & 0 \\ R_1K + R_2 & T \end{bmatrix}^{-1} \Sigma^-$$
$$= [G_{11}K + G_{12} \quad G_{13}]\Sigma^- = GHM(G^*; [K \quad 0]).$$

We thus finish the proof.   □

(J, J')-lossless factorization [22, 90] is a general notion that includes the well-known inner-outer factorization and the spectral factorization. If a matrix $G(s) \in RL^{\infty}_{(m+r)\times(p+q)}$ is represented as a product $G(s) = \Theta(s)\Pi(s)$, where $\Theta(s) \in RL^{\infty}_{(m+r)\times(p+q)}$ is (J_{mr}, J_{pq})-lossless and $\Pi(s)$ is unimodular in $RH^{\infty}_{(m+q)\times(p+q)}$, then $G(s)$ is said to have a (J_{mr}, J_{pq})-lossless factorization. It is established [22, 90] that the problem of regular H_{∞} control can be reduced to finding a special class of (J, J')-lossless factorization of the CSR. On the bases of Definition 14.3.1 and Theorems 14.3.1 and 14.3.3, concerning the solvability of the nonstandard H_{∞} control problem, we are now ready to propose the following important result.

Theorem 14.3.3. *Assume that the GCSR $G^* = GCHAIN(P)$ has no zeros or poles on the $j\omega$-axis. If the GCSR G^* has a $(J_{mr}, J_{(p+q)r})$-lossless factorization*

$$G^*(s) = \Theta(s)\Pi(s), \tag{14.32}$$

where $\Pi(s)$ has the form

$$\Pi(s) = \left[\begin{array}{cc} \Pi_{11}(s) & 0 \\ \Pi_{21}(s) & \Pi_{22}(s) \end{array} \right], \tag{14.33}$$

where $\Pi_{11}(s)$, $\Pi_{21}(s)$, and $\Pi_{22}(s)$ are of $(p+q) \times (p+q)$, $r \times (p+q)$, and $r \times r$, respectively, and if the output uniqueness condition

$$[G_{11}HM(\Pi_{11}^{-1}; S) + G_{12} \quad G_{13}] \tag{14.34}$$

$$\times (I - [G_{21}HM(\Pi_{11}^{-1}; S) + G_{22} \quad G_{23}]^- \tag{14.35}$$

$$\times [G_{21}HM(\Pi_{11}^{-1}; S) + G_{22} \quad G_{23}]) = 0 \tag{14.36}$$

is satisfied, where S is an arbitrary matrix in $BH^{\infty}_{(p+q)\times(p+q)}$, the $\{1\}$-inverses can be arbitrary, then the nonstandard d H_{∞} control problem is solvable for the original plant (14.1), the desirable H_{∞} controllers are given by

$$K = HM(\Pi_{11}^{-1}; S), \quad \forall S \in BH^{\infty}_{(p+q)\times(p+q)} \tag{14.37}$$

and the closed-loop transfer function is $\Phi = HM(\Theta; S)$.

Proof. According to the reformulation of the problem, to prove this statement it is sufficient to see that any desirable controller given by Eq. (14.37) satisfies the output uniqueness condition (14.34) and

$$\Phi = GHM(G^*; [K \quad 0]) \in BH^{\infty} \tag{14.38}$$

if G^* has a $(J_{mr}, J_{(p+q)r})$-lossless factorization given by Eq. (14.32) with $\Pi(s)$ being the specific form described in Eq. (14.33). □

As far as the output uniqueness condition (14.34) is concerned, Theorem 14.2.2 tells us that it is a must for the generalized homographic transformations to exist.

From Lemma 4.4 in [90] if the matrix $\Theta(s)$ is (J, J')-lossless, then there exists a lossless matrix P_1 such that $\Theta(s)$ is a CSR of P_1, i.e., $\Theta(s) = CHAIN(P_1)$. Subsequently, if $K \in BH^{\infty}$, taking into account Theorem 4.15 in [90], then we find that

$$GHM(\Theta; [K \quad 0]) = GHM(CHAIN(P_1); [K \quad 0]) \tag{14.39}$$
$$= HM(CHAIN(P_1); K) \in BH^\infty.$$

Based on the above observations, by using Theorems 14.3.1 and 14.3.3, it follows that

$$GHM(G^*, [K \quad 0]) = GHM\left(\Theta\begin{bmatrix} \Pi_{11}(s) & 0 \\ \Pi_{21}(s) & \Pi_{22}(s) \end{bmatrix}; [HM(\Pi_{11}^{-1}; S) \quad 0]\right)$$

$$= GHM\left(\Theta\begin{bmatrix} \Pi_{11}(s) & 0 \\ \Pi_{21}(s) & \Pi_{22}(s) \end{bmatrix}\begin{bmatrix} \Pi_{11}^{-1} & 0 \\ 0 & I \end{bmatrix}; [S \quad 0]\right)$$

$$= GHM\left(\Theta\begin{bmatrix} I & 0 \\ \Pi_{21}(s)\Pi_{11}^{-1} & \Pi_{22}(s) \end{bmatrix}; [S \quad 0]\right)$$

$$= GHM(\Theta; [S \quad 0])$$

$$= GHM(CHAIN(P_1); [S \quad 0])$$

$$= HM(CHAIN(P_1); S) \in BH^\infty \quad \forall S \in BH^\infty_{(p+q) \times (p+q)}.$$

As far as the issue of internal stability is concerned in the above equaling process and considering Definition 14.3.1, one can see that the H_∞ controller only cancels out all the stable poles and zeros from the (J, J')-lossless factorization. The internal stability is therefore left invariant.

Due to the fact that one needs to use Theorem 4.15 of [90] in proposing the above theorem, we have assumed that GCSR $G^* = GCHAIN(P)$ has no zeros or poles on the $j\omega$-axis. However, this condition is not as restrictive as the standard assumption (A) that we are trying to relax in this approach. To see this, take the following simple example.

Example 14.3.1. Consider the plant

$$P = \begin{bmatrix} P_{11} & P_{12} \\ P_{21} & P_{22} \end{bmatrix}$$

$$= \begin{bmatrix} 1/(s+1) & 0 & 0 & 0 \\ 1 & 0 & 0 & 1 \\ 0 & 0 & 0 & 1/(s+2) \end{bmatrix},$$

where

$$P_{21} = \begin{bmatrix} 1 & 0 & 0 \\ 0 & 0 & 0 \end{bmatrix}$$

is not of full row rank. The standard assumption (A) is not satisfied by the above plant. It is seen that

$$P_{21}^\dagger = \begin{bmatrix} 1 & 0 \\ 0 & 0 \\ 0 & 0 \end{bmatrix}.$$

Further calculations yield the GCSR matrix

$$
GCHAIN(P) = \begin{bmatrix} P_{12} - P_{11}P_{21}^{\dagger}P_{22} & P_{11}P_{21}^{\dagger} & P_{11}(I - P_{21}^{\dagger}P_{21}) \\ -P_{21}^{\dagger}P_{22} & P_{21}^{\dagger} & I - P_{21}^{\dagger}P_{21} \end{bmatrix}
$$

$$
= \begin{bmatrix} -1/(s+1) & 1/(s+1) & 0 & 0 & 0 & 0 \\ -1 & 1 & 0 & 0 & 0 & 0 \\ 0 & 0 & 0 & 0 & 1 & 0 \\ 0 & 0 & 0 & 0 & 0 & 1 \end{bmatrix},
$$

which has no zeros or poles on the imaginary axis.

The invariant zeros or poles of GCSR $G^* = GCHAIN(P)$ can be understood in the usual way [5] for a GCSR matrix is still a rational matrix though it is essentially a generalized transfer function [116] of the plant. In the case that the GCSR matrix holds any invariant zero or pole on the imaginary axis one can choose a controller to create a pole or zero in the same point. Then the invariant zero or pole is canceled via pole-zero cancelation. Because of our requirement of internal stability this is only possible for zero or pole in the open left half plane. For zero or pole in the open right half plane it is clearly not possible. In the case of an invariant zero or pole on the imaginary axis we can achieve this cancelation proximately by creating a pole or a zero in the left half plane that is very close to the imaginary axis. A treatment in this manner is given in [126].

It should also be noted that the above triangular structure of the unimodular factor $\Pi(s)$ in Eq. (14.33) is similar to that of Theorem 7.7 in [90] for the four block cases. In [22], a procedure based on J-lossless conjugation and then formulated in terms of the solutions of two relevant algebraic Riccati equations has already developed, which brings any rational functional matrix $G(s)$ into its (J, J')-lossless factorization, i.e., $G(s) = \Theta_1(s)\Pi_1(s)$, where $\Theta_1(s)$ is (J, J')-lossless and $\Pi_1(s)$ is unimodular, though $\Pi_1(s)$ is not necessarily in a triangular form. This procedure can also be applied to find the factorization required in the above theorem in the following manner. One first obtains the above (J, J')-lossless factorization

$$
G^*(s) = \Theta_1(s)\Pi_1(s)
$$

by using Kimura's approach, and then computing the LU decomposition of $\Pi_1(s)$ by the *row − echelon factoring* method to obtain the required triangular form. In more detail, considering $\Pi_1(s)$ is a unimodular matrix, one finds a special LU decomposition such that $\Pi_1(s) = L(s)U(s)$, where the row-echelon factoring matrix $L(s)$ can be expected to be orthogonal, and $U(s)$ is in the required triangular form. Note that, in this case the matrix $\Theta(s) = \Theta_1(s)L(s)$ is still (J, J')-lossless for it satisfies [22]

$$
\Theta^T(-s)J_{mr}\Theta(s) = J_{(p+q)r}, \quad \forall s
$$
$$
\Theta^*(s)J_{mr}\Theta(s) \leq J_{(p+q)r}, \quad \forall Re[s] \geq 0.
$$

For the techniques of traditional LU decomposition and a modified LU decomposition of a rational function matrix, we can refer to [127]. For numerical symbolic computation, Maple has provided a routine *LUdecomp* for this purpose. A remaining interesting topic worthy of further research, however, is to show how this kind of special triangular structure can be directly linked to certain algebraic Riccati equations.

The above theorem is interesting in that it generalizes the result proposed by Kimura [22, 90] from the regular case to the nonstandard case. It establishes a close tie between the nonstandard H_∞ control problem and (J, J')-lossless factorization by displaying the fact that, just as in the regular problem, if the unimodular part of a GCSR is completely canceled out by the controller and a certain output uniqueness condition is satisfied, then this nonstandard problem is solvable.

14.4 Conclusions

We have presented a GCSR approach to the nonstandard H_∞ control problem. Certain sufficient conditions for the solvability of this problem are established via a (J, J')-lossless factorization of GCSR. These results thus present extensions to Kimura's results on the CSR approach to H_∞ control. The suggested approach is believed to be workable in the practical control setting due to the fact that some computational formulas and procedures of the Moore-Penrose inverse of a matrix are available through the work of Klema and Laub [128], which can be applied to obtain the GCSR, and the J-lossless conjugation approach [22, 90] has provided a powerful tool for computing the required (J, J')-lossless factorization of the GCSR. Future research would focus on the issue of how these solution conditions in terms of the GCSR of the plant can be directly linked to the relevant algebraic Riccati equations in a state-space scheme.

Internet congestion control: A linear multivariable control look

Congestion control mechanisms in the Internet represent one of the largest deployed artificial feedback systems.

(from [42])

The working mechanism of today's Internet is a feedback system, which has the responsibilities of managing the allocation of bandwidth resources among competing traffic flows and controlling the congestion. Being sharply different from the traditional telephony network, where resources are allocated by the network core using an admission policy, the Internet's resources are allocated in a real-time manner, mainly by the end computer and communication systems themselves. The response is required to meet the desire to accommodate a diversity of demands, from "mice" being made of a few packets, to long "elephants" being greedy for whatever bandwidth is available, and to avoid a centralized allocation mechanism while performing all the manipulations in a distributed manner. The fact that end-systems must control their throughput with little information about the overall network facilitates the use of feedback; such mechanisms have been incorporated since the late 1980s [129] (the well-known Jacobson's congestion avoidance algorithm) into the transport control protocol (TCP) layer of the Internet protocol (IP) stack. For a survey of these algorithms, see [42].

Congestion control and resource allocation mechanisms in today's wireless and wire-lined networks, including the Internet, have already represented many challenges in design, as they continues to expand in size, diversity, reaching scope, integration, and convergence. Having a deep understanding of how their fundamental resource is controlled, and how the allocated resource is contributing to congestion control and service quality, becomes extremely important. Computer networks including the Internet are initially designed on a heuristic, intricate basis of many control mechanisms microscopically on the packet level, they are understood afterwards macroscopically in the flow level.

The notable approaches to address the congestion control and resource allocation issues from a macroscopic perspective are the primal algorithm, the dual algorithm, and the primal-dual algorithm. All these algorithms result from the network utility maximization model. The significance in these algorithms are in terms of the modeling of the congestion measuring and source actions, which is presented in the following:

- Links can feedback to data sources information on the congestion in the resources being used. It is assumed that each network link measures its congestion by a scalar variable (termed price) and that sources have information on the aggregate price of links in their path.

A Generalized Framework of Linear Multivariable Control. http://dx.doi.org/10.1016/B978-0-08-101946-7.00015-9

- After gaining congestion information, flows can take action by using their control mechanisms to react to the congestion along the path to adjust their sending rate.

These assumptions are implicitly presented in many variants of today's TCP protocols. The price signal being used in these protocols can be, for example, loss probability and queueing delay. Also, it is the natural setting for exploring alternative protocols based on more explicit price signaling; for example, the random marking and random dropping actions of packet.

In the primal-dual congestion control algorithm, there are two key blocks: the source rates are adjusted to be adapted to aggregate prices in the TCP algorithm and the link prices are updated in terms of link utilization. Mathematically, we can model the TCP/AQM systems by the following general forms:

$$\dot{X}(t) = F(X(t), Q(t)) \times Q(t), \tag{15.1}$$

$$\dot{P}(t) = G(P(t), Y(t)) \times Y(t). \tag{15.2}$$

The first equation in the above is the mechanism of source rate vector X updating in terms of the aggregate price vector Q, while the second equation in the above is responsible for generating the link price vector P in terms of the aggregate flow rate Y. Note that $F(X(t), Q(t))$ and $G(P(t), Y(t))$ are matrix functions, which are involved into generating the source rate vector and the link price vector. Being deviated from the model presented by Low et al. [42] and Wang [130], herein we are considering a more general communication model, where there are possibly multicast communications.

A multicast (one-to-many or many-to-many distribution [131]) is group communication, where information is addressed to a group of destination computers simultaneously. Group communication has two forms: application layer multicast and network assisted multicast. The network assisted multicast makes it possible for the source to efficiently send to a specific group in a single transmission. Copies are automatically created in other network elements, such as routers, switches, and cellular network base stations, but only to network segments that currently contain members of the group. One typical example of multicast communications is the IP multicast. IP multicast is a technique for one-to-many communication over an IP infrastructure in a network. The destination nodes send join and leave messages, for example in the case of Internet television when the user changes from one TV channel to another. IP multicast scales to a larger receiver population by not requiring prior knowledge of the identity of number of receivers. Multicast uses network infrastructure efficiently by requiring that the source send a packet only once, even if it needs to be delivered to a large number of receivers. The nodes in the network take care of replicating the packet to reach multiple receivers only when it is necessary.

In the situation of multicast, there are interactions (coupled components) in the source rate vector X and in the link price vector P, as one source may issue more than one flow and the link may set more than one price value for a flow. In the usual model presented in, for example, [42, 130], an inherited assumption is that any source only issues one flow. This usual model does not consider the multicast flows and multiple pricing mechanism. Therefore in our model presented in Eqs. (15.1), (15.2)

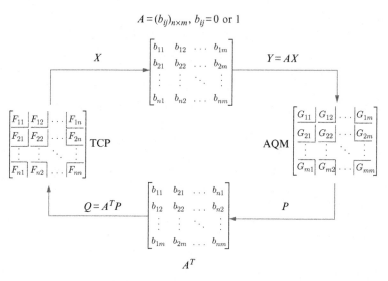

Fig. 15.1 The mechanism of the source rate updating and link price updating in the Internet.

there may be interactions among the components of X and P. This is the reason why we have written the source rate updating and the link price updating mechanisms into the matrix equation forms rather than the individual and independent equations as in Eqs. (15.1), (15.2).

The complete system, which is coupled and interacted by Eqs. (15.1), (15.2), determines both the equilibrium and dynamic characteristics of the TCP/AQM network [130]. However, the equilibrium point is not necessarily unique and it will depend on the initial settings of the system. Therefore the dynamic characteristics of the network may have a diversity. This issue is, however, very complex and deserves further study.

The mechanism of the source rate updating and link price updating, the relation between the individual source rate and link aggregate rate, which is related to the routing matrix, and the relation between the individual link price and the aggregated flow price, which is also related to the routing matrix, are demonstrated in Fig. 15.1.

As a distributed algorithm [132] to allocate network resources among competing users, congestion control consists of two components: a source algorithm that dynamically adjusts the rate in response to the congestion in its path, and a link algorithm that implicitly or explicitly, updates a congestion measure and sends it back to the sources that use that link. In the current Internet, the source algorithm is carried out by TCP, and the link algorithm by active queue management (AQM) schemes. The TCP algorithms include TCP Reno [129], TCP Vegas [133, 134], High-speed TCP [135], Scalable-TCP [136], FAST TCP [137–139], and Generalized FAST TCP [132, 140]; while the AQM algorithms include Drop-Tail [141], RED [142], BLUE [143], REM/PI [144], and AVQ [139, 145]. The working mechanism of the TCP algorithms and the AQM algorithms in the Internet are displayed in Fig. 15.2.

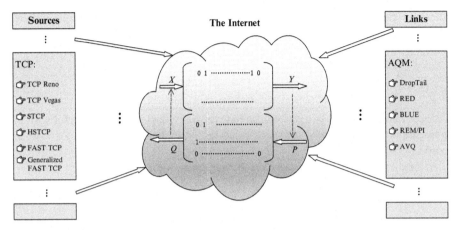

Fig. 15.2 The interactions between the transport control protocol algorithms and the active queue management algorithms in the Internet.

Different protocols use different metrics to measure congestion. For example, TCP Reno [129] and its variants use loss probability as congestion measure, while TCP Vegas [133, 134], FAST TCP [137, 139], and the generalized Fast TCP [140] use queueing delay.

It is established in [146, 147] that, the overall operations of TCP congestion control algorithms work very similarly to a time-delayed closed-loop control system. The feedback control theory for time-delayed systems have been successfully used to enhance the performance of the TCP/RED system in a very logical manner in [146, 147]. Other developments on stability of TCP/AQM systems along the line of control theory can be found in [148–153].

The main aim of this chapter is to give an outline of the Internet congestion control from the perspective of linear multivariable control system. We bring this subject into a control domain by looking at the transferring relations between the dynamics of individual flow rates and the dynamics of aggregate flow rates, the transferring relations between link price and flow aggregate price from a multivariable control point of views including the Smith-McMillan forms. Finally, we look at the issue of feedback control design to relate the TCP source flow rate with the aggregate flow price, which is in a proportional (P) controller structure. These analyses in the flow level then constitute a theoretical basis for TCP protocol design on the packet level.

15.1 The basic model of Internet congestion control

This section presents the basic model of Internet congestion control.

Consider a network that consists of a set $L = \{1, \ldots, M\}$ of unidirectional links of capacity $c_l, l \in L$. The network is shared by a set $I = \{1, \ldots, N\}$ of sources. Let x_i be the rate of flow i and let $X(t) = \{x_i(t), i \in I\}$ be the rate vector. Let $R = (R_{li}, i \in I, l \in L)$ be the routing matrix, where $R_{li} = 1$ if flow i traverses link l, and 0 otherwise. By introducing the link rate vector $Y(t) = \{y_l(t), l \in L\}$, the link price vector $P(t) = \{p_l(t), l \in L\}$, and the source price vector $Q(t) = \{q_i(t), i \in I\}$, we then have

$$Y(t) = RX(t), \tag{15.3}$$

$$Q(t) = R^T P(t). \tag{15.4}$$

It is noted that the above two equations relate the individual source rate vector with the aggregate flow (at the links) rate vector and the individual link price vector with the source aggregate price vector.

At time t, we assume link l observes the aggregate source rate

$$y_l(t) = \sum_{i=1}^{N} R_{li} x_i(t - \tau_{li}^f), \tag{15.5}$$

and source i observes the aggregate price in its path

$$q_i(t) = \sum_{l=1}^{M} R_{li} p_l(t - \tau_{li}^b), \tag{15.6}$$

where τ_{li}^f is the forward feedback delay from source i to link l, and τ_{li}^b is the backward feedback delay from link l to source i. For simplicity, we assume that the feedback delays τ_{li}^f and τ_{li}^b are constants. Further, we consider propagation delay and neglect the queueing delay, thus treating round trip delay for any link l on the path of source i, i.e., $\tau_i = \tau_{li}^f + \tau_{li}^b$, as a constant.

15.2 Internet congestion control: A multivariable control look

In this section, by describing the congestion control network model as a time-delayed multivariable control system, we propose a method to analyze the transfer functions between the individual source rate and the link aggregate rate, between the link price and the source aggregate price. Based on this, we look at the issue of feedback control design to relate the TCP source flow rate with the aggregate flow price, which is in a proportional (P) controller structure. This can further be utilized to analyze the important properties, such as stability, of the network systems.

Stability of the algorithm in Internet congestion control is especially important when it is implemented within a network. For example, to have an efficient throughput, to bound the delay and to minimize packet loss, the rate of data flow through its links should tend toward an equilibrium value, rather than continually oscillating between having bandwidth spare and being completely overloaded. Given the fact that the control system structure of Eqs. (15.5), (15.6) like the Smith-McMillan form governs their system performance, e.g., the stability, let us take a closer look.

Taking the Laplace transform of Eqs. (15.5), (15.6), respectively, and writing the delay routing matrices in the frequency domain as

$$R_f(s) = (R_{li}^f(s), i \in I, l \in L),$$

where

$$R_{li}^f(s) = \begin{cases} e^{-\tau_{li}^f s}, & \text{if flow } i \text{ traverses link } l, \\ 0, & \text{otherwise,} \end{cases}$$

and

$$R_b(s) = (R_{li}^b(s), i \in I, l \in L),$$

where

$$R_{li}^b(s) = \begin{cases} e^{-\tau_{li}^b s}, & \text{if flow } i \text{ traverses link } l, \\ 0, & \text{otherwise,} \end{cases}$$

we then have

$$Y(s) = R_f(s)X(s), \tag{15.7}$$

$$Q(s) = (R_b(s))^T P(s), \tag{15.8}$$

where $X(s)$, $Y(s)$, $P(s)$, and $Q(s)$ denote the Laplace transform of the vectors $X(t)$, $Y(t)$, $P(t)$, and $Q(t)$, respectively, and T denotes the transport of the matrix.

15.3 Padé approximations to the system (15.7) and (15.8)

Polynomials are not such a good class of functions if one wants to approximate functions with singularities because polynomials are entire functions without singularities. They are only useful up to the first singularity. Rational functions are the simplest ones with singularities. The idea is that the poles of the rational functions will move to the singularities of the function and hence the domain of convergence can be enlarged, and singularities of it may be discovered using the poles of the rational approximations.

The $[m, n]$ Padé approximation of a function f in a point a is the rational function $\frac{Q_m}{P_n}$, with Q_m a polynomial of *degree* $\leq m$ and P_n a polynomial of *degree* $\leq n$, for which we have the following interpolation condition at a

$$f(z) - \frac{Q_m(z)}{P_n(z)} = O((z - a)^{m+n+1}), \quad z \to a. \tag{15.9}$$

The computation of the polynomials P_n and Q_m is not so straightforward in the above condition since one first has to compute the Taylor expansion of $\frac{Q_m}{P_n}$ and then equate the first $m + n + 1$ Taylor coefficients to the first $m + n + 1$ Taylor coefficients of f. Usually the Padé approximation is defined by linearizing the interpolation condition as

$$P_n(z)f(z) - Q_m(z) = O((z-a)^{m+n+1}), \quad z \to a. \tag{15.10}$$

For Padé approximation near infinity to a function of the form

$$f(z) = \sum_{k=0}^{\infty} \frac{c_k}{z^{k+1}},$$

we let $m = n - 1$ and the interpolation condition is

$$P_n f(z) - Q_{n-1}(z) = O(z^{-n-1}), \quad z \to \infty.$$

There is a degree of freedom since we can multiply both sides of Eq. (15.10) by a constant. Usually we normalize this by taking P_n monic, that is it is in the form $x^n + \cdots$, when this is possible. This can only be done if P_n is of the exact degree n. If we take P_n monic, then we can determine the n unknown coefficients a_k $(k = 1, \ldots, n)$ in

$$P_n =: \sum_{k=0}^{n} a_k(z-a)^{n-k}, \quad a_0 = 1, \tag{15.11}$$

by putting the coefficients of $(z-a)^k$ for $k = m+1, m+2, \ldots, m+n$ in the Taylor expansion of $P_n f$ equal to zero. The polynomial Q_m then corresponds to the Taylor polynomial of degree m of $P_n f$.

Hereafter we present another approach. Suppose f is analytic in a domain Ω that contains a. Again we take a contour Γ inside Ω encircling a once in the positive direction. Divide both sides of Eq. (15.10) by $(z-a)^{m+k+2}$ and integrate, to find

$$\frac{1}{2\pi i} \int_{\Gamma} \frac{P_n(z)f(z)}{(z-a)^{m+k+2}} dz - \frac{1}{2\pi i} \int_{\Gamma} \frac{Q_m(z)}{(z-a)^{m+k+2}} dz$$

$$= \sum_{j=m+n+1}^{\infty} b_{nj} \frac{1}{2\pi i} \int_{\Gamma} (z-a)^{j-m-k-2} dz,$$

where the $b_{n,j}$s are the coefficients in the expansion of $P_n f - Q_m$ around a. The integral involving Q_m is zero for $k \geq 0$ since it is proportional to the $(m+k+1)$th derivative of Q_m, which is zero for $k \geq 0$. The summation on the right-hand side has a contribution only when $j = m+k+1$, but when $0 \leq k \leq n-1$ then $j \leq m+n$, and such indices do not appear in the summation. Hence the right hand side also vanishes for $k \leq n-1$. Therefore Eq. (15.10) tells us that

$$\frac{1}{2\pi i} \int_{\Gamma} \frac{P_n(z)}{(z-a)^{m+k+2}} f(z)dz = 0, \quad k = 0, 1, \ldots, n-1.$$

If we use the expansion (15.11) then this gives

$$\sum_{j=0}^{n} a_j \frac{1}{2\pi i} \int_{\Gamma} (z-a)^{n-j-m-k-2} f(z)dz = 0, \quad k = 0, 1, 2, \ldots, n-1$$

and then

$$\frac{1}{2\pi i} \int_\Gamma (z-a)^{n-j-m-k-2} f(z)dz = c_{m-n+k+j+1}.$$

Therefore we obtain the following system of equations

$$\begin{pmatrix} c_{m-n+1} & c_{m-n+2} & \cdots & c_{m+1} \\ c_{m-n+2} & c_{m-n+3} & \cdots & c_{m+2} \\ \vdots & \vdots & \cdots & \vdots \\ c_m & c_{m+1} & \cdots & c_{m+n} \end{pmatrix} \begin{pmatrix} a_0 \\ a_1 \\ \vdots \\ a_n \end{pmatrix} = \begin{pmatrix} 0 \\ 0 \\ \vdots \\ 0 \end{pmatrix}.$$

There is one degree of freedom here since we have $n+1$ unknowns and n (homogeneous) equations. The choice $a_0 = 1$ (if possible) gives the monic polynomial P_n, but sometimes other normalization will be used.

15.3.1 Padé approximation to the transfer functions of the time-delay system (15.7) and (15.8)

For the transfer function of a time-delay system

$$H(S) = e^{-\tau s}, \tag{15.12}$$

which is irrational (it has no s-polynomial in the numerator and in the denominator), in some situations as in analyzing the frequency response of this sort of control system containing a time-delay, it is necessary to substitute $e^{-\tau s}$ with the following Padé approximation

$$e^{-\tau s} \approx \frac{1 - k_1 s + k_2 s^2 + \cdots \pm k_n s^n}{1 + k_1 s + k_2 s^2 + \cdots + k_n s^n}, \tag{15.13}$$

where n is the order of the approximation.

In particular, to find a second order Padé approximation ($n = 2$) for the time-delay transfer function $H(s)$, we have

$$k_1 = \frac{\tau}{2}, \quad k_2 = \frac{\tau^2}{12}. \tag{15.14}$$

Subsequently, the second order Padé approximation to the transfer function $H(s)$, denoted by $H^P(s)$, comes out to be

$$H^P(s) = \frac{1 - k_1 s + k_2 s^2}{1 + k_1 s + k_2 s^2} = \frac{1 - \frac{\tau}{2}s + \frac{\tau^2}{12}s^2}{1 + \frac{\tau}{2}s + \frac{\tau^2}{12}s^2}. \tag{15.15}$$

With regard to the Padé approximation to the time-delay system (15.7) and (15.8), so far we are able to establish the following result.

Theorem 15.3.1. *The Padé approximations to the time-delay system (15.7) and (15.8) are given by*

$$Y(s) = R_f^P(s)X(s), \tag{15.16}$$

$$Q(s) = (R_b^P(s))^T P(s), \tag{15.17}$$

where $R_f^P(s)$ and $(R_b^P(s))^T$ are the respectively Padé approximations to the transfer function matrices (with delays) $R_f(s)$ and $(R_b(s))^T$. They are given by

$$R_f^P(s) = ((R_{li}^f(s))^P, i \in I, l \in L),$$

where

$$(R_{li}^f(s))^P = \begin{cases} \frac{1 - k_1 s + k_2 s^2 + \cdots \pm k_n s^n}{1 + k_1 s + k_2 s^2 + \cdots + k_n s^n}, & \text{if flow } i \text{ traverses link } l, \\ 0, & \text{otherwise}, \end{cases}$$

and

$$R_b^P(s) = ((R_{li}^b(s))^P, i \in I, l \in L),$$

where

$$(R_{li}^b(s))^P = \begin{cases} \frac{1 - k_1' s + k_2' s^2 + \cdots \pm k_n' s^n}{1 + k_1' s + k_2' s^2 + \cdots + k_n' s^n}, & \text{if flow } i \text{ traverses link } l, \\ 0, & \text{otherwise}. \end{cases}$$

In the above, all the parameters k_i $(i = 1, 2, \ldots, n)$ and k_i' $(i = 1, 2, \ldots, n)$ can be determined with relations to the order of the approximation and the delays.

It should be noted that the above theorem is very important for us to explore the system structure of Internet congestion control, in that it has actually reduced the original transfer functions from irrational to rational. This enables us to work on the rational function matrices given in Eqs. (15.16), (15.17) by using the generalized linear multivariable control approaches.

15.3.2 A proportional controller

Now we design the following proportional control to relate the source rate with the experienced aggregate price of that source

$$x_i(t) = \psi_i q_i(t), \quad i \in I, \tag{15.18}$$

where $0 \le \psi_i \le 1$ is the control gain.

We have the following formulations to calculate the congestion price. For a given flow that traverses a set of link L_i, if all the packet loss probabilities $p_l(t)$ at all links are small enough, the end-to-end packet loss probabilities $q_i(t)$ satisfy

$$q_i(t) = 1 - \Pi_{l \in L_i}(1 - p_l(t)) \approx \Sigma_{l \in L_i} p_l(t).$$

The aggregate queuing delay for that flow is given by

$$q_i(t) = \Sigma_{l \in L_i} R_{li} \frac{b_l}{c_l},$$

where b_l is the backlog at link l and c_l is the leaking capacity of the link.

From a TCP protocol point of view, the controller (15.18) suggests that if q_i represents congestion price (either queuing delay or packet loss probability) in the sources path, the source rate should be a monotonically decreasing function of it. Specifically, as a larger queuing delay or packet loss probability suggests a higher degree of congestion, the source rate should be decreased proportionally to the congestion degree parameter (the aggregate queuing delay or the aggregate packet loss probability). Alternatively, if a source is likely to experience slight congestion along its path, i.e., the aggregate queuing delay or the aggregate packet loss probability is smaller, one should increase the sending rate for this source.

Combining all the above N equations (15.18) together, we have the following matrix equation

$$X(t) = \Psi Q(t), \tag{15.19}$$

where $\Psi = diag\{\psi_1, \psi_2, \ldots, \psi_N\}$. By taking the Laplace transform of the above equation and then substituting this Laplace transformed form into Eq. (15.16), with Eq. (15.17) being considered, we arrive at the following relation between $Y(s)$ and $P(s)$

$$Y(s) = R_f^P(s)\Psi(R_b^P(s))^T P(s). \tag{15.20}$$

The above equation is the Padé approximation to the transfer function from the link price vector to the link aggregate rate vector. We will work on it to gain insights into its frequency structure later on by using a simple network example.

15.4   Analyses into system structure of congestion control of a simple network in frequency domain

Let us consider a simple network with three links and four flows as shown in Fig. 15.3. In this network, there are four flows: $f1$ traverses link $L1$, $f2$ traverses link $L2$, $f3$ traverses link $L3$, and $f4$ traverses link $L1$, link $L2$, and link $L3$. Flows $f1$, $f2$, and $f3$ represent one-hop short flow, which is dominant in the Internet: most peer-to-peer applications involve one-link flows between a data source and a receiver (downloader); while flow $f4$ represents long flows which traverse more than one hop. The routing structure of this network is described by the following routing matrix

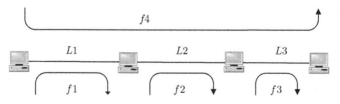

Fig. 15.3 A simple but representative network: three links and four flows.

$$R = \begin{bmatrix} 1 & 0 & 0 & 1 \\ 0 & 1 & 0 & 1 \\ 0 & 0 & 1 & 1 \end{bmatrix}.$$

Letting x_i denote the source sending rate of flow fi, ($i = 1, 2, 3, 4$). The individual flow rate vector is then

$$X(t) = \begin{bmatrix} x_1(t) \\ x_2(t) \\ x_3(t) \\ x_4(t) \end{bmatrix}.$$

Letting y_l ($l = 1, 2, 3$) denote the aggregate flow rate at link l, the aggregate flow rate vector is then denoted by

$$Y(t) = \begin{bmatrix} y_1(t) \\ y_2(t) \\ y_3(t) \end{bmatrix}.$$

Each link is suggesting a price (either delay or packet loss ratio), one has the link price vector

$$P(t) = \begin{bmatrix} p_1(t) \\ p_2(t) \\ p_3(t) \end{bmatrix}.$$

The total price for flow fi is the sum of all the prices for all links it traverses and is denoted by q_i. We then denote the aggregate price vector as

$$Q(t) = \begin{bmatrix} q_1(t) \\ q_2(t) \\ q_3(t) \\ q_4(t) \end{bmatrix}.$$

The aggregate flow rate vector and the individual rate vector is associated by the routing matrix, and the aggregate price vector and the individual price vector are also associated by the transpose of the routing matrix. They are mathematically described by

$$Y(t) = RX(t), \tag{15.21}$$

$$Q(t) = R^T P(t). \tag{15.22}$$

In particular, at time t, the links observe the following corresponding aggregate source rate

$$y_1(t) = x_1(t - \tau_{11}^f) + x_4(t - \tau_{14}^f), \tag{15.23}$$

$$y_2(t) = x_2(t - \tau_{22}^f) + x_4(t - \tau_{24}^f), \tag{15.24}$$

$$y_3(t) = x_3(t - \tau_{33}^f) + x_4(t - \tau_{34}^f), \tag{15.25}$$

where τ_{li}^f $(l = 1,2,3; i = 1,2,3,4)$ denotes the feed-forward delay of the data flow f_i by the traveling link l. This delay is determined by the propagation delay and the forward queuing delay at the link. The propagation delay is determined by the distance between the source and the destination while the forward queuing delay is determined by the queue length and the leaking capacity at the link.

Similarly, the sources observe the following corresponding aggregate price along their path

$$q_1(t) = p_1(t - \tau_{11}^b), \tag{15.26}$$

$$q_2(t) = p_2(t - \tau_{22}^b), \tag{15.27}$$

$$q_3(t) = p_3(t - \tau_{33}^b), \tag{15.28}$$

$$q_4(t) = p_1(t - \tau_{14}^b) + p_2(t - \tau_{24}^b) + p_3(t - \tau_{34}^b), \tag{15.29}$$

where τ_{li}^b $(l = 1,2,3; i = 1,2,3,4)$ denotes the feed-backward delay of flow f_i. These backward delays are actually experienced by the acknowledgment (ack) packets for their corresponding data flows. They are all related to the distance between the destination and the source and the capacity (in the opposite direction) correspondingly.

By taking the Laplace transform of Eqs. (15.23), (15.24), (15.25), and denoting the Laplace transformation of $X(t)$ and $Y(t)$ as $X(s)$ and $Y(s)$, respectively, one has

$$Y(s) = R_f^P(s)X(s),$$

where the transfer function $R_f^P(s)$ is given by

$$R_f^P(s) = \begin{bmatrix} e^{-\tau_{11}^f s} & 0 & 0 & e^{-\tau_{14}^f s} \\ 0 & e^{-\tau_{22}^f s} & 0 & e^{-\tau_{24}^f s} \\ 0 & 0 & e^{-\tau_{33}^f s} & e^{-\tau_{34}^f s} \end{bmatrix}.$$

The second order Padé approximation to the above transfer function is

$$\mathbf{R_f^P(s)} = \begin{bmatrix} \dfrac{12-6\tau_{11}^f s+(\tau_{11}^f)^2 s^2}{12+6\tau_{11}^f s+(\tau_{11}^f)^2 s^2} & 0 & 0 & \dfrac{12-6\tau_{14}^f s+(\tau_{14}^f)^2 s^2}{12+6\tau_{14}^f s+(\tau_{14}^f)^2 s^2} \\ 0 & \dfrac{12-6\tau_{22}^f s+(\tau_{22}^f)^2 s^2}{12+6\tau_{22}^f s+(\tau_{22}^f)^2 s^2} & 0 & \dfrac{12-6\tau_{24}^f s+(\tau_{24}^f)^2 s^2}{12+6\tau_{24}^f s+(\tau_{24}^f)^2 s^2} \\ 0 & 0 & \dfrac{12-6\tau_{33}^f s+(\tau_{33}^f)^2 s^2}{12+6\tau_{33}^f s+(\tau_{33}^f)^2 s^2} & \dfrac{12-6\tau_{34}^f s+(\tau_{34}^f)^2 s^2}{12+6\tau_{34}^f s+(\tau_{34}^f)^2 s^2} \end{bmatrix}.$$

If we set all the forward delays as $\tau_{11}^f = 1, \tau_{22}^f = 2, \tau_{33}^f = 6, \tau_{14}^f = 3, \tau_{24}^f = 4, \tau_{34}^f = 9$, by denoting

$$d_1^f = 12 - 6\tau_{11}^f s + (\tau_{11}^f)^2 s^2 = 12 - 6s + s^2$$

$$d_2^f = 12 - 6\tau_{22}^f s + (\tau_{22}^f)^2 s^2 = 12 - 12s + 4s^2$$

$$d_3^f = 12 - 6\tau_{33}^f s + (\tau_{33}^f)^2 s^2 = 12 - 36s + 36s^2$$

$$d_4^f = 12 - 6\tau_{14}^f s + (\tau_{14}^f)^2 s^2 = 12 - 18s + 9s^2$$

$$d_5^f = 12 - 6\tau_{24}^f s + (\tau_{24}^f)^2 s^2 = 12 - 24s + 16s^2$$

$$d_6^f = 12 - 6\tau_{34}^f s + (\tau_{34}^f)^2 s^2 = 12 - 54s + 81s^2$$

$$b_1^f = 12 + 6\tau_{11}^f s + (\tau_{11}^f)^2 s^2 = 12 + 6s + s^2$$

$$b_2^f = 12 + 6\tau_{22}^f s + (\tau_{22}^f)^2 s^2 = 12 + 12s + 4s^2$$

$$b_3^f = 12 + 6\tau_{33}^f s + (\tau_{33}^f)^2 s^2 = 12 + 36s + 36s^2$$

$$b_4^f = 12 + 6\tau_{14}^f s + (\tau_{14}^f)^2 s^2 = 12 + 18s + 9s^2$$

$$b_5^f = 12 + 6\tau_{24}^f s + (\tau_{24}^f)^2 s^2 = 12 + 24s + 16s^2$$

$$b_6^f = 12 + 6\tau_{34}^f s + (\tau_{34}^f)^2 s^2 = 12 + 54s + 81s^2$$

then we have the following approximated transfer function

$$R_f^P(s) = \begin{bmatrix} \frac{d_1^f}{b_1^f} & 0 & 0 & \frac{d_4^f}{b_4^f} \\ 0 & \frac{d_2^f}{b_2^f} & 0 & \frac{d_5^f}{b_5^f} \\ 0 & 0 & \frac{d_3^f}{b_3^f} & \frac{d_6^f}{b_6^f} \end{bmatrix}.$$

Using $d^f(s)$ to denote the least common multiple (**lcm**) of the denominators of all the elements of $R_f^P(s)$, one has

$$d^f(s) = b_1^f \times b_2^f \times b_3^f \times b_4^f \times b_5^f \times b_6^f.$$

Subsequently the Smith form of $R_f^P(s)$ is $S_f(s)$, which is given by

$$S_f(s) = \begin{bmatrix} 1 & 0 & 0 & 0 \\ 0 & b_1^f \times b_2^f \times b_3^f \times b_4^f \times b_5^f \times b_6^f & 0 & 0 \\ 0 & 0 & b_1^f \times b_2^f \times b_3^f \times b_4^f \times b_5^f \times b_6^f & 0 \end{bmatrix}.$$

Accordingly the Smith-McMillan form of $R_f^P(s)$ is $M_f(s)$

$$M_f(s) = \begin{bmatrix} \frac{1}{b_1^f \times b_2^f \times b_3^f \times b_4^f \times b_5^f \times b_6^f} & 0 & 0 & 0 \\ 0 & 1 & 0 & 0 \\ 0 & 0 & 1 & 0 \end{bmatrix}$$

$$= \begin{bmatrix} \frac{1}{(s+3+\sqrt{3}i)\times(s+3-\sqrt{3}i)} \times \frac{1}{\left(s+\frac{3+\sqrt{3}i}{2}\right)\times\left(s+\frac{3-\sqrt{3}i}{2}\right)} & & \\ \times \frac{1}{\left(s+\frac{3+\sqrt{3}i}{6}\right)\times\left(s+\frac{3-\sqrt{3}i}{6}\right)} \times \frac{1}{\left(s+\frac{3+\sqrt{3}i}{3}\right)\times\left(s+\frac{3-\sqrt{3}i}{3}\right)} & & \\ \times \frac{1}{\left(s+\frac{3+\sqrt{3}i}{4}\right)\times\left(s+\frac{3-\sqrt{3}i}{4}\right)} \times \frac{1}{\left(s+\frac{3+\sqrt{3}i}{9}\right)\times\left(s+\frac{3-\sqrt{3}i}{9}\right)} & 0 & 0 & 0 \\ 0 & & & 1 & 0 & 0 \\ 0 & & & 0 & 1 & 0 \end{bmatrix}.$$

Similarly, by taking the Laplace transform of Eqs. (15.26), (15.27), (15.28), (15.29), and denoting the Laplace transformation of $P(t)$ and $Q(t)$ as $P(s)$ and $Q(s)$, respectively, one has

$$Q(s) = (R_f^P(s))^T X(s),$$

where the transfer function $(R_f^P(s))^T$ is given by

$$(R_b^P(s))^T = \begin{bmatrix} \frac{12-6\tau_{11}^b s+(\tau_{11}^b)^2 s^2}{12+6\tau_{11}^b s+(\tau_{11}^b)^2 s^2} & 0 & 0 \\ 0 & \frac{12-6\tau_{22}^b s+(\tau_{22}^b)^2 s^2}{12+6\tau_{22}^b s+(\tau_{22}^b)^2 s^2} & 0 \\ 0 & 0 & \frac{12+6\tau_{33}^b s+(\tau_{33}^b)^2 s^2}{12+6\tau_{33}^b s+(\tau_{33}^b)^2 s^2} \\ \frac{12-6\tau_{14}^b s+(\tau_{14}^b)^2 s^2}{12+6\tau_{14}^b s+(\tau_{14}^b)^2 s^2} & \frac{12-6\tau_{24}^b s+(\tau_{24}^b)^2 s^2}{12+6\tau_{24}^b s+(\tau_{24}^b)^2 s^2} & \frac{12-6\tau_{34}^b s+(\tau_{34}^b)^2 s^2}{12+6\tau_{34}^b s+(\tau_{34}^b)^2 s^2} \end{bmatrix}.$$

If the backward delays are set as $\tau_{11}^b = 11, \tau_{22}^b = 5, \tau_{33}^b = 7, \tau_{14}^b = 14, \tau_{24}^b = 8, \tau_{34}^b = 12$, by denoting

$$a_1^b = 12 - 6\tau_{11}^b s + (\tau_{11}^b)^2 s^2 = 12 + 66s + 121s^2$$
$$a_2^b = 12 - 6\tau_{22}^b s + (\tau_{22}^b)^2 s^2 = 12 + 30s + 25s^2$$
$$a_3^b = 12 - 6\tau_{33}^b s + (\tau_{33}^b)^2 s^2 = 12 + 42s + 49s^2$$
$$a_4^b = 12 - 6\tau_{14}^b s + (\tau_{14}^b)^2 s^2 = 12 + 84s + 196s^2$$
$$a_5^b = 12 - 6\tau_{24}^b s + (\tau_{24}^b)^2 s^2 = 12 + 48s + 64s^2$$
$$a_6^b = 12 - 6\tau_{34}^b s + (\tau_{34}^b)^2 s^2 = 12 + 72s + 144s^2$$
$$b_1^b = 12 + 6\tau_{11}^b s + (\tau_{11}^b)^2 s^2 = 12 + 66s + 121s^2$$
$$b_2^b = 12 + 6\tau_{22}^b s + (\tau_{22}^b)^2 s^2 = 12 + 30s + 25s^2$$

$$b_3^b = 12 + 6\tau_{33}^b s + (\tau_{33}^b)^2 s^2 = 12 + 42s + 49s^2$$
$$b_4^b = 12 + 6\tau_{14}^b s + (\tau_{14}^b)^2 s^2 = 12 + 84s + 196s^2$$
$$b_5^b = 12 + 6\tau_{24}^b s + (\tau_{24}^b)^2 s^2 = 12 + 48s + 64s^2$$
$$b_6^b = 12 + 6\tau_{34}^b s + (\tau_{34}^b)^2 s^2 = 12 + 72s + 144s^2,$$

we then have

$$(R_b^P(s))^T = \begin{bmatrix} \frac{a_1^b}{b_1^b} & 0 & 0 \\ 0 & \frac{a_2^b}{b_2^b} & 0 \\ 0 & 0 & \frac{a_3^b}{b_3^b} \\ \frac{a_4^b}{b_4^b} & \frac{a_5^b}{b_5^b} & \frac{a_6^b}{b_6^b} \end{bmatrix}.$$

Further, if $d^b(s)$ is the least common multiple (lcm) of the denominators of all the elements in $(R_b^P(s))^T$, that is

$$d^b(s) = b_1^b \times b_2^b \times b_3^b \times b_4^b \times b_5^b \times b_6^b,$$

then the Smith form of $(R_b^P(s))^T$ (denoted by $S_b(s)$) is given by

$$S_b(s) = \begin{bmatrix} 1 & 0 & 0 \\ 0 & b_1^b \times b_2^b \times b_3^b \times b_4^b \times b_5^b \times b_6^b & 0 \\ 0 & 0 & b_1^b \times b_2^b \times b_3^b \times b_4^b \times b_5^b \times b_6^b \\ 0 & 0 & 0 \end{bmatrix}.$$

Therefore the Smith-McMillan form of $(R_b^P(s))^T$ (denoted by $M_b(s)$) is given by

$$M_b(s) = \begin{bmatrix} \frac{1}{b_1^b \times b_2^b \times b_3^b \times b_4^b \times b_5^b \times b_6^b} & 0 & 0 \\ 0 & 1 & 0 \\ 0 & 0 & 1 \\ 0 & 0 & 0 \end{bmatrix}$$

$$= \begin{bmatrix} \frac{1}{\left(s+\frac{3+\sqrt{3}i}{11}\right)\times\left(s+\frac{3-\sqrt{3}i}{11}\right)} \times \frac{1}{\left(s+\frac{3+\sqrt{3}i}{5}\right)\times\left(s+\frac{3-\sqrt{3}i}{5}\right)} \\ \times \frac{1}{\left(s+\frac{3+\sqrt{3}i}{7}\right)\times\left(s+\frac{3-\sqrt{3}i}{7}\right)} \times \frac{1}{\left(s+\frac{3+\sqrt{3}i}{14}\right)\times\left(s+\frac{3-\sqrt{3}i}{14}\right)} \\ \times \frac{1}{\left(s+\frac{3+\sqrt{3}i}{8}\right)\times\left(s+\frac{3-\sqrt{3}i}{8}\right)} \times \frac{1}{\left(s+\frac{3+\sqrt{3}i}{12}\right)\times\left(s+\frac{3-\sqrt{3}i}{12}\right)} \quad 0 \quad 0 \\ 0 \qquad\qquad\qquad\qquad\qquad\qquad 1 \quad 0 \\ 0 \qquad\qquad\qquad\qquad\qquad\qquad 0 \quad 1 \\ 0 \qquad\qquad\qquad\qquad\qquad\qquad 0 \quad 0 \end{bmatrix}.$$

Now we design the following proportional control to relate the source rate with the experienced aggregate price of that source

$$x_i(t) = \psi_i q_i(t), \quad i = 1, 2, 3, 4,$$

where $0 \leq \psi_i \leq 1$ is the control gain. The transfer function from $P(s)$ to $Y(s)$ is then given by

$$R_f^P(s)\Psi((R_f^P(s))^T) = \begin{bmatrix} \frac{d_1^f}{b_1^f} & 0 & 0 & \frac{d_4^f}{b_4^f} \\ 0 & \frac{d_2^f}{b_2^f} & 0 & \frac{d_5^f}{b_5^f} \\ 0 & 0 & \frac{d_3^f}{b_3^f} & \frac{d_6^f}{b_6^f} \end{bmatrix} \begin{bmatrix} \psi_1 & 0 & 0 & 0 \\ 0 & \psi_2 & 0 & 0 \\ 0 & 0 & \psi_3 & 0 \\ 0 & 0 & 0 & \psi_4 \end{bmatrix} \begin{bmatrix} \frac{a_1^b}{b_1^b} & 0 & 0 \\ 0 & \frac{a_2^b}{b_2^b} & 0 \\ 0 & 0 & \frac{a_3^b}{b_3^b} \\ \frac{a_4^b}{b_4^b} & \frac{a_5^b}{b_5^b} & \frac{a_6^b}{b_6^b} \end{bmatrix}$$

$$= \begin{bmatrix} \psi_1 \frac{d_1^f}{b_1^f}\frac{a_1^b}{b_1^b} + \psi_4 \frac{d_4^f}{b_4^f}\frac{a_4^b}{b_4^b} & \psi_4 \frac{d_4^f}{b_4^f}\frac{a_5^b}{b_5^b} & \psi_4 \frac{d_4^f}{b_4^f}\frac{a_6^b}{b_6^b} \\ \psi_5 \frac{d_4^f}{b_4^f}\frac{a_4^b}{b_4^b} & \psi_2 \frac{d_2^f}{b_2^f}\frac{a_2^b}{b_2^b} + \psi_4 \frac{d_5^f}{b_5^f}\frac{a_5^b}{b_5^b} & \psi_4 \frac{d_5^f}{b_5^f}\frac{a_6^b}{b_6^b} \\ \psi_4 \frac{d_6^f}{b_6f}\frac{a_4^b}{b_4^b} & \psi_4 \frac{d_6^f}{b_6^f}\frac{a_5^b}{b_5^b} & \psi_3 \frac{d_3^f}{b_3^f}\frac{a_3^b}{b_3^b} + \psi_4 \frac{d_6^f}{b_6^f}\frac{a_6^b}{b_6^b} \end{bmatrix}$$

$$= \begin{bmatrix} \frac{\psi_1 d_1^f a_1^b b_4^f b_4^b + \psi_4 d_4^f a_4^b b_1^f b_1^b}{b_1^f b_1^b b_4^f b_4^b} & \frac{\psi_4 d_4^f a_5^b}{b_4^f b_5^b} & \frac{\psi_4 d_4^f a_6^b}{b_4^f b_6^b} \\ \frac{\psi_5 d_4^f a_5^b}{b_4^f b_5^b} & \frac{\psi_2 d_2^f a_2^b b_5^f b_5^b + \psi_4 d_5^f a_5^b b_2^f b_2^b}{b_2^f b_2^b b_5^f b_5^b} & \frac{\psi_4 d_5^f a_6^b}{b_5^f b_6^b} \\ \frac{\psi_4 d_6^f a_4^b}{b_6^f b_4^b} & \frac{\psi_4 d_6^f a_5^b}{b_6^f b_5^b} & \frac{\psi_3 d_3^f a_3^b b_6^f b_6^b + \psi_4 d_6^f a_6^b b_3^f b_3^b}{b_3^f b_3^b b_6^f b_6^b} \end{bmatrix}$$

In the above all the elements are given by

$$b_1^f b_1^b b_4^f b_4^b = \left(s + \frac{3 + \sqrt{3}i}{\tau_{11}^f}\right) \times \left(s + \frac{3 - \sqrt{3}i}{\tau_{11}^f}\right) \times \left(s + \frac{3 + \sqrt{3}i}{\tau_{11}^b}\right) \times \left(s + \frac{3 - \sqrt{3}i}{\tau_{11}^b}\right)$$

$$\times \left(s + \frac{3 + \sqrt{3}i}{\tau_{14}^f}\right) \times \left(s + \frac{3 - \sqrt{3}i}{\tau_{14}^f}\right) \times \left(s + \frac{3 + \sqrt{3}i}{\tau_{14}^b}\right) \times \left(s + \frac{3 - \sqrt{3}i}{\tau_{14}^b}\right)$$

$$= (s + 3 + \sqrt{3}i) \times (s + 3 - \sqrt{3}i) \times \left(s + \frac{3 + \sqrt{3}i}{11}\right) \times \left(s + \frac{3 - \sqrt{3}i}{11}\right)$$

$$\times \left(s + \frac{3 + \sqrt{3}i}{3}\right) \times \left(s + \frac{3 - \sqrt{3}i}{3}\right) \times \left(s + \frac{3 + \sqrt{3}i}{14}\right) \times \left(s + \frac{3 - \sqrt{3}i}{14}\right)$$

$$b_4^f b_5^b = \left(s + \frac{3 + \sqrt{3}i}{\tau_{14}^f}\right) \times \left(s + \frac{3 - \sqrt{3}i}{\tau_{14}^f}\right) \times \left(s + \frac{3 + \sqrt{3}i}{\tau_{24}^b}\right) \times \left(s + \frac{3 - \sqrt{3}i}{\tau_{24}^b}\right)$$

$$= \left(s + \frac{3 + \sqrt{3}i}{3}\right) \times \left(s + \frac{3 - \sqrt{3}i}{3}\right) \times \left(s + \frac{3 + \sqrt{3}i}{8}\right) \times \left(s + \frac{3 - \sqrt{3}i}{8}\right)$$

$$b_4^f b_6^b = \left(s + \frac{3 + \sqrt{3}i}{\tau_{14}^f}\right) \times \left(s + \frac{3 - \sqrt{3}i}{\tau_{14}^f}\right) \times \left(s + \frac{3 + \sqrt{3}i}{\tau_{34}^b}\right) \times \left(s + \frac{3 - \sqrt{3}i}{\tau_{34}^b}\right)$$

$$= \left(s + \frac{3 + \sqrt{3}i}{3}\right) \times \left(s + \frac{3 - \sqrt{3}i}{3}\right) \times \left(s + \frac{3 + \sqrt{3}i}{12}\right) \times \left(s + \frac{3 - \sqrt{3}i}{12}\right)$$

$$b_4^f b_5^b = \left(s + \frac{3 + \sqrt{3}i}{\tau_{14}^f}\right) \times \left(s + \frac{3 - \sqrt{3}i}{\tau_{14}^f}\right) \times \left(s + \frac{3 + \sqrt{3}i}{\tau_{24}^b}\right) \times \left(s + \frac{3 - \sqrt{3}i}{\tau_{24}^b}\right)$$

$$= \left(s + \frac{3 + \sqrt{3}i}{3}\right) \times \left(s + \frac{3 - \sqrt{3}i}{3}\right) \times \left(s + \frac{3 + \sqrt{3}i}{8}\right) \times \left(s + \frac{3 - \sqrt{3}i}{8}\right)$$

$$b_2^f b_2^b b_5^f b_5^b = \left(s + \frac{3 + \sqrt{3}i}{\tau_{22}^f}\right) \times \left(s + \frac{3 - \sqrt{3}i}{\tau_{22}^f}\right) \times \left(s + \frac{3 + \sqrt{3}i}{\tau_{22}^b}\right) \times \left(s + \frac{3 - \sqrt{3}i}{\tau_{22}^b}\right)$$

$$\times \left(s + \frac{3 + \sqrt{3}i}{\tau_{24}^f}\right) \times \left(s + \frac{3 - \sqrt{3}i}{\tau_{24}^f}\right) \times \left(s + \frac{3 + \sqrt{3}i}{\tau_{24}^b}\right) \times \left(s + \frac{3 - \sqrt{3}i}{\tau_{24}^b}\right)$$

$$= \left(s + \frac{3 + \sqrt{3}i}{2}\right) \times \left(s + \frac{3 - \sqrt{3}i}{2}\right) \times \left(s + \frac{3 + \sqrt{3}i}{5}\right) \times \left(s + \frac{3 - \sqrt{3}i}{5}\right)$$

$$\times \left(s + \frac{3 + \sqrt{3}i}{4}\right) \times \left(s + \frac{3 - \sqrt{3}i}{4}\right) \times \left(s + \frac{3 + \sqrt{3}i}{8}\right) \times \left(s + \frac{3 - \sqrt{3}i}{8}\right)$$

$$b_5^f b_6^b = \left(s + \frac{3 + \sqrt{3}i}{\tau_{24}^f}\right) \times \left(s + \frac{3 - \sqrt{3}i}{\tau_{24}^f}\right) \times \left(s + \frac{3 + \sqrt{3}i}{\tau_{34}^b}\right) \times \left(s + \frac{3 - \sqrt{3}i}{\tau_{34}^b}\right)$$

$$= \left(s + \frac{3 + \sqrt{3}i}{4}\right) \times \left(s + \frac{3 - \sqrt{3}i}{4}\right) \times \left(s + \frac{3 + \sqrt{3}i}{12}\right) \times \left(s + \frac{3 - \sqrt{3}i}{12}\right)$$

$$b_6^f b_4^b = \left(s + \frac{3 + \sqrt{3}i}{\tau_{34}^f}\right) \times \left(s + \frac{3 - \sqrt{3}i}{\tau_{34}^f}\right) \times \left(s + \frac{3 + \sqrt{3}i}{\tau_{14}^b}\right) \times \left(s + \frac{3 - \sqrt{3}i}{\tau_{14}^b}\right)$$

$$= \left(s + \frac{3 + \sqrt{3}i}{9}\right) \times \left(s + \frac{3 - \sqrt{3}i}{9}\right) \times \left(s + \frac{3 + \sqrt{3}i}{14}\right) \times \left(s + \frac{3 - \sqrt{3}i}{14}\right)$$

$$b_6^f b_5^b = \left(s + \frac{3 + \sqrt{3}i}{\tau_{34}^f}\right) \times \left(s + \frac{3 - \sqrt{3}i}{\tau_{34}^f}\right) \times \left(s + \frac{3 + \sqrt{3}i}{\tau_{24}^b}\right) \times \left(s + \frac{3 - \sqrt{3}i}{\tau_{24}^b}\right)$$

$$= \left(s + \frac{3 + \sqrt{3}i}{9}\right) \times \left(s + \frac{3 - \sqrt{3}i}{9}\right) \times \left(s + \frac{3 + \sqrt{3}i}{8}\right) \times \left(s + \frac{3 - \sqrt{3}i}{8}\right)$$

$$b_3^f b_3^b b_6^f b_6^b = \left(s + \frac{3 + \sqrt{3}i}{\tau_{33}^f}\right) \times \left(s + \frac{3 - \sqrt{3}i}{\tau_{33}^f}\right) \times \left(s + \frac{3 + \sqrt{3}i}{\tau_{11}^b}\right) \times \left(s + \frac{3 - \sqrt{3}i}{\tau_{11}^b}\right)$$

$$\times \left(s + \frac{3 + \sqrt{3}i}{\tau_{34}^f}\right) \times \left(s + \frac{3 - \sqrt{3}i}{\tau_{34}^f}\right) \times \left(s + \frac{3 + \sqrt{3}i}{\tau_{34}^b}\right) \times \left(s + \frac{3 - \sqrt{3}i}{\tau_{34}^b}\right)$$

$$= \left(s + \frac{3 + \sqrt{3}i}{6}\right) \times \left(s + \frac{3 - \sqrt{3}i}{6}\right) \times \left(s + \frac{3 + \sqrt{3}i}{7}\right) \times \left(s + \frac{3 - \sqrt{3}i}{7}\right)$$

$$\times \left(s + \frac{3 + \sqrt{3}i}{9}\right) \times \left(s + \frac{3 - \sqrt{3}i}{9}\right) \times \left(s + \frac{3 + \sqrt{3}i}{12}\right) \times \left(s + \frac{3 - \sqrt{3}i}{12}\right)$$

Therefore one finds that all the poles of the matrix $R_f^P(s)\Psi((R_f^P(s))^T)$ are

$$-3 + \sqrt{3}i, \quad -3 - \sqrt{3}i$$

$$-\frac{3 + \sqrt{3}i}{11}, \quad -\frac{3 - \sqrt{3}i}{11},$$

$$-\frac{3 + \sqrt{3}i}{2}, \quad -\frac{3 - \sqrt{3}i}{2},$$

$$-\frac{3 + \sqrt{3}i}{5}, \quad -\frac{3 - \sqrt{3}i}{5},$$

$$-\frac{3 + \sqrt{3}i}{6}, \quad -\frac{3 - \sqrt{3}i}{6},$$

$$-\frac{3 + \sqrt{3}i}{7}, \quad -\frac{3 - \sqrt{3}i}{7},$$

$$-\frac{3 + \sqrt{3}i}{3}, \quad -\frac{3 + \sqrt{3}i}{3},$$

$$-\frac{3 - \sqrt{3}i}{14}, \quad -\frac{3 + \sqrt{3}i}{14},$$

$$-\frac{3 + \sqrt{3}i}{4}, \quad -\frac{3 + \sqrt{3}i}{4},$$

$$-\frac{3 - \sqrt{3}i}{8}, \quad -\frac{3 + \sqrt{3}i}{8},$$

$$-\frac{3 + \sqrt{3}i}{9}, \quad -\frac{3 + \sqrt{3}i}{9},$$

$$-\frac{3 - \sqrt{3}i}{12}, \quad -\frac{3 + \sqrt{3}i}{12}.$$

Observing from the above analyses, one finds that all the poles of the open-loop transfer functions from $X(s)$ to $Y(s)$ and from $P(s)$ to $Q(s)$, and the closed-loop transfer function from $P(s)$ to $Y(s)$ are lying in the left half complex plane. This suggests that all the three considered systems are bounded-input bounded-output (BIBO) stable. However, note that we have taken the second-order Padé approximation. The stability of the systems may be specific to the particular Padé approximations.

15.5   Conclusions and further discussions

Control theory provides a powerful mathematical tool for describing dynamic systems. It lends itself to modeling congestion control—the interacting function of TCP and AQM is a perfect example of a typical "closed loop" system that can be described in control theoretic terms. However, control theory has had to be extended to model the interactions between multiple control loops like the multicast communication system in the Internet, or even a huge number of control loops with extremely complex interactions inside the loops and among the loops. The latter is referred to as the complex network, in which the interactions among the elements of the network are neither purely regular nor purely random.

In a stable system, for any bounded input over any amount of time, the output will also be bounded. For congestion control, what is actually meant by global stability is typically asymptotic stability: a mechanism should converge to a certain state irrespective of the initial state of the network. Local stability means that if the system is perturbed from its stable state it will quickly return toward the locally stable state. In a distributed feedback system, equilibrium and stability properties determine system performance. Regarding the Internet, the protocols are designed and implemented first at the packet level, and the dynamics (performance) are then understood at the flow level. Flow dynamics at the equilibrium determine performance, for example the throughput, fairness, etc. One key issue in yielding satisfactory system dynamics is the stability, which helps us to bound queueing delay and delay jitter, to bring the buffer queue length to the desired point and thus to obtain many benefits. From this perspective, studies of flow level properties can be a good guide to packet level design of TCP Protocols. Successful examples are FAST TCP [137, 139] and the generalized Fast TCP [140], both of which achieve stability in terms of window updating.

In designing the Internet protocol, some other basic mechanisms from control theory are useful as guidelines. For instance, in terms of multivariable control theory a controller should only be fed a system state that determines its output. A (low-pass) filter function should be used in order to pass only states to the controller that are expected to last long enough for its action to be meaningful. Action should be carried out whenever such feedback arrives, as it is a fundamental principle of control that the control frequency should ideally be equal to the feedback frequency. Faster actions lead to oscillations and instability while slower reactions make the system tardy. Following the control theory, only rate increase needs to be inversely proportional to round-trip delay (RTD) (whereas window-based control converges on a target rate inversely proportional to RTD). A congestion control mechanism can therefore converge on a rate that is independent of RTD, as long as its dynamics depend on RTD. This basic and key rule is implemented in, for example, FAST TCP.

Control theory, particularly multivariable control, is believed to be playing an even more important rule in dealing with the following issues that are crucial in congestion control in the Internet:

- **ACK feedback**: The acknowledgment function is used in the Internet to notify the source the receipt of data packets by the destination. Acknowledgments for data sent, or lack of acknowledgments, are used by senders to infer network conditions between the TCP sender and the receiver. A TCP sender uses a timer to recognize lost segments. Duplicate acknowledgment is the basis for the fast retransmit mechanism, which works as follows: after receiving a packet (e.g., with sequence number 1), the receiver sends an acknowledgment by adding 1 to the sequence number (i.e., acknowledgment number 2), which means that the receiver has received the packet number 1 and it expects packet number 2 from the sender. Let's assume that three subsequent packets have been lost. In the meantime the receiver receives packet numbers 5 and 6. After receiving packet number 5, the receiver sends an acknowledgment, but still only for sequence number 2. When the receiver receives packet number 6, it sends yet another acknowledgment value of 2. Because the sender receives more than one acknowledgment with the same sequence number (2 in this example) this is called a duplicate acknowledgment. If an acknowledgment is not received for a particular segment within a specified time (a function of the estimated round-trip delay time), the sender will assume the segment was lost in the network, and will retransmit the segment. Herein, one identifies one key question: in order to optimize the network performance, what is the optimal frequency of feedback: only in case of congestion events, per RTD, per packet, per session etc.? Frequency analyses by using multivariable control can be helpful in understanding this issue. However, it is subject to further research.

- **Heterogeneity**: The Internet encompasses a large variety of heterogeneous IP networks that are run by a multitude of technologies, which results in a tremendous variety of link and path characteristics: capacity can be either small in very slow speed radio links (several kbps), or high in high-speed optical links (several gigabit per second). Latency can range from much less than a millisecond (in networks of local interconnects) to over a second (in satellite links). Even higher latencies can occur in space communication. Subsequently, both the available bandwidth and the end-to-end delay in the Internet may vary over many orders of magnitude, and it is likely that the range of parameters will further increase in the future. Interactions between the sources and destinations in such a high degree of heterogeneity can be modeled by large scale interconnected systems.

- **Delays**: The interactions between the source and its destination always experience delays. For the data packets sent by the source to the destination will have a forward delay while ack packets feedbacked to the source will also experience a backward delay. These two sorts of delay are determined by the propagating distance between the source and its destination, the queue length and the leaking capacity at the route. The importance of the ack packets lies with their role in conveying the congestion information to the source by using the measurement of either queuing delay or packet loss ratio at the link. It can be observed that when the ack packet reaches the source, there have already been a certain number of data packets sent out from the source that are traveling toward the destination. These "in flight" data packets are blind of the congestion situation at the route. The essential difficulty of congestion control lies with these "in flight" data packets. The product of bandwidth and delay is equal to the total amount of "in flight" data packets. As this product becomes larger, congestion controlling becomes more difficult. The time delay control theory has been successfully applied [146, 147] in analyzing the TCP/AQM systems with delayed interactions. Insightful study is still needed to thoroughly look at the dynamics of the interactions between the source and its destination in terms of the forward and backward delays in the Internet under various TCP protocols (at source) and AQM schemes (at destination); for which the time delay control theory is believed to be able to play a key role in understanding the Internet's dynamics with consideration to delays.

- **From a dumbbell network to the Internet**: Currently in networking community, a simple network with dumbbell topology and with aggregate traffic is being widely used in the modeling of the AQM schemes, scheduling and differentiated services. Multivariable control theory is believed to be able to provide theoretical bases for understanding the characteristics of congested links, range of round-trip times and traffic characterization (distribution of transfer sizes, reverse-path traffic effects on congestion elsewhere, etc.) looking from a dumbbell network to the whole Internet. Herein, it is envisaged that the whole Internet can be decomposed into a large number of dumbbell networks. By studying the dynamics of the dumbbell networks, one may gain an overall understanding of the dynamics of the whole Internet.

Conclusions and further research

16

In this book a number of structural and behavioral problems have been investigated that are based on the polynomial matrix description (PMD) theory, behavioral approach and chain-scattering representation approach. Several new and interesting results have been established that it is hoped, will lead to further research.

Chapter 1 is an introduction of the work contained in this book. The main motivations and the main contributions are briefly introduced. These can be divided into six distinct components, namely the theoretical analysis of the solution of regular PMDs, the associated results of behavior theory, the generalizations to chain-scattering representations, the generalized notions that are related to $\mathcal{H}_\infty$ control theory, the generalized $\mathcal{H}_\infty$ control model, formulation and the solution, and, as an example of applications of linear multivariable control, a system structure look into Internet congestion control in frequency domain.

Chapters 2 through 5 covers the mathematical preliminaries and the basic background in linear control systems.

Chapter 6 briefly introduces the background and preliminary results in PMD theory, behavioral theory, and chain-scattering representation approaches, which are needed in the representation of our contributions.

Chapter 7 addresses an important subject in system analysis and design: the determination of finite and infinite frequency structure of a rational matrix. A novel method has been developed that determines the finite and infinite frequency structure of any rational matrix. Section 7.3 establishes a natural relationship between the rank information of the Toeplitz matrices and the multiplicities of the corresponding irreducible elementary divisors (IREDs) in its Smith form. Based on this, Sections 7.4–7.6 present the corresponding methods to determine the Smith form of a polynomial matrix, to determine the Smith-McMillan form at infinity of a rational matrix and the Smith-McMillan form of a rational matrix, respectively. The proposed methods are believed to be neater and more numerically attractive when compared with the classical methods and the methods of Van Dooren et al. [1], Verghese and Kailath [2], and Pugh et al. [3].

Chapter 8 discusses the complete solution of a regular PMD. Two main contributions have been made therein, one is the complete solution of a regular PMD, which takes into account the nonzero initial conditions of the state as well as the nonzero initial conditions of the input, the other is the results on the set of impulsive free initial conditions. Specifically, Theorem 8.3.1 presents the complete solution of a regular PMD, which is an extension of that given by Vardulakis [13] to the case where the

A Generalized Framework of Linear Multivariable Control. http://dx.doi.org/10.1016/B978-0-08-101946-7.00016-0

initial conditions of the pseudo-state and the input are not zero. Further, by using by using this theorem, reformulation of the solution of a regular PMD in terms of the regular derivatives of $u(t)$ is given in Theorem 8.4.1. System behaviors of a regular PMD are decomposed into the slow and fast solution components. This approach therefore provides an efficient method to analyze the impulse free initial conditions of a regular PMD.

Chapter 9 presents a resolvent decomposition, which is a refinement of both results obtained by Gohberg et al. [17] and Vardulakis [13]. It is formulated in terms of the notions of the finite Jordan pairs, infinite Jordan pairs, and the generalized infinite Jordan pairs that were defined by Gohberg et al. [17] and Vardulakis [13]. This refined resolvent decomposition captures the essential feature of the system structure, the redundant information that is included in the resolvent decomposition of Gohberg et al. [17] is deleted through certain transformations, thus the resulting resolvent decomposition inherits the advantages of both the results of Gohberg et al. [17] and those of Vardulakis [13]. In the proposed approach the matrices Z, Z_∞ in Eq. (9.3) are formulated explicitly, which means that this method is more constructive. The main idea in this proposed approach is to calculate an elementary matrix P, which is very easy to obtain, to delete the redundant information, then to propose the refined resolvent decomposition. This elementary matrix has the effect of deleting the redundant information in two ways. First, it deletes the redundant information in those blocks in the infinite Jordan pair of Gohberg et al. [17] that corresponds to the infinite zeros and brings them into the correct sizes. Second, it deletes the whole blocks in the infinite Jordan pair of Gohberg et al. [17] that correspond to the infinite poles and the whole blocks that are not dynamically important. This elementary matrix serves to transform the partitioned block matrix in Z_∞ that corresponds to the redundant information into zero, the resulting refined resolvent decomposition is thus of minimal dimensions. As well, by using this elementary matrix the mechanism of decoupling in the solution of Gohberg et al. [17] is explained clearly. This refined resolvent decomposition facilitates computation of the inverse matrix of $A(s)$ due to the fact that the dimensions of the matrices used are of minimal. Rather calculating the inverse matrix $A^{-1}(s)$ with an overly large dimension, it is suggested that the easily available elementary matrix P be calculated first to delete all the redundant information and then obtain the inverse matrix with the minimal dimension. Once the refined resolvent decomposition is obtained, the generalized infinite Jordan pair and the elementary matrix P do not need to be involved in the calculation of the solution of the regular PMDs. This represents another merit to this method, which is algorithmically attractive when applied in actual computation.

Based on this proposed resolvent decomposition, a complete solution of a PMD follows that reflects the detailed structure of the *zero state response* and the *zero input response* of the system. The complete impulsive properties of the system are also displayed in our solution. Such impulsive properties is not properly displayed in the solution of Gohberg et al. [17]. Although in the homogeneous case it is considered in Vardulakis [13], such complete impulsive properties of the system for the general nonhomogeneous regular PMDs are not available from Vardulakis [13].

In Chapter 10, a novel approach is presented to determine the complete solution of regular PMDs, which is based on one kind of linearization [17] of the regular polynomial matrix $A(s)$, the so-called *generalized companion matrix*. In this chapter certain properties of this companion form have been established, and a special *resolvent decomposition* of $A(s)$ has been proposed that is based on the Weierstrass canonical form of this companion form. The solution of a regular PMD has then been formulated from this resolvent decomposition. An obvious advantage of the approach adopted therein is that it avoids the polynomial matrix transformation necessary to obtain the finite and infinite Jordan pairs of $A(s)$, and only requires the constant matrix transformation to obtain the Weierstrass canonical form of the generalized companion form, which is less sensitive than the former in computational terms. Since numerically efficient algorithms to generate the canonical form of a matrix pencil are well developed (see, e.g., [19, 20]), the formula proposed there is more attractive in computational terms than the previously known results.

Theorem 10.2.1 relates the infinite frequency structure of a polynomial matrix with the associated Weierstrass canonical matrix structure in a very natural way. Thus together with Propositions 10.2.1 and 10.2.2 one can immediately, from the Weierstrass canonical form of the generalized companion matrix, give characterizations of the finite and infinite frequency structure of the polynomial matrix $A(s)$. These results can be considered as the generalizations of the associated properties from the matrix pencil case to the polynomial matrix case. These results are also interesting from the numerical computation point of view. They suggest an alternative way to find the finite and infinite frequency structure of a regular polynomial matrix. Since the finite and infinite frequency structure of a regular polynomial matrix are completely characterized by the Weierstrass canonical form of its companion matrix, the only thing we need to do is to transform the easily formed companion matrix (a regular matrix pencil) into its Weierstrass canonical form by some constant matrix transformation. Compared to the classical methods, which use the polynomial matrix transformations, the method proposed here will be less sensitive to data perturbations and rounding errors.

From Gohberg et al. [17] and Vardulakis [13], we can see that the *resolvent decomposition* of the regular polynomial matrix $A(s)$ plays a key role in formulating the solution of the regular PMD. The generalized companion form of $A(s)$ enables us to establish the new resolvent decomposition in Theorem 10.2.2, which is based on the Weierstrass canonical form of $C_A(s)$ and is different from that of Gohberg et al. [17] and that of Vardulakis [13]. Theorem 10.3.1 has thus proposed the complete of a regular PMD on the basis of this new resolvent decomposition. This proposed solution procedure is, therefore, more attractive in actual computation than the classical solution methods (Gohberg et al. [17] and that of Vardulakis [13]).

In Chapter 11, the chain-scattering representation proposed by Kimuura for the regular case has been extended to the general case where no rank condition is assumed on P_{21} and P_{12}. The conditions under which the generalized chain-scattering representation (GCSR) and the dual generalized chain-scattering representation (DGCSR) exist have been developed from the point of view of input-output consistency. The generalized chain-scattering matrices have been formulated into parameterized form by using the generalized inverse of matrices. The essential cascade structure of them has been clarified in a general setting. Some algebraic system properties of these general matrices that are relevant to control system design requirements, especially $\mathcal{H}^\infty$ control problem, have been developed. In particular, Section 11.2 has presented the chain-scattering representation from the point of view of input-output consistency, and the formulations of the GCSR and the DGCSR have been parameterized. Section 11.3 has investigated two fundamental algebraic properties the chain-scattering representation, i.e., the cascade structure and the duality in a general setting. These results are extensions to the regular cascade connection in classical circuit theory. In Section 11.4, the realizations of GCSR and DGCSR have been obtained.

Chapter 12 has presented a new notion of realization of behavior. It has been shown that realization of behavior generalizes the classical concept of realization of transfer function matrix. The basic idea in this approach is to find an autoregressive moving-average (ARMA) representation for a given frequency behavior description, such that the known frequency behavior is completely recovered to the corresponding dynamical behavior. From this point of view, realization of behavior is seen to be a converse procedure to the latent variable eliminating process that was studied by Willems [27]. Such a realization approach is believed to be highly significant in modeling a dynamical system in some real cases where the system behavior is conveniently described in the frequency domain. Since no numerical computation is needed, the realization of behavior is believed to be particularly suitable for situations in which the coefficients are symbolic rather than numerical.

Based on this idea, the behavior structures of the GCSRs and the DGCSRs have been clarified. In Theorem 12.3.2 it has been shown that any GCSR or any DGCSR develops the same (frequency) behavior. Subsequently in Theorems 12.3.3–12.3.6, the corresponding ARMA representations are proposed and are proved to be realizations of behavior for any GCSR and any DGCSR. More specifically, two Rosenbrock PMDs are found to be the realizations of behavior for any GCSR $GCHAIN$ $(P^*; (P_{21}^*)^-)$ and any DGCSR $DGCHAIN$ $(P^*, (P_{12}^*)^-)$. Once these ARMA representations are proposed, one can further find the corresponding first-order system representations by using the method of [30] or other well-developed realization approaches such as Kailath [32]. These results are thus interesting in that they provide a natural linkage between the new chain-scattering approach and the well-developed Rosenbrock PMD theory or the developing behavior theory.

Chapter 13 is concerned two key notions, i.e., well-posedness and internal stability, which are closely related to control system designs, such as $\mathcal{H}_\infty$ control and $\mathcal{H}_2$ control. By Looking into the standard $\mathcal{H}_\infty$ control scheme in detail, we have presented certain generalizations to the classical concepts of well-posedness and internal stability, which are termed as *externally internal well-posedness* and *externally internal stability*. The main significance of these generalizations lies in the fact that the issue of well-posedness can be unifiedly clarified from a general setting, the *input-output stability* of the closed-loop system can be characterized in terms of the generalized input-output transfer functions. Such generalizations also throw a new light on the impact of the input signal information on system control design.

It is shown that externally internal well-posedness and externally internal stability generalize the classical notions of well-posedness and internal stability respectively.

The generality of externally internal well-posedness is seen from the fact that the resulting closed-loop system can be "ill-posed" provided that the related input signals are appropriate and the matrices P_{21} and P_{12} do not necessarily satisfy the full rank assumptions. The generality of externally internal stability is achieved by its facility for allowing the system to be internally unstabilized and for allowing the system not necessarily to satisfy the relevant rank assumptions. Especially when these restrictions on the system are satisfied, system externally internal well-posedness is essentially equivalent to system well-posedness and externally internal stability is equivalent to internal stability.

Our approach was exposed by using the matrix generalized inverses, which in turn led to a generalization of *linear fractional transformation* and an extended definition of $\mathcal{H}_\infty$ control problem. The *extended* $\mathcal{H}_\infty$ control problem is cast in such a general setting that it allows one to cancel two restrictions that the known approaches in $\mathcal{H}_\infty$ control theories usually impose on systems. One is that the considered system must be internally stabilized, the other is that the system must satisfy the relevant rank assumption on P_{21} and P_{12}. The solution to this extended problem is thus enlarged from that of the regular $\mathcal{H}_\infty$ control problem, which enables one to search the appropriate controllers in a wider area, the resulting controllers might be subsequently more flexible in achieving the control aims.

There are several directions in which we believe our research can be further usefully pursued. One of the interesting topics deserving further research would be that of a complete solution to a *nonregular* PMD and the set of impulsive free initial conditions. A nonregular PMD is described by

$$A(\rho)\beta(t) = B(\rho)u(t), \tag{16.1}$$

where $A(\rho)$ is a nonsquare and singular polynomial matrix. Even though some characterizations of the solution space to Eq. (16.1) have been available from [154], a complete solution that takes into account both the initial conditions of $\beta(t)$ and $u(t)$ is, however, not available at this stage. Furthermore, the impulsive property of such a system can be much more complicated than that of a regular PMD, the main difficulty one may encounter in dealing with these issues is that the above representation cannot be reduced to a full row rank representation via any left unimodular operation because the impulsive behavior of such a system does not remain invariant under unimodular transformation. Obviously, the solutions to Eq. (16.1) are not unique due to the nonregularity of $A(\rho)$. However, $A(\rho)$ provides some structural invariant, which should be reflected in the solutions. All these issues still call for a general treatment.

Characterizing the minimal behavior realization is also a topic for further research. Introduction of a latent variable is a fundamental idea in the behavior realization approach and it would be interesting, therefore, if minimality of realization of behavior could be defined and accomplished via a specific introduction process. These results would definitely complete the behavior realization approach that is proposed in this book.

The chain-scattering representation together with the theory of J-lossless factorization seems to be the most compact theory for $\mathcal{H}_\infty$ control. The chain-scattering representation approach proposed by Kimura inherits certain limitations and its applicability is subsequently restricted. The GCSR, which is proposed in this book, seems to have canceled those limitations to some extent and the applicability seems to have been broadened. The application of our approach to analysis and synthesis problems in general are also subjects for further research.

Chapter 13 has led to an extended $\mathcal{H}_\infty$ control problem, which generalizes the standard problem in two senses, i.e., the resulting closed-loop system can be "ill-posed" and the full rank assumptions on the plant can be relaxed. In future research it may be worthwhile to investigate such "ill-posed" controller structure along the lines of this approach and to solve the extend $\mathcal{H}_\infty$ control problem in a general setting.

Chapter 14 defines and discusses the nonstandard H_∞ control problem from a GCSR approach perspective and points out its solvability. This seems to open up a widely extended area in engineering practice, for which the usual H_∞ control theory is not applicable. Recently, the ill-conditioned plants have received considerable attention in the literature (see, e.g., [155–161]), but unfortunately so far no direct systematic design methodology has been yielded. There should be a relationship between the "ill-posed" system and "ill-conditioned" plants and would be interesting to explore the control system designing philosophy (see, for example [162–188]) for the $\mathcal{H}_\infty$ control of ill-posed system. Therefore a generalized H_∞ control design framework for stable multivariable plants, which suffer an ill-posed-ness, is calling for in future research efforts.

Bibliography

[1] P.M. Van Dooren, D. Patrick, J. Vandewalle, On the determination of the Smith-McMillan form of a rational matrix from its Laurent expansion, IEEE Trans. Circ. Syst. CAS-26 (3) (1979) 180–189.

[2] G. Verghese, T. Kailath, Rational matrix structure, IEEE Trans. Autom. Control 26 (2) (1981) 434–439.

[3] A.C. Pugh, E.R.L. Jones, O. Dcmianczuk, G.E. Hayton, Infinite frequency structure and a certain matrix Laurent expansion, Int. J. Control 50 (5) (1989) 1973–1805.

[4] F.R. Gantmacher, The Theory of Matrices, Chelsea, New York, 1971.

[5] H.H. Rosenbrock, State Space and Multivariable Theory, Nelson, London, 1970.

[6] N.A. Emami, P. Van Dooren, Computation of zeros of linear multivariable systems, Automatica 18 (82) (1982) 415–430.

[7] S.L. Campbell, Singular Systems of Differential Equations 1, Pitman, London, 1980.

[8] S.L. Campbell, Singular Systems of Differential Equations 2, Pitman, London, 1982.

[9] S.L. Campbell, Linear systems of differential equations with singular coefficients, SIAM J. Math. Anal. 8 (6) (1977) 1057–1066.

[10] D. Cobb, On the solution of linear differential equations with singular coefficients, J. Differ. Equ. 46 (3) (1982) 310–323.

[11] G. Verghese, B.C. Levy, T. Kailath, A generalized state space for singular systems, IEEE Trans. Autom. Control AC-26 (4) (1981) 811–830.

[12] G. Fragulis, A closed formula for the determination of the impulsive solutions of linear homogeneous matrix differential equations, IEEE Trans. Autom. Control 38 (11) (1993) 1688–1695.

[13] A.I.G. Vardulakis, Linear Multivarialble Control: Algebraic Analysis and Synthesis Methods, Wiley, New York, 1991.

[14] E. Yip, R. Sineovec, Solvability, controllability and observability of continuous descriptor systems, IEEE Trans. Autom. Control 26 (3) (1981) 702–707.

[15] K. Ozcaldiran, Control of descriptor systems, Ph.D. thesis, School of Electrical Engineering, Georgia Institute of Technology, Atlanta, GA, 1985.

[16] G.F. Fragulis, A.I.G. Vardulakis, Reachability of polynomial matrix descriptions (PMDs), Circ. Syst. Signal Process 14 (6) (1995) 787–815.

[17] I. Gohberg, P. Langaster, I. Rodman, Matrix Polynomial, Academic Press, New York, 1982.

[18] G.E. Hayton, A.C. Pugh, P. Fretwell, Infinite elementary divisors of a matrix polynomial and its implications, Int. J. Control 47 (1) (1988) 53–64.

[19] D. Jordan, L.F. Godbout, On the computation of the canonical pencil of a linear system, IEEE Trans. Autom. Control 22 (1) (1977) 112–114.

[20] P. Van Dooren, The computation of Kronecker's canonical form of a singular pencil, Linear Algebra Appl. 27 (1979) 103–140.

[21] V. Belevitch, Classical Network Theory, Holden-Day, San Francisco, 1968.

[22] H. Kimura, Chain-scattering representation, J-lossless factorization and H_∞ control, J. Math. Syst. Estimation Control 5 (1995) 203–255.

[23] W. Kongprawechnon, H. Kimura, J-lossless conjugation and factorization for discrete-time systems, Int. J. Control 65 (5) (1996) 867–884.

[24] H. Kimura, Conjugation, interpolation and model-matching in H_∞, Int. J. Control 49 (1) (1989) 269–307.

[25] J.A. Ball, J.W. Helton, M. Verma, A factorization principle for stabilization of linear control systems, Int. J. Robust Nonlinear Control 1 (4) (1991) 229–294.

[26] J.C. Willems, From time series to linear system: Part 1. Finite dimensional linear time invariant systems: Part 2. Exact modeling: Part 3: Approximate modelling, Automatica 22 (1986) 561–580.

[27] J.C. Willems, Paradigms and puzzles in the theory of dynamical systems, IEEE Trans. Autom. Control 36 (3) (1991) 259–294.

[28] A.C. Antoulas, J.C. Willems, A behavioural approach to linear exact modeling, IEEE Trans. Autom. Control 38 (12) (1993) 1776–1800.

[29] M. Kuijper, J.M. Schumacher, Input-output structure of linear differential/algebraic systems, IEEE Trans. Autom. Control 38 (3) (1993) 404–414.

[30] J. Rosenthal, J.M. Schumacher, Realization by inspection, IEEE Trans. Autom. Control 42 (9) (1997) 1257–1263.

[31] A.C. Pugh, L. Tan, A generalized chain-scattering representation and its algebraic system properties, Report Number A316, Department of Mathematical Sciences, Loughborough University, UK, 1998.

[32] T. Kailath, Linear Systems, Prentice-Hall, Englewood Cliffs, NJ, 1980.

[33] H. Kwakernaak, R. Sivan, Modern Signals and Systems, Prentice-Hall, Englewood Cliffs, NJ, 1991.

[34] J.T.H. Chan, Mimo model-matching controller for minimum phase systems: a computational design with input-output data, Control Comput. 24 (1996) 12–26.

[35] G. Zames, Feedback and optimal sensitivity: model reference transformations, multiplicative seminorms, and approximate inverses, IEEE Trans. Autom. Control 26 (2) (1981) 301–320.

[36] B.A. Francis, A Course in H-infinity Control, in: Lecture Notes in Control and Information Sciences, vol. 88, Springer-Verlag, Berlin, 1987.

[37] J.C. Doyle, K. Klover, P.P. Khargonekar, B.A. Francis, State-space solutions to standard H_2 and H_∞ control problems, IEEE Trans. Autom. Control 34 (1989) 831–847.

[38] H. Kwakernaak, M.J. Grimble, V. Kucera (Eds.), Frequency domain solution of the standard H-infinity Control problem, in: Polynomial Methods for Control Systems Design, Springer-Verlag, Berlin, 1996, pp. 57–105.

[39] J.C. Doyle, B. Francis, A. Tannenbaum, Feedback Control Theory, Macmillan Publishing Co., New York, 1992.

[40] M. Vidyasagar, Control System Synthesis: A Factorization Approach, MIT Press, Cambridge, MA, 1985.

[41] H. Kwakernaak, Robust control and H-infinity optimization, Automatica 29 (2) (1993) 255–273.

[42] S.H. Low, F. Paganini, J.C. Doyle, Internet congestion control, IEEE Control Syst. Mag. 22 (1) (2002) 28–43.

[43] T. Kardi, Linear Algebra Tutorial, 2001, http://people.revoledu.com/kardi/toturial/LinearAlgebra/.

[44] E.Y. Shapiro, H.E. Decarli, Generalized inverse of a matrix: the minimization approach, AIAA J. 14 (10) (1976) 1483–1484.

[45] T.S. Shores, Applied Linear Algebra and Matrix Analysis, Springer Science Business Media, LLC, New York, NY, 2007.

[46] G. Strang, Linear Algebra and Its Applications, fourth ed., Wellesley-Cambridge Press, Cambridge, MA, 2005.

[47] S.J. Leon, Linear Algebra With Applications, seventh ed., Pearson Prentice Hall, Upper Saddle River, NJ, 2006.

[48] R.A. Horn, C.R. Johnson, Topics in Matrix Analysis, Cambridge University Press, Cambridge, 1994.

[49] S. Lang, Linear Algebra (Undergraduate Texts in Mathematics), third ed., Springer, New York, 2004.

[50] M. Marcus, H. Minc, A Survey of Matrix Theory and Matrix Inequalities, Dover Publications, New York, NY, 2010.

[51] F.E. Udwadia, P. Phohomsiri, Recursive determination of the generalized Moore–Penrose M-Inverse of a matrix, J. Optim. Theory Appl. 127 (3) (2005) 639–663.

[52] S.L. Campbell, C.D. Meyer, Generalized Inverses of Linear Transformations, Dover, New York, NY, 1991.

[53] R. Penrose, A generalized inverse for matrices, Proc. Cambridge Philos. Soc. 51 (4) (1955) 406–413.

[54] A. Ben-Israel, T.N.E. Greville, Generalized Inverses: Theory and Applications, Wiley, New York, 1974.

[55] E. Polak, An algorithm for reducing a linear time-invariant differential system to state form, IEEE Trans. Autom. Control 11 (3) (1966) 577–579.

[56] A. Ilchmann, I. Nurnberger, W. Schmale, Time-varying polynomial matrix systems, Int. J. Control 40 (2) (1984) 329–362.

[57] D.F. Delchamps, State Space and Input-Output Linear Systems, Springer-Verlag, New York, 1988.

[58] G. Liu, K. Lu, Polynomial matrix description and structural controllability of composite system, in: Proceedings of 2009 International Conference on Information Engineering and Computer Science, 2009, pp. 1–4.

[59] C.H. Fang, A new approach for calculating doubly-coprime matrix fraction descriptions, IEEE Trans. Autom. Control 37 (1) (1992) 138–141.

[60] C.A. Desoer, J.D. Schulman, Zeros and poles of matrix transfer functions and their dynamical interpretation, IEEE Trans. Circ. Syst. 21 (1) (1974) 3–8.

[61] E.W. Kamen, Poles and zeros of linear time-varying systems, Linear Algebra Appl. 98 (1988) 263–289.

[62] C.B. Schrader, M.K. Sain, Research in system zeros: a survey, Int. J. Control 50 (4) (1989) 1407–1433.

[63] O.M. Grasselli, S. Longhi, Zeros and poles of linear periodic multivariable discrete-time systems, Circ. Syst. Signal Process. 7 (3) (1988) 361–380.

[64] B.D.O. Anderson, V.V. Kucera, Matrix fraction construction of linear compensators, IEEE Trans. Autom. Control 30 (11) (1985) 112–1114.

[65] G. Zhang, A. Lanzon, On poles and zeros of input-output and chain-scattering systems, Syst. Control Lett. 55 (4) (2006) 314–320.

[66] B. Bekhiti, A. Dahimene, B. Nail, K. Hariche, A. Hamadouche, On Block roots of matrix polynomials based MIMO control system design, in: Proceedings of 4th International Conference on Electrical Engineering (ICEE), 2015, pp. 1–6.

[67] A. Sandmann, A. Ahrens, S. Lochmann, Resource allocation in SVD-assisted optical MIMO systems using polynomial matrix factorization, in: Proceedings of 16th ITG Symposium on Photonic Networks, 2015, pp. 1–7.

[68] F.D. Freitas, S. Luis Varricchio, A power system dynamical model in the matrix polynomial description form, in: 2013 IEEE Power & Energy Society General Meeting, Vancouver, BC, 2013, pp. 1–5.

[69] S. Barnett, Polynomials and Linear Control Systems, Marcel Dekker, New York, 1983.

[70] J. Wilson, Rugh, Linear System Theory, in: T. Kailath (Ed.), Prentice Hall Information and System Science Series, Prentice Hall, Upper Saddle River, NJ, 1996.

[71] W.A. Wolovich, Linear Multivariable Systems, Springer-Verlag, New York, 1974.

[72] H. Blomberg, R. Ylinen, Algebraic Theory for Multivariable Linear Systems, in: Mathematics in Science and Engineering, vol. 166, Academic Press, London, 1983.

[73] M. Hou, A.C. Pugh, G.E. Hayton, A test for behavioural equivalence, IEEE Trans. Autom. Control 45 (2000) 2177–2182.

[74] M. Hou, A.C. Pugh, G.E. Hayton, General solution to systems in polynomial matrix form, Int. J. Control 73 (2000) 733–743.

[75] B.D.O. Anderson, J.B. Moore, Optimal Filtering, Prentice-Hall, Englewood Cliffs, NJ, 1979.

[76] R. Bellman, Stability Theory of Differential Equations, McGraw-Hill, New York, 1953.

[77] W.A. Coppel, Stability and Asymptotic Behavior of Differential Equations, Heath, Boston, 1965.

[78] J.L. Willems, Stability Theory of Dynamical Systems, John Wiley, New York, 1970.

[79] B.R. Barmish, New Tools for Robustness of Linear Systems, Macmillan, New York, 1994.

[80] C. Chicone, Stability theory of ordinary differential equations, in: Mathematics of Complexity and Dynamical Systems, Springer-Verlag, New York, 2011, pp. 1653–1671.

[81] H. Nasiri, M. Haeri, How BIBO stability of LTI fractional-order time delayed systems relates to their approximated integer-order counterparts, IET Control Theory Appl. 8 (8) (2014) 598–605.

[82] K.A. Varma, D.K. Mohanta, M.J.B. Reddy, Applications of type-2 fuzzy logic in power systems: a literature survey, in: Proceedings of 12th IEEE International Conference on Environment and Electrical Engineering (EEEIC), 2013, pp. 282–286.

[83] Y. Nishimura, K. Tanaka, Y. Wakasa, Almost sure asymptotic stabilizability for deterministic systems with wiener processes, in: Proceedings of 20th Mediterranean Conference on Control & Automation (MED), 2012, pp. 410–415.

[84] M. Tiwari, A. Dhawan, A survey on stability of 2-D discrete systems described by Fornasini-Marchesini first model, in: Proceedings of 2010 International Conference on Power, Control and Embedded Systems (ICPCES), 2010, pp. 1–4.

[85] G. Chesi, LMI techniques for optimization over polynomials in control: a survey, IEEE Trans. Autom. Control 55 (11) (2010) 2500–2510.

[86] B.D.O. Anderson, R.R. Bitmead, C.R. Johnson, P.V. Kokotovic, R.L. Kosut, I.M.Y. Mareels, L. Praly, B.D. Riedle, Stability of Adaptive Systems: Passivity and Averaging Analysis, The MIT Press, Cambridge, MA/London, 1986.

[87] A.I.G. Vardulakis, D. Limebeer, N. Karcanias, Structure and Smith-McMillan form of a rational matrix at infinity, Int. J. Control 35 (4) (1982) 701.

[88] J.C. Willems, On interconnections, control, and feedback, IEEE Trans. Autom. Control 42 (3) (1997) 326–339.

[89] J.W. Polderman, J.C. Willems, Introduction to Mathematical Systems Theory: A Behavioural Approach, in: Texts in Applied Mathematics, vol. 26, Springer, New York, 1997.

[90] H. Kimura, Chain-scattering Approach to H-infinity Control, Birkhäuser, Boston, 1997.

[91] M.R. Redheffer, On a certain linear fractional transformation, J. Math. Phys. 39 (1) (1960) 269–286.

[92] N. Karcanias, G. Kalogeropoulos, On the 'Segre' Wegr characteristics of right (left) regular pencils, Int. J. Control 44 (1986) 991–1015.

[93] H. Eliopoulou, N. Karcanias, Geometric properties of Segre characteristic at infinity of singular pencils, in: H. Kimura (Ed.), MTNS-91 Proceedings, MITA Press, 1991, pp. 109–111.

[94] N. Karcanias, G. Kalogeropoulos, The prime and generalised null space of right regular pencils, Circ. Syst. Signal Process. 14 (4) (1995) 495–524.

[95] H. Eliopoulou, N. Karcanias, The fundamental subspace sequences of matrix pencils, Circ. Syst. Signal Process. 17 (5) (1998) 559–574.

[96] O.H. Bosgra, A.I.J. Van Der Weiden, Realizations in generalized state space form for polynomial system matrices, and the definition of poles, zeros and decoupling zeros at infinity, Int. J. Control 33 (3) (1981) 393–411.

[97] G.E. Haylon, A.B. Walker, A.C. Pugh, Matrix pencil equivalents of a general polynomial matrix, Int. J. Control 49 (6) (1989) 1979–1987.

[98] A.C. Pugh, G.E. Hayton, P. Fretwell, Transformations of matrix pencils and implications in linear systems theory, Int. J. Control 54 (2) (1987) 529–543.

[99] C.B. Moler, G.W. Stewart, An algorithm for the generalized matrix eigenvalue problem, SIAM J. Numer. Anal. 10 (2) (1973) 241–256.

[100] R.C. Ward, The combination shift QZ algorithm, SIAM J. Numer. Anal. 12 (6) (1975) 835–853.

[101] D. Vafiadis, N. Karcanias, Generalized state space realizations from matrix fraction descriptions, IEEE Trans. Autom. Control 40 (6) (1995) 1134–1137.

[102] A.I.G. Vardulakis, E.N. Antoniou, N. Karampetakis, On the solution of polynomial matrix descriptions of free linear multivariable systems, Int. J. Control 72 (3) (1999) 215–228.

[103] L. Tan, The disturbance localization problem for singular systems, Acta Math. Sci. 15 (3) (1995) 241–246.

[104] L. Tan, On disturbance localization in singular systems with direct feedthrough, Int. J. Syst. Sci. 26 (11) (1995) 2235–2244.

[105] P. Kokotovic, H.K. Khalil, J. O'Reilly, Singular Perturbation Methods in Control: Analysis and Design, Academic Press, New York, 1986.

[106] A.H. Zemanian, Distribution Theory and Transform Analysis, McGraw-Hill Inc., New York, 1965.

[107] A.C. Pugh, The McMillan degree of a polynomial system matrix, Int. J. Control 24 (1) (1976) 129–135.

[108] A. Fossard, Commande des Systems Multidimensionnels, Dunod, France, 1972.

[109] G.C. Verghese, Infinite frequency behaviour in dynamical systems, Ph.D. dissertation, Department of Electronic Engineering, Stanford University, 1978.

[110] A.I.G. Vardulakis, G. Fragulis, Infinite elementary divisors of polynomial matrices and impulsive solutions of linear homogeneous matrix differential equations, Circ. Syst. Signal Process 8 (3) (1989) 357–373.

[111] I. Gohberg, P. Lancaster, I. Rodman, Spectral analysis or matrix polynomials, Res. Paper No. 313, University of Calgary, Calgary, 1976.

[112] C.R. Cao, S.K. Mitra, Generalized Inverse of Matrices and its Applications, John Wiley, New York, 1971.

[113] P. Dewilde, H. Dym, Lossless chain-scattering matrices and optimum linear prediction: the vector case, Int. J. Circ. Theory Appl. 9 (2) (1981) 135–175.

[114] D. Vafiadis, N. Karcanias, First-order realizations of autoregressive equations, Int. J. Control 72 (12) (1999) 1043–1053.

[115] D.G. Luenberger, Time invariant descriptor systems, Automatica 14 (5) (1978) 473–480.

[116] M. Hou, A.C. Pugh, G.E. Hayton, Generalized transfer functions and input-output equivalence, Int. J. Control 68 (5) (1997) 1163–1178.

[117] M. Green, D.J.N. Limebeer, Linear Robust Control, Prentice-Hall, Englewood Cliffs, NJ, 1995.

[118] M.G. Safonov, E.A. Jonckheere, M. Verma, D.J.N. Limebeer, Synthesis of positive real multivariable feedback systems, Int. J. Control 45 (3) (1987) 817–842.

[119] D. Vafiadis, N. Karcanias, Model matching under external and input-output equivalence, Circ. Syst. Signal Process. 16 (4) (1997) 429–483.

[120] A.A. Stoorvogel, The singular H^∞ control with dynamic measurement feedback, SIAM J. Control Optim. 39 (1991) 355–360.

[121] M. Green, K. Glover, D. Linebeer, D. Doyle, A J-spectral factorization approach to H_∞ control, SIAM J. Control Optim. 28 (1990) 1350–1371.

[122] A.A. Stoorvogel, A. Sabell, B.M. Chen, A reduced order observer based controller design for H^∞ optimization, IEEE Trans. Autom. Control 39 (1994) 355–360.

[123] A.A. Stoorvogel, The H^∞ Control Problem: A State Space Approach, Prentice Hall, Upper Saddle River, NJ, 1992.

[124] X. Xin, L. Guo, C. Feng, Reduced-order controllers for continuous and discrete-time singular H^∞ control problems based on LMI, Automatica 32 (1996) 1581–1585.

[125] L. Tan, A.C. Pugh, A note on the solution of regular PMDs, Int. J. Control 72 (14) (1999) 1235–1248.

[126] S. Hara, T. Sugie, R. Kondo, Descriptor form solution for H-infinity control problem with $j\omega$-axis zeros, Automatica 28 (1992) 55–70.

[127] A.G. Akritas, Elements of Computer Algebra with Applications, Wiley, New York, 1989.

[128] V.C. Klema, A.J. Laub, The singular value decomposition: its computation and some applications, IEEE Trans. Autom. Control 25 (1980) 164–176.

[129] V. Jacobson, Congestion avoidance and control, in: Proceedings of ACM SIGCOMM'88, 1988.

[130] J. Wang, A theoretical study of Internet congestion control: equilibrium and dynamics, Ph.D. thesis, California Institute of Technology, Pasadena, California, USA, 2005.

[131] H. Lawrence, Introduction to Data Multicasting, Althos Publishing, Fuquay Varina, NC, 2008.

[132] L. Tan, C. Yuan, M. Zukerman, A price-based Internet congestion control scheme, IEEE Commun. Lett. 12 (4) (2008) 331–333.

[133] L.S. Brakmo, S.W. Omalley, L.L. Peterson, TCP Vegas: new techniques for congestion detection and avoidance, in: Proceedings of 1994 SIGCOMM, 1994.

[134] S.H. Low, L.L. Peterson, L. Wang, Understanding Vegas: a duality model, J. ACM 49 (2) (2002) 207–235.

[135] S. Floyd, High-Speed TCP for large congestion windows, IETF, 2002, Internet Draft. Available from: http://www.ietf.org/internet-drafts/draft-floyd-tcp-highspeed-00.txt.

[136] T. Kelly, Scalable TCP: improving performance in high-speed wide area networks, ACM SIGCOMM Comput. Commun. Rev. 33 (2) (2003) 83–91.

[137] C. Jin, D. Wei, S.H. Low, FAST TCP: motivation, architecture, algorithms, performance, in: Proceedings of IEEE INFOCOM 2004, 7–11 March 2004, vol. 4, Hong Kong, pp. 2490–2501.

[138] C. Jin, D.X. Wei, S.H. Low, G. Buhrmaster, J. Bunn, D.H. Choe, R.L.A. Cottrell, J.C. Doyle, W. Feng, O. Martin, H. Newman, F. Paganini, S. Ravot, S. Singh, FAST TCP: from theory to experiments, IEEE Netw. 19 (1) (2005) 4–11.

[139] L. Tan, C. Yuan, M. Zukerman, FAST TCP: fairness and queueing issues, IEEE Commun. Lett. 9 (8) (2005) 762–764.

[140] C. Yuan, L. Tan, L.L.H. Andrew, W. Zhang, M. Zukerman, A generalized FAST TCP scheme, Comput. Commun. 31 (14) (2008) 3242–3249.

[141] D.E. Comer, Internetworking with TCP/IP, fifth ed., Prentice Hall, Upper Saddle River, NJ, 2005.

[142] S. Floyd, V. Jacobson, Random early detection (RED) gateways for congestion avoidance, IEEE/ACM Trans. Netw. 1 (4) (1993) 397–413.

[143] W.C. Feng, D.D. Kandlur, D. Saha, K.G. Shin, BLUE: A New Class of Active Queue Management Algorithms, Computer Science Technical Report CSE-TR-387-99, University of Michigan, 2013.

[144] S. Athuraliya, V.H. Li, S.H. Low, Q. Yin, REM: active queue management, IEEE Netw. 15 (3) (2001) 48–53.

[145] A. Lakshmikantha, C.L. Beck, R. Srikant, Robustness of real and virtual queue-based active queue management schemes, IEEE/ACM Trans. Netw. 13 (1) (2005) 81–92.

[146] L. Tan, W. Zhang, G. Peng, G. Chen, Stability of TCP/RED systems in AQM routers, IEEE Trans. Autom. Control 51 (8) (2006) 1393–1398.

[147] W. Zhang, L. Tan, C. Yuan, G. Chen, F. Ge, Internet primal-dual congestion control: stability and applications, Control Eng. Pract. 21 (1) (2013) 87–95.

[148] F. Ge, L. Tan, R.A. Kennedy, Stability and throughput of FAST TCP traffic in bidirectional connections, IET Commun. 4 (6) (2010) 639–644.

[149] F. Ge, L. Tan, M. Zukerman, Throughput of FAST TCP in asymmetric networks, IEEE Commun. Lett. 12 (2) (2008) 158–160.

[150] L. Tan, W. Zhang, C. Yuan, On parameter tuning for FAST TCP, IEEE Commun. Lett. 11 (5) (2007) 458–460.

[151] L. Tan, Y. Yang, C. Lin, N. Xiong, M. Zukerman, Scalable parameter tuning for AVQ, IEEE Commun. Lett. 9 (1) (90–92) 2005.

[152] C.N. Houmkozlis, G.A. Rovithakis, End-to-End Adaptive Congestion Control in TCP/IP Networks, CRC Press Taylor & Francis Group, Boca Raton, 2012.

[153] L. Tan, Y. Yang, W. Zhang, M. Zukerman, On control gain selection in dynamic-RED, IEEE Commun. Lett. 9 (1) (2005) 81–83.

[154] N.P. Karampetakis, A.I.G. Vardulakis, A.C. Pugh, On the solution space of continuous time AR representation, Mathematical Sciences Report, Department of Mathematical Sciences, Loughborough University, UK, 1994.

[155] S. Skogestad, I. Postlethwaite, Multivariable Feedback Control: Analysis and Design, Wiley, New York, 2005.

[156] S. Skogestad, K. Havre, The use of RGA and condition number as robustness measures, Comput. Chem. Eng. 20 (1996) 1005–1010.

[157] J.S. Freudenberg, D.P. Looze, Relations between properties of multivariable feedback systems at different loop-breaking points: Part I, in: IEEE Conference on Decision and Control, 1986, p. 250.

[158] J.S. Freudenberg, D.P. Looze, Relations between properties of multivariable feedback systems at different loop-breaking points: Part II, in: American Control Conference, 1986, pp. 771–777.

[159] J.S. Freudenberg, Analysis and design for ill-conditioned plants, in: American Control Conference, 1988, pp. 372–377.

[160] J.S. Freudenberg, Directionality, coupling, and multivariable loopshaping, in: IEEE Conference on Decision and Control, 1988, pp. 399–340.

[161] J. Chen, J.S. Freudenberg, C.N. Nett, The role of the condition number and the RGA in robustness analysis, Automatica 30 (6) (1994) 1029–1035.

[162] A.C. Pugh, N.P. Karampetakis, A.I.G. Vardulakis, G.E. Hayton, A fundamental notion of equivalence for linear multivariable system, IEEE Trans. Autom. Control 39 (5) (1994) 1141–1145.

[163] A.C. Pugh, N.P. Karampetakis, S. Mahmood, G.E. Hayton, Admissible initial conditions for regular PMDs, in: 34th IEEE CDC, Louisiana, USA, 1995, pp. 307–308.

[164] B.D.O. Anderson, J.B. Moore, Linear Optimal Control, Prentice-Hall, Englewood Cliffs, NJ, 1971.

[165] B. McMillan, Introduction to formal realisability theory, Bell Syst. Tech. J. 31 (2) (1952) 541–600.

[166] D. Cobb, Feedback and pole placement in descriptor variable systems, Int. J. Control 33 (6) (1981) 1135–1146.

[167] D. Cobb, A further interpretation of inconsistent initial conditions in descriptor variable systems, IEEE Trans. Autom. Control. AC-28 (9) (1983) 920–922.

[168] D.G. Luenberger, Dynamic systems in descriptor form, IEEE Trans. Autom. Control AC-22 (3) (1977) 312–321.

[169] D.C. Youla, H. Jabr, J.J. Bongiorno, Modern Wiener-Hopf design of optimal controllers, Part 2: The multivariable case, IEEE Trans. Autom. Control 21 (3) (1976) 319–338.

[170] F. Lewis, Descriptor systems: expanded descriptor equations and Markov parameters, IEEE Trans. Autom. Control AC-28 (5) (1983) 623–627.

[171] F. Lewis, A survey of linear singular systems, Circ. Syst. Signal Process 5 (1) (1986) 3–36.

[172] G. Verghese, T. Kailath, Impulsive behaviour in dynamical systems: structure and significance, in: Proceedings of 4th International Symposium on Mathematical Theory Networks Systems, Delft, The Netherlands, 1979, pp. 162–168.

[173] H.H. Rosenbrock, Structural properties of linear dynamical systems, Int. J. Control 20 (2) (1974) 191–202.

[174] H. Kwakernaak, A polynomial approach to minimax frequency domain optimization of multivariable feedback systems, Int. J. Control 44 (1) (1986) 117–156.

[175] H. Kwakernaak, M. Sebek, Polynomial J-spectral factorization, IEEE Trans. Autom. Control. 39 (2) (1994) 315–328.

[176] H. Kimura, Y. Lu, R. Kawatani, On the structure of H-infinity control systems and related extensions, IEEE Trans. Autom. Control 36 (6) (1991) 653–667.

[177] I. Gohberg, L. Rodman, On spectral analysis of nonmonic matrix and operator polynomial, I. Reduction to monic polynomials, Israel J. Math. 30 (1 and 2) (1978) 133–151.

[178] K. Glover, D. McFarlane, Robust stabilization of normalized co-prime factor plant descriptions with H-infinity-bounded uncertainty, IEEE Trans. Autom. Control 34 (8) (1989) 821–830.

[179] L. Dai, Singular Control Systems, Springer-Verlag, Berlin, 1989.

[180] N.P. Karampetakis, A.I.G. Vardulakis, On the reduction of a polynomial matrix description of a linear multivariable system to generalised state space form, in: EURACO Workshop on Recent Results in Robust and Adaptive Control, Florence, Italy, 1995.

[181] N.P. Karampetakis, A.I.G. Vardulakis, On the solution of ARMA representations, in: IEEE Mediterranean Symposium on New Directions in Control and Automation, vol. 1, Cyprus, 1995, pp. 156–163.

[182] S. Mahmood, N.P. Karampetakis, A.C. Pugh, Reachability, controllability and observability properties of regular PMDs, Mathematics Report A. 269, Loughborough University, 1996.

[183] L. Tan, Structural and behavioral analyses to linear multivariable control systems, Ph.D. thesis, Department of Mathematical Science, Loughborough University, UK, 1999.

[184] A.C. Pugh, L. Tan, A generalised chain-scattering representation and its algebraic system properties, IEEE Trans. Autom. Control 45 (2000) 1002–1007.

[185] V. Kucera, Discrete Linear Control: The Polynomial Equation Approach, Wiley, New York, 1979.

[186] P. Van Dooren, G. Verghese, T. Kailath, Properties of the system matrix of a generalized state-space system, in: Proceedings of IEEE Conference on Decision and Control, 1978, pp. 173–175.

[187] W. Rudin, Principles of Mathematical Analysis, McGraw-Hill Book Company, London, 1987.

[188] Y. Wang, Generalized input-output equations and nonlinear realizability, Int. J. Control 64 (4) (1996) 615–629.

Index

Note: Page numbers followed by *f* indicate figures and *t* indicate tables.

Printed in the United States
By Bookmasters